CORALS IN A CHANGING WORLD

Edited by **Carmenza Duque Beltran**
and **Edisson Tello Camacho**

Corals in a Changing World
http://dx.doi.org/10.5772/65203
Edited by Carmenza Duque Beltran and Edisson Tello Camacho

Contributors
Lene Buhl-Mortensen, Pål Buhl-Mortensen, Edwin Hernandez-Delgado, Alex E. Mercado-Molina, Samuel E. Suleiman-Ramos, Mary Ann Lucking, Michela Angiolillo, Simonepietro Canese, Patrick Schubert, Thomas Wilke, Alejandra Rojas-Molina, Norma Olguín-López, César Ibarra, Víctor Hugo Hernández-Elizárraga, Carolina Gutiérrez-Chávez, Sonia Barba-Herrera, Angel Torres-Valcarcel, Monica Puyana, Edisson Tello Camacho, Carmenza Duque, Leonardo Castellanos, Pamela Hallock, Lucy Bartlett, Vanessa Brinkhuis, R.R. Ruzicka, M.A. Colella, Kathleen Semon Lunz, Erin H. Leone, Luwei Fan

Published by InTech
Janeza Trdine 9, 51000 Rijeka, Croatia

Publishing Process Manager Romina Skomersic
Technical Editor SPi Global
Cover InTech Design team

First published March, 2018
Printed in Croatia
Legal deposit, Croatia: National and University Library in Zagreb

Additional hard copies can be obtained from orders@intechopen.com

Corals in a Changing World, Edited by Carmenza Duque Beltran and Edisson Tello Camacho
p. cm.
Print ISBN 978-953-51-3909-6
Online ISBN 978-953-51-3910-2

For EU product safety concerns:
IN TECH d.o.o., Prolaz Marije Krucifikse Kozulić 3, 51000 Rijeka, Croatia,
info@intechopen.com or visit our website at intechopen.com.

Meet the editors

Dr. Carmenza Duque Beltrán is currently an emeritus professor at the Department of Chemistry of the National University of Columbia, Colombia. She was an executive director of the Corporation Center of Excellence in Marine Science (CEMarin), Colombia, and the vice president of the Colombian Academy of Exact, Physical and Natural Sciences. She graduated with a degree in Chemistry in 1972 from the National University of Columbia and completed her Dr. Sc. degree from the Tokyo Institute of Technology, Japan, in 1980 and postgraduate studies from the Stanford University, United States, in 1982 and the Würzburg University, Germany, in 1988 and 1991. Since then, she has dedicated her life to the education of graduate and postgraduate students in analytical chemistry and natural product chemistry and her scientific research in the same field. In 1993, she became a full professor at the National University of Columbia, Colombia. She is one of the founders of the postgraduate school of the Department of Chemistry at the National University of Columbia, where she proposed and initiated two master's and doctorate programs. She has published nearly 120 scientific papers in high-impact journals, edited, and coedited 4 books dealing with her scientific research results and with the scientific policy of Colombia. She has also authored 20 chapters in scientific books and presented 2 patents and 2 prototypes as technological research products.

Edisson Tello Camacho, Dr. Sc. chemist and masters in Chemical Science from the National University of Columbia, Columbia, in 2005 and 2008, respectively. In 2013, he received his master's degree in Chemical Science from the National University of Columbia with an emphasis on natural products, bioprospecting, and organic synthesis. In 2012 and 2013, he developed a research internship in the synthesis of natural products at the University of Nottingham, United Kingdom. Currently, he is an assistant professor at the Department of Chemical Engineering of the La Sabana University, Columbia, and supports the bioscience PhD program at the same university. He is also the leader of the Bioprospecting Research Group at the La Sabana University, where he focuses his research on marine natural product chemistry, searching for cytotoxic and antimicrobial compounds and working in an organic synthesis. He has published nearly 15 scientific papers and 1 chapter dealing with his scientific research results in high-impact journals. He received awards such as the "Early Career Chemist Award" from the American Chemical Society (ACS) in 2015, a Laureate PhD Thesis in 2013, and a Meritorious Magister Thesis in 2008.

Contents

Preface

The 12 chapters of the book "Corals in a Changing World" written by authors from different countries such as the USA, Puerto Rico, Germany, Colombia, Italy, China, Mexico, and Spain are grouped into three thematic sections: diversity of corals, corals in changing environments, and industrial products from corals.

The first section is dedicated to the present studies on coral biodiversity. One of its three chapters carried out by innovative technological tools for coral characterization such as remotely operated vehicles (ROVs) shows the richness of the Mediterranean deep-sea environments. The second chapter describes the cold-water coral habitats in the North Atlantic, and the third chapter shows the use of Raman spectroscopic and modern microscopic techniques to clearly identify the main commercial species of precious corals.

The second section contains a brief overview of the actual status of corals and, in some cases, the resilience state of coral reef systems with the hope that timely information will help managers and decision makers to implement sustainable management measures to alleviate coral detriment. Furthermore, one of the book chapters presents interesting lessons learned from the community-based coral aquaculture and reef rehabilitation program in Puerto Rico, from a multidisciplinary standpoint. The corals will survive, as it is assured in one of the chapters in this section of the book, but their communities will no longer be the same; therefore, the author states that we need to work on the acquisition of scientific knowledge, so that the managers could implement best management practices and reef rehabilitation strategies. In this section, the book also discusses the use of well-maintained coral microcosms to provide a good basis for performing experiments with natural fluctuations.

The last section of the book discusses corals, as a source of new bioactive compounds useful to humans, in fields, namely, new drugs, molecular tools, fine chemicals, cosmetics, nutraceuticals, and agrochemical industries. Some study cases on structure-activity relationship (SAR) as a strategy to maximize the activity initially detected in compounds isolated from octocorals are particularly discussed as a part of the global strategy to make more efficient the process of sustainable production and industrial application of these compounds.

In conclusion, as the book points out, coral reefs are recognized as unique habitats characterized by numerous structuring species, generating complex ecosystems, which act as important oases of biodiversity in the oceans. For these reasons, and due to the fragility of these ecosystems under changes (threats) either natural or anthropogenic very frequently introduced in their environment, it is evident that the key to understanding their future requires an insightful comprehension of the actual status of corals and continuous following up to better understand the causes and effects of the coral deterioration, their survival mechanisms, and their ecological importance. Surely, the knowledge provided and the results proposed by each of the chapters of "Corals in a Changing World" will be useful to design strategies and policies to restore coral reefs and to promote their conservation and an adequate sustainable commercial use.

Finally, the editors express their deep gratitude to all the scientists and contributors who decided to be a part of this important editorial project. It is also important to mention our indebt and appreciation to InTechOpen Science/Open Minds Publisher and Ms. Romina Skomersic, the Publishing Process Manager, for the continuous effort and support in the task of publishing this book.

Emeritus Professor Carmenza Duque Beltran
Universidad Nacional de Colombia
Bogotá, Colombia

Edisson Tello Camacho
Universidad de La Sabana
Chía, Colombia

Section 1

Introduction

Chapter 1

Introductory Chapter: Introduction to Corals in a Changing World

Carmenza Duque Beltran and
Edisson Tello Camacho

Additional information is available at the end of the chapter

http://dx.doi.org/10.5772/intechopen.73868

1. Introduction

1.1. Corals and coral reefs

Corals have taken about 200–300 million years of evolution to reach the way we currently know them. They are mostly polypoid colonial marine invertebrates belonging to the phylum Cnidaria in the subclasses, Hexacorallia (having six tentacles and mesenteries or multiple mesenteries thereof) and Octocorallia (with eight tentacles and eight mesenteries), both included in the class Anthozoa [1, 2]. They are sessile organisms, which have developed unique form of symbiosis to be successful in complex marine environments. This complex biological assemblage is known to be composed of the coral, its endolithic algae (zooxanthellae), and the associated community of microorganisms including bacteria, archaea, viruses, and fungi. The plant host contributes with nitrogenous waste and receives photosynthetic products from the symbiont in return. Additionally, the association between plants and animals also contributes to the brilliant colors of corals.

1.2. Geographical distribution

Those marine animals composed by millions of tiny polyps have evolved in such extraordinary way that they construct through excretion of calcium carbonate, truly underwater cities (coral reefs) being homes for numerous invertebrate and fish species. The number of species harboring these coral reefs is so high that many people consider them as hotspots of biodiversity. Corals are distributed worldwide in the oceans in shallow and deep waters, but reef-building corals are only limited to shallow waters in tropical and subtropical areas; their

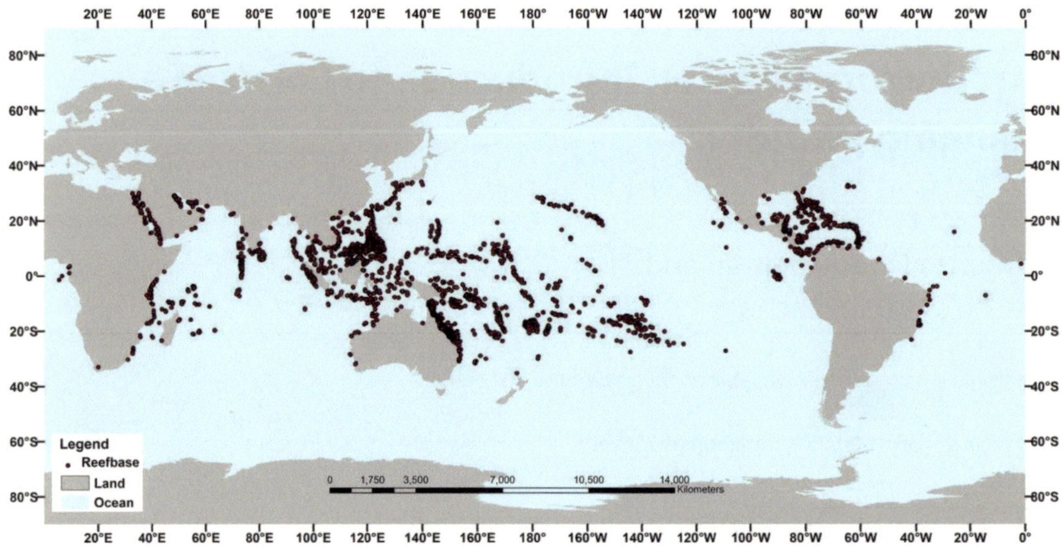

Figure 1. Major coral reef regions of the world as presented in the website: https://aamboceanservice.blob.core.windows.net/oceanservice-prod/education/kits/corals/media/coralreefmap.jpg [accessed 2018-01-08].

symbionts, particularly the photosynthetic algae, called zooxanthella, the main producer of energy, needs light and warm temperatures (generally no lower than 20°C) to provide up to 98% of the nutritional needs of this unique marine organism. In addition, coral reefs are the only living structure to be visible from space (**Figure 1**) being within the jurisdiction of more than 100 countries and occupying more than 600,000 km^2 of tropical oceans.

1.3. Benefits that corals provide and current threats

Furthermore, coral reefs are important ecosystems that harbor many aquatic animals, food sources of superior organisms (they give almost to one quarter of all ocean species, food, and shelter), and commercial and sports fishing for humans. They also provide substantial ecological and economic services to coastal communities. However, they are highly susceptible to natural and anthropogenic threats.

Although in the last two decades, deterioration in coral environments worldwide has been documented, due mainly to the interaction of anthropogenic and climatic stress factors through observation of changes in coral composition, coral bleaching, mass mortality of reef organisms, reduction in the number of coral species, and coral reef coverage, among others; large-scale information on the current state of the reefs is not up to date. As a result, there is no certainty about the actual degree of their status or the degree of recovery or if, on the contrary, the degradation has continued. Nowadays, it is of general consensus that to provide scientific knowledge has become an international priority to make possible their conservation and the reduction of the damages suffered.

http://dx.doi.org/10.5772/intechopen.73868

Thus, not only due to the undeniable importance of corals but also due to its continuous anthropogenic exploitation, climate change, and other threats such as strong El Niño Oscillation events, ocean acidification, and ecological aspects that have caused the increase of algae and cyanobacteria competitors in affected coral reefs, we have considered of strategic importance to present this book. Furthermore, knowledge on corals is now an international priority to implement effective management and proper conservation measures.

Author details

Carmenza Duque Beltran[1]* and Edisson Tello Camacho[2]

*Address all correspondence to: cduqueb@unal.edu.co

1 Universidad Nacional de Colombia, Bogotá, Colombia

2 Universidad de La Sabana, Chía, Colombia

References

[1] Daly M, Brugler MR, Cartwright P, Collins AG, Dawson MN, Fautin DG, France SC, McFadden CS, Opresko DM, Rodriguez E, Romano SL, Stake JL. The phylum Cnidaria: A review of phylogenetic patterns and diversity 300 years after Linnaeus. Zootaxa. 2007;**1668**:127-182

[2] McFadden CS, Sánchez JA, France SC. Molecular phylogenetic insights into the evolution of Octocorallia: A review. Integrative and Comparative Biology. 2010;**50**:389-410. DOI: 10.1093/icb/icq056

Section 2

Coral Diversity

Chapter 2

Cold Temperate Coral Habitats

Lene Buhl-Mortensen and Pål Buhl-Mortensen

Additional information is available at the end of the chapter

http://dx.doi.org/10.5772/intechopen.71446

Abstract

Cold-water coral habitats are constituted by a great variety of anthozoan taxa, with reefs and gardens being homes for numerous invertebrates and fish species. In the cold temperate North Atlantic, some coral habitats such as *Lophelia pertusa* reefs, and *Primnoa*/*Paragorgia* dominated coral gardens occur on both sides of the Atlantic over a wide latitudinal range. Other habitats, as some dominated by species of Isididae and Chrysogorgidae seem to have a more local/regional distribution. In this chapter, we describe the habitat characteristics of cold-water coral reefs, soft and hard-bottom coral gardens, and sea pen meadows with their rich associated fauna illustrated with numerous photos.

Keywords: cold-water corals, associated fauna, coral garden, coral reef, Scleractinia, Alcyonacea, gorgonians, Antipatharia

1. Main subtitles of chapter

- Cold water coral reefs; *Lophelia*
- Hard-bottom coral gardens:

  Sections cover different communities with key coral species (e.g., Alcyonacea (Gorgonians), Antipatharia) and their associated fauna.
- Soft-bottom coral gardens:

  Sections cover different communities with key coral species (e.g., Alcyonacea (Gorgonians), Antipatharia) and their associated fauna.
- Sea pen meadows:

  Two sections covering shallow and deep water meadows and their associated fauna.

2. Introduction

"Coral reefs," "Coral gardens" (reef-forming Scleractinian corals and aggregations of gorgonians, black corals, and sea pens) and "sea pen and burrowing megafauna communities" are habitats classified by the Oslo Paris Convention for the Protection of the Marine Environment of the Northeast Atlantic (OSPAR) as "Threatened and/or declining" [1]. Sea pen and burrowing megafauna communities are also of key conservation importance as defined under Annex V of the 1992 OSPAR Convention [2, 3]. Coral gardens are sensitive to physical disturbance impacts caused by bottom trawling and activities related to the petroleum industry [4–6]. Bottom trawling is known to be one of the most destructive ways of fishing and causes reductions in habitat complexity, changes in species composition, and reductions in biodiversity [7–9]. These threats highlight that it is crucial to assess the ecological importance of these deepwater communities, to develop sound scientific advice for management of cold-water ecosystems [10, 11].

Corals include species from various taxonomic groups (including Scleractinia, Zoanthidea, Antipatharia, Gorgonians, Pennatulacea, and Stylasteridae). According to Roberts et al. [12], 65% of a total of 5160 coral species occur deeper than 50 m. Several of the groups (Gold corals, Antipatharia, Octocorallia, and Stylasterida) are represented by more species in deepwater than shallow.

Studies on cold-water corals have expanded dramatically in recent years and their role as habitat providers for a rich fauna has been shown in several studies [11–18].

Cold-water corals are found around the world, most commonly in the depth range of 200–1500 m [19–33]. Shallower, upper depth limits (up to 40 m) are found in Norway, New Zealand, Chile, and British Columbia and demonstrate that it is not the depth per se, but the vertical range of intermediate and deep water masses that controls the bathymetric distribution of these corals.

Corals typically create habitats elevated above the surrounding seabed (up to several meters) and occur on bottoms with mixed substrata in areas with relatively strong currents (**Table 1**). They offer a variety of microhabitats with different current speeds, food sources, and substrates. Most corals have an arborescent morphology with branches reaching out of the near-bottom boundary into the faster flowing water above. Corals have a complex 3-D architecture and provide substrata of different ages, due to their continuous growth and decay. Sheltered cavities within a colony often contain organic-rich sediments, while the outer parts provide a high water flow with elevated rates of food supply and little sedimentation.

The relative abundance of food at the shelf-slope transition argues against food limitation in this zone and focuses attention upon physical factors [34] to understand the distribution of cold-water corals.

Internal waves on continental margins can induce resuspension and even an upward transport of particles in periods of strong wind [35]. Here, biological structures such as corals can provide shelter and protection for some organisms against strong currents and predators and, at the same time, offer a reliable supply of detrital food within their interstices. Higher structures that reach into laminar currents above the more turbulent near-bottom currents may provide other food sources (e.g., zooplankton) [15].

Coral habitats	Habitat provision	Main key species	Associated fauna	Longevity of key species (years)
Cold-water coral reefs	Hard substrate, shelter, Elevated feeding position	*Lophelia pertusa, Oculina varicosa, Madrepora oculata, Solenosmilia variabilis, Gonicorella dumosa, Enallopsammia profunda*	Mobile: Fish, crustaceans, brittle stars, polychaetes. Sessile: Actiniarians, gorgonians, sponges, bryozoans, polychaetes	>1000
Hard-bottom coral garden	Shelter, elevated feeding position	*Paragorgia arborea, Primnoa* spp., *Paramuricea* spp., *Callogorgia verticillata, Antipathes* spp., *Bathypathes* spp., *Leiopathes* spp.	Mobile: Fish, crustaceans, brittle stars. Sessile: Actiniarians, parasitic crustaceans, polychaetes	50–300
Soft-bottom coral garden	Elevated feeding position, Shelter	*Acanella arbuscula, Isidella lofotensis, Radicipes gracilis*	Mobile: Crustaceans, polychaetes	50–100
Sea pen meadow	Elevated feeding position, shelter	*Funiculina quadrangularis, Kophobelemnon steølliferum, Virgularia mirabilis, Pennatula phosforea, Halipteris* spp., *Umbellula encrinus*	Sessile: Crustaceans, brittle stars	15–80

Table 1. Coral habitats.

In many ways, the biological habitat structures provide more food particles and other vital resources compared to the framing habitat. Often propagules and larvae are present in the deep-sea demersal plankton, but suitable firm substratum is lacking. Thus, organisms that provide an elevated position on a stable substratum represent a scarce habitat, contributing substantially to the species richness of their respective environments [15, 36].

It is known that coral habitats house a large assemblage of crustaceans, mollusks, ophiuroids, sponges, and fish [11, 15, 16, 37–42]. These organisms can use corals as a substrate for attachment and shelter [43], for feeding [44, 45] and parasitism [16]. The mobile fauna is particularly difficult to document [11, 15, 17, 41], but with the recent development of underwater video equipment, it has become possible to inspect coral colonies for even rather small associated organisms.

3. Cold-water coral reefs *Lophelia pertusa*

L. pertusa is common along the European margin and develops reefs in several places where the environmental conditions are right. The Northeast Atlantic can be divided into three main reef provinces based on geography and environmental similarities: (1) The Nordic occurrences, including Sweden, Norway, Faroe Island, and Iceland, (2) Irish-British margins, and (3) Franco-Iberian margin [46]. *L. pertusa* is also found along the mid-Atlantic ridge, but living reefs have not been confirmed and live coral is only represented by small scattered colonies [47]. However, large patches of coral rubble indicate proliferation of reefs occurred in the past. Changing ocean currents resulting from the disappearing glaciers during the last deglaciation probably changed the environmental settings in a negative way for the coral.

On the North American margin, reefs are much less common than at the other side of the Atlantic. Larger reef structures in the eastern USA are only found from North Carolina and southward, and into the Mexican Gulf. The reason for this is not clear, but temperature variation caused by North Atlantic Drift, better known as the Gulf Stream, may prevent long-term stable conditions for reefs to develop in the northeastern USA-Canadian margin. A single reef occurrence in the mouth of the Laurentian Channel in Atlantic Canada is an exception to this [48]. This reef occurs in the southward flowing warm water that has passed the southern coast of Greenland after branching off the Gulf Stream south of Iceland. The recently discovered reef off southwest Greenland [49] occurs in the same water.

3.1. Environment

L. pertusa can use all kinds of hard substrate as a foundation, even human-made structures such as legs of oil platforms in the North Sea. In Norwegian waters, *Lophelia* reefs are found on finer sediments mixed with gravel as well as directly on bedrock. Occurrences on bedrock are found in fjords and coastal areas. Further from the coast, on the continental shelf and slope, bedrock is rare and morainic material with gravel and boulder on banks and edges of troughs is the most common reef foundation substrate [50]. Strong tidal currents, together with seasonal changes in temperature and wave energy, influence the habitats and generate large sand waves [51]. In these settings, the large biogenic structures formed by the reef-building *Lophelia* are found in the upper range [28, 32, 52, 53].

3.2. Morphology

At a local scale, the morphology of organisms shapes the environment by modifying the hydrodynamics and providing shelter, pockets with trapped particles, and other microhabitats. A coral reef can be defined as an aggregation of coral skeletons completely covering the substrate underneath (**Figure 1**). Colonial scleractinians need hard substrate for settlement. This substrate can be a shell or a pebble, and as soon as one colony is present, it provides a new hard substrate for subsequent colonization. Coral colonies may grow at one site for hundreds of years. During that time, it transforms the seabed to a complete cover of coral skeleton fragments through alternating growth, death, and fragmentation. When the bottom beneath the colonies consists of a layer of dead skeleton, the area can be termed a "coral reef." Corals growing on a steep surface may not develop reefs, but are rather called coral gardens. When the coral grows large, and break up, due to their own weight, skeletal fragments will not accumulate at the site but fall deeper, outside the favorable environment.

Cold-water coral reefs typically have a circular or elongated outline with a maximum length of c. 1000 m. At the Norwegian continental shelf, it is estimated that there are around 6000 *Lophelia* reefs [52]. Many of these reefs are several 100 m long and occur in clusters (reef complexes) up to 35 km long (Røst reef) [54]. Their area, however, covers less than 0.1% of the total

http://dx.doi.org/10.5772/intechopen.71446

Figure 1. The cold-water coral reef habitat. (A) Summit of cold-water coral reef. Live *Lophelia pertusa* (white branches) together with a cod. (B) Lower part of *Lophelia* reef. Typically dominated by dead *Lophelia* being a substrate for gorgonian corals and other associated organisms. (C) Richness of old parts of a *Lophelia* reef with gorgonians that are overgrown with hydroids and sponges. Two redfish are taking shelter below the old *Paragorgia*. Red laser dots indicates a scale of 10 cm.

area of the depth zone where they occur. The reefs commonly have vertical zones, with living coral at the top and skeletal fragments at increasing stages of decay toward the bottom of the reef [55]. The reefs may have different shapes depending on currents and seabed topography.

3.3. Provision of habitat

Reefs represent large and complex structures that significantly increase habitat heterogeneity. Framing habitats is varied and offers a wide range of substrates, but the complexity of these large structures represents an increased variety of microhabitats that elevate local species diversity [16, 17, 42, 56]. The associated organisms of cold-water coral reefs are comprised mainly of species that occur on other hard-bottom substrates, and their relationships with the coral are facultative.

Three successive habitat zones can be observed when crossing a reef, namely (1) the coral rubble zone, bordering the framing habitats, consists of small pieces of skeleton, followed by (2) the coral block zone dominated between the foot and the top of the reefs, with mixed coral substrates dominated by larger dead blocks, which lead to (3) the top of the reef where live colonies proliferate.

Within coral colonies of the live reef, four distinct microhabitats can be recognized, namely (i) live coral tissue, (ii) surfaces of dead corals often slightly covered with detritus, (iii) cavities inside coral skeletons, and (iv) open space between coral branches.

Most coral-associated species are facultative symbionts without a direct relationship with the living corals and can survive in similar microhabitats on bottoms without corals [56]. The endosymbionts (mainly copepods) are an exception to this. Rather than the presence of live coral, it is the hard substrate, and thus the diverse microhabitats provided by dead coral skeletons, that facilitates the high biodiversity associated with reef-forming, cold-water corals [56] (and references therein) [42].

3.4. Associated fauna

There is a great species diversity of animals found together with cold-water, reef-forming corals [32, 56–59]; however, there are no examples of associated species with an obligate relationship between dead coral skeletons. The highest diversity of associated species is found in the zone with dead coral block [56, 59]. Here, the skeletons are exposed and occur with a higher three-dimensional complexity than in the rubble zone surrounding the reefs.

The live tissue of cold-water scleractinians seems to prevent attachment of sessile epibiotic species. Even among the few species that are commonly found intimately associated with living coral polyps, there are few examples of obligate relationships [15] (**Figure 2**). However, many of these species are rarely found in other habitats. The polychaetes *Eunice norvegica* and *Harmothoe oculinarum* are two good examples: *E. norvegica* lives in a close relationship with *L. pertusa*. The coral embed the parchment-like tube of the polychaete in its skeleton. After some years of skeletal growth, the tube of *E. norvegica* may contain several openings, each one close to a polyp, where it can search for food spills [60]. *E. norvegica* spends time searching for food, cleaning the coral's surface for organic particles, and removing organisms invading its territory or over-growing the coral [59]. The strategically located tube openings allow easy access to food trapped by *L. pertusa*. The polynoid polychaete *H. oculinarum* is a commensal that can be found inside the tube of *E. norvegica*. The highest diversity of associated species is found in the zone with dead coral block [56, 59]. Here, the skeletons are exposed and occur with a higher three-dimensional complexity than in the rubble zone surrounding the reefs.

Figure 2. Close associates with *Lophelia pertusa*. A: Living part of a *Lophelia* reef with sponges. B: *Eunice norvegica* is a common polychaet living intimately with the polyps of *Lophelia* that rarely occur in other habitats. C: The scleractinian *Madrepora oculata* with the basket star *Gorgonocephalus* sp. (lower right). D: The squat lobster *Munidopsis serricornis* on the branches of *Lophelia*.

4. Hard-bottom coral gardens

4.1. Alcyonarian corals

The alcyonarian corals (soft corals) utilize a wide range of substrates, including semiconsolidated mudstone. Alcyonarian corals, in particular Nephtheidae, have a wide geographical and bathymetric distribution [33, 47]. The colonies are rather small (<30 cm) but may occur in relatively high densities (>500 colonies per 100 m^2) [33]. The extent of patches of this coral group seems to be larger than for gorgonians. There are few known relationships documented with other invertebrates. The association between ophiuroids and nephtheids has been reported at various locations [61–63]. Mortensen [62] observed juveniles of the basket star, *Gorgonocephalus eucnemis*, parasitic on *Eunephtia*, and Fedotov [61] reports juveniles of *Gorgonocephalus* on colonies and within the polyps of *Drifa glomerata*. The foraminiferan *Planispirinoides bucculentus* has been observed on *Duva florida* off Nova Scotia [40]. Compared to the rigid structures of scleractinians and gorgonians, the soft coral represents an unstable substrate not suitable for attached species.

4.2. Gorgonian corals

Gorgonian corals provide habitats within and between colonies, when they occur in stands. The density of colonies within stands is typically higher for smaller species than for larger species [64]. This is illustrated by coral gardens off the Aleutian Islands (Alaska) [64–66] and Nova Scotia (Canada) [64], where smaller gorgonians and stylasteridae are found in densities of up to 200–400 colonies per 100 m^2. The larger gorgonian *Paragorgia arborea* occur less dense with a maximum of 49 colonies per 100 m^2. The gorgonian stands extend horizontally from 10 to 100 m [64]. In addition to accessing faster-flowing water above the bottom, colonies orient themselves toward the main current to maximize the amount of water passing the polyps [67]. The advantages of this morphologically enhanced feeding may also be utilized by suspension feeding and epizoic animal attached to the colony. In addition, the suspension feeders may also derive nutrition from detritus or microorganisms commonly found trapped in the mucus secreted by the gorgonians [13].

4.2.1. *Associated fauna*

The gorgonian-associated fauna is dominated by crustaceans, particularly amphipods. This is true for tropical gorgonians [68], which also host parasitic copepods, but deepwater gorgonian taxa exhibit a richer echinoderm fauna, including ophiuroids adapted to capturing particles in the elevated flows provided by the corals [39, 69]. The cold-water gorgonians have fewer decapods (crabs and shrimps) and gastropods than warm-water corals [13, 68]. In a study focusing on the associated fauna of cold-water gorgonians, Buhl-Mortensen and Mortensen [17] found that only a few specialized and obligate symbionts are connected to the live parts of corals. Among these are three highly specialized parasitic copepod species, presumably feeding on their hosts' coenenchyme [13]. The number of obligate symbionts is higher for gorgonians than for antipatharians, alcyonarians, and scleractinians. In their review of available literature, Buhl-Mortensen and Mortensen [17] reported 24 species having obligate relationships with 21 gorgonian host species. For comparison, only seven obligate symbionts have been reported for four scleractinians.

4.2.2. *Habitat provision*

Primnoa resedaeformis and *P. arborea* are the most abundant and widely distributed large gorgonians in the North Atlantic [25, 30, 33, 64, 70], where they can form stands or "coral gardens" (**Figure 3**). They are among the largest cold-water gorgonian corals, reaching a height of 50–250 cm. These corals offer two different microhabitats for associated species: (i) the clean and living surface of coral tissue in the younger parts of the colony and (ii) pockets of detritus and exposed skeleton in the older parts.

4.2.3. *Associated fauna*

P. arborea and *P. resedaeformis* host a rich fauna, dominated by suspension feeders using the coral as substratum or mobile animals using it as a refuge against predators [15–17] (**Figure 4**). The fauna composition differs for the two corals, but consists mainly of species also occurring in other habitats. However, a few highly specialized parasites have been identified associated with each of the species [15–17]. The abundance and species richness of the associates are significantly correlated with the host morphology, such as number of

Figure 3. Hard-bottom coral gardens. Three species that form hard-bottom coral gardens are the "bubble gumme" coral *Paragorgia arborea*, on top, *Primnoa resedaeformis*, in the middle, and *Paramuricea placomus*. *P. arborea* can reach several meters in height and widths, while the two other species normally reaches a height of 20–70 cm.

branches and area of exposed skeleton. Even though the cold-water gorgonians support fewer obligate associations, several of the associated species are rare in other habitats and seem to prefer gorgonian species [15].

Close inspection of *P. arborea* using video [71] shows that crustaceans are the most common group of associates. Amphipods belonging to the family Stegocephalidae were found on 26% of the colonies, and other common crustacean associates were shrimps and hermit crabs.

Figure 4. Habitats hard-bottom coral gardens. The larger sea trees can form forest-like habitats (upper photo). Here, we find a mixture of coral species. Shrimps are hiding between the branches, and the large "Basket star" *Gorgonocephalus* sp. is feeding from their elevated position on the branches of *Paragorgia arborea*. (photo lower left). It is the oldest parts of the colonies that hosts the riches associated fauna (photo lower right). Her we often find hydroids, crinoids and even hermit crabs. Bellow the old colony is a sponge.

Galls of the endoparasitic copepod *Gorgonophilus canadensis,* which is highly adapted to its host *P. arborea* [16], has been observed on both sides of the North Atlantic. In addition to parasitic copepods, the ophiuroid *Gorgonocephalus* is one of very few examples of host-specific associates. It uses the elevated position offered by *P. arborea* to collect particles (detritus or plankton) from the water passing by.

4.3. Antipatharia

North Atlantic black corals (Antipatharia) appear to be restricted to open ocean areas with *Antipathes erinaceus*, *Distichopathes* sp., *Phanopathes* sp., and *Stauropathes punctata* only recorded on Josephine seamount, the Azores, and Cape Verde Islands [72]. Around the Azores, *Antipathella wollastoni* is the most common species in deep infralittoral and circalittoral grounds (>20 m) and is known to form dense stands. In bathyal areas, the black corals, *Leiopathes* spp., are common between 200 and 600 m and can grow to a height of 2.5 m [73, 77].

http://dx.doi.org/10.5772/intechopen.71446

4.3.1. Associated fauna

Black corals have many associated epifauna, most commonly serpulid worms, bryozoans, and ascidians, and the number of epibionts of Antipathella subpinnata shows an increase with the age of the corals and decrease with the depth [74].

5. Soft-bottom coral gardens

5.1. Gorgonian corals

Most gorgonians are confined to hard bottoms, except for some species of Isididae and Chrysogorgiidae, such as *Isidella lofotensis, Acanella arbuscula,* and *Radicipes gracilis,* which attach to sandy and muddy bottoms with root-like holdfasts (**Figure 5**).

Figure 5. Soft-bottom coral gardens. The bamboo coral *I. lofotensis* is in the Norwegian Hardangerfjord and in the Norwegian through forming coral gardens at a depth of 200–300 m depth (upper photo). This species is only known from Norwegian waters. In deep and cold North Atlantic waters the gorgonian *Radicipes* "pigtail coral" forms coral gardens.

Soft-bottom bamboo coral gardens are found on both sides of the North Atlantic, represented by two species with different geographic ranges. The geographic barrier represented by the Greenland-Scotland Ridge separates the Nordic Seas and the North Atlantic. To the south of this, *A. arbuscula* occur sometimes with sea pens and the solitary scleractinian cup coral *Flabellum alabastrum* [33, 75], while on the other side of the barrier, in the North Sea and the Norwegian Sea, *I. lofotensis* occur in troughs and fjords [76]. In the western North Atlantic, *A. arbuscula* is found in the upper and middle bathyal (200–1000 m depth), while in the eastern North Atlantic, it occurs deeper (1800–2700 m depth) in the lower bathyal zone [69].

5.1.1. Associated fauna

In the western North Atlantic, the brittle *star Ophiomuseim lymani* is often found on *A. arbuscula* and assemblage. From the eastern side, Buhl-Mortensen and Mortensen [40] found that the polynoid polychaeta *Eunoe spinulosa* was strongly associated to this coral. Very little is known about the associated fauna of *Isidella* and *Radicipes*.

6. Sea pen medows

Sea pens are slender anthozoans reaching 0.1–2 m above the bottom accessing the elevated bottom currents. They provide predation shelter and good position for particle collection away from the slower current in the near-bottom boundary layer. Although the number of studies is limited, sea pens appear to have fewer associated organisms compared to scleractinians and gorgonians. *Funiculina quadrangularis* is a species with greatest conservation importance in the greater North Sea and Celtic Sea areas [78]. It can become a little more than 2 m tall, with approximately one quarter of the lower part of the structure embedded in the sediment [78]. Predators on sea pens include nudibranchs, which have been observed preying on sea pen polyps. The nudibranch *Armina loveni* is a specialized predator on the sea pen *Virgularia mirabilis*. It is infrequently recorded but known to occur from Norway to Western France. In Puget Sound (western USA), a related species, *Armina californica,* is one of the predators on *Ptilosarcus gurneyi* [79]. Many specimens of *V. mirabilis* lack the uppermost part of the colony, a feature that has been attributed to predation by fish.

6.1. Associated fauna

The associated fauna of sea pens is poor compared to gorgonian corals. In a study of >1000 sea pens from Norway [71], only 4% of the colonies had fauna on them, but 15% had organisms sitting near the colony. The squat lobster *Munida* sp. was found close to 8% of the 584 *Kophobelemnon stelliferum* colonies studied (**Figures 6** and **7**). It appears to use the sea pen as a base station for scavenging, active hunting, and sheltering against predators. Sea pens

Figure 6. Sea pen meadows. *K. stelliferum*, upper photo, is together with *Pennatula phosphorea*, two common and relatively small sea pens (10–20 cm) forming "sea pen meadows" in the North Atlantic. There are few associates living on them, but organisms are often found to hide below a sea pen, likely as a protection against predators. Below *K. stelliferum*, we see two *Munida* squat lobsters and sitting on the *P. phosphorea* is a shrimp and an ophiuroid.

have stinging cells and often emits light that could scare away the potential predators of the organisms seeking shelter under these [80]. The only associated species found on *F. quadrangularis* was the ophiuroid *Asteronyx loveni* that has a close relationship with its host that is also reported from the west coast of Scotland [81]. The sea pen provides a suitable feeding platform in an elevated position for particle collection away from the slower current in the near-bottom boundary layer [39]. *A. loveni* catches small pelagic animals, mainly copepods, for food, but polyps and mucus of from sea pens have also been reported as stomach content [39]. It has been suggested by Buhl-Mortensen and Mortensen [40] that the relationship with the host could be mutualistic as, *A. loveni* could keep colonies clear of sediment and therefore, making them less vulnerable to smothering. Associated fauna has only in very few occasions been found on the sea pens *Pennatula phosphorea* and *V. mirabilis* [71].

Figure 7. Sea pen meadows, *Funiculina quadrangularis*, upper photo, and *Umbellula encrinus* are among the largest sea pens (1–2.5 m) in the North Atlantic. They form sea pen meadows at larger depths and firmer substrate than the smaller sea pens. *U. encrinus* occur at 900 m in arctic waters off Norway. These large sea pens have few associated organisms, but there is a strong and likely mutualistic relation between the ophiuroid *Asteronyx loveni* and *F. quadrangularis*. In upper photo (a compilation of three photos), we see how the ophiuroid stretching outs its arms uses its elevated position to catch particles from the water.

7. Conclusion

Clearly, cold-water coral habitats in the North Atlantic represent several different ecosystems, with different species compositions and habitat characteristics. Coral garden is a heterogeneous habitat covering contrasting environments and a wide range of anthozoan taxa. A main division of this habitat relating to substrate (hard vs. soft seabed) is relevant to provide better consistency of habitat definitions. Further subdivisions are presented in this chapter. Main problem of management of cold-water coral habitats is still a lack of knowledge. Mapping of these habitats cannot rely only on bycatch records from the fishing fleet. Directed mapping of identified priority areas should be carried out before new industries move into deeper oceanic waters. Climate change (increased temperature, changing current patterns, and other indirect effects) represents a pressure that could compromise the coral livelihood differentially in different areas. There is a risk that the combined effect of human impact and climate change will cause greater negative effects in some places than anticipated today.

Author details

Lene Buhl-Mortensen* and Pål Buhl-Mortensen

*Address all correspondence to: lenebu@imr.no

Institute of Marine Research, Bergen, Norway

References

[1] OSPAR. Descriptions of habitats on the initial OSPAR list of threatened and/or declining species and habitats. OSPAR Convention for the Protection of the Marine Environment of the NorthEast Atlantic. 2004;**7**:7

[2] Convention for the protection of the marine environment of the north-East Atlantic (OSPAR Convention) of 22nd September 1992 Entered into Force on 25th March 1998

[3] Curd A. Background Document for Seapen and Burrowing Megafauna Communities. OSPAR comission publication, Biodiversity Series. 2010;**481**:26

[4] Troffe PM, Levings CD, Piercey GE, Keong V. Fishing gear effects and ecology of the sea-whip (Halipteris willemoesi (Cnidaria: Octocorallia:Pennatulacea)) in British Columbia, Canada: Preliminary observations. Aquatic Conservation: Marine and Freshwater Ecosystems. 2005;**15**:523-533

[5] Malecha PW, Stone RP. Response of the seawhip *Halipteris willemoesi* to simulated trawl disturbance and its vulnerability to subsequent predation. Marine Ecology Progress Series. 2009;**388**:197-206

[6] Tyler-Walters H, Rogers SI, Marshall CE, Hiscock K. A method to assess the sensitivity of sedimentary communities to fishing activities. Aquatic Conservation: Marine and Freshwater Ecosystems. 2009;**19**:285-300

[7] Watling L, Norse EA. Disturbance of the seabed by mobile fishing gear: A comparison to forest clear cutting. Conservation Biology. 1998;**12**:1180-1197

[8] Chuenpagdee R, Morgan LE, Maxwell SM, Norse EA, Pauly D. Shifting gears: Assessing collateral impacts of fishing methods in US waters. Frontiers in Ecology and the Environment. 2003;**1**:517-524

[9] Buhl-Mortensen L, Ellingsen KE, Buhl-Mortensen P, Skaar KL, Gonzalez-Mirelis G. Trawling disturbance on megabenthos and sediment in the Barents Sea: Chronic effects on density, diversity, and composition. ICES Journal of Marine Science. 2016;**73**(Supplement 1):98-114

[10] Greathead CF, Donnan DW, Mair JM, Saunders GR. The seapens *Virgularia mirabilis, Pennatula phosphorea* and *Funiculina quadrangularis*: Distribution and conservation issues in Scottish waters. Journal of the Marine Biological Association of the UK. 2007;**87**:1095-1103

[11] Buhl-Mortensen L, Vanreusel A, Gooday AJ, Levin LA, Priede IG, Buhl-Mortensen P, Gheerardyn H, King NJ, Raes M. Biological structures as a source of habitat heterogeneity and biodiversity on the deep ocean margins. Marine Ecology-An Evolutionary Perspective. 2010;**31**:21-50

[12] Roberts JM, Wheeler AJ, Freiwald A, Cairns SD. Cold-water Corals: The Biology and Geology of Deep-Sea Coral Habitats. Cambridge University Press. 2009; 334 pp

[13] Patton WK. Studies on animal symbionts of gorgonian coral *Leptogorgia virgulata* (Lamarck). Bulletin of Marine Science. 1972;**22**:419-431

[14] Krieger KJ, Wing BL. Megafauna associations with deep water corals (*Primnoa* spp.) in the Gulf of Alaska. Hydrobiologia. 2002;**471**:83-90

[15] Buhl-Mortensen L, Mortensen PB. Crustaceans associated with the deep-water gorgonian corals *Paragorgia arborea* (L., 1758) and *Primnoa resedaeformis* (Gunn., 1763). Journal of Natural History. 2004;**38**:1233-1247

[16] Buhl-Mortensen L, Mortensen PB. *Gorgonophilus canadensis* n. Gen.,sp. (Copepoda: Lamippidae), a gall forming endoparasite in the octocoral *Paragorgia arborea* (L., 1758) from the northwest Atlantic. Symbiosis. 2004;**37**:155-168

[17] Buhl-Mortensen L, Mortensen PB. Distribution and diversity of species associated with deep-sea gorgonian corals off Atlantic Canada. In: Freiwald A, Roberts JM, editors. Cold-water Corals and Ecosystems. Berlin Heidelberg: Springer-Verlag; 2005. p. 849-879

[18] Carvalho S, Curdia J, Pereira F, Guerra-Garcia JM, Santos MN, Cunha MR. Biodiversity patterns of epifaunal assemblages associated with the gorgonians *Eunicella gazella* and *Leptogorgia lusitanica* in response to host, space and time. Journal of Sea Research. 2014;**85**:37-47

[19] Broch H. Die Alcyonarien des Trondhjemsfjordes I. Alcyonacea. Det Kongelige Norske Videnskabers Selskabs Skrifter. 1912;**1911**(7):1-48

[20] Broch H. Die Alcyonarien des Trondhjemsfjordes II. Gorgonacea. Det Kongelige Norske Videnskabers Selskabs Skrifter. 1912;**1912**(**2**):1-48

[21] Broch H. Riffkorallen im Nordmeer einst und jetzt. Die Naturwissenschaften. 1922;**37**:1-3

[22] Broch H. Oktokorallen des nördlichsten Pazifischen Ozeans. Det norske Videnskaps-Akademi i Oslo I. Matematisk-Naturvidenskabelig klasse. 1935;**1935**:1-53

[23] Jungersen HFE. The Alcyonaria, Antipatharia and Madreporaria. Conceptus Faunae Groenlandicae. Meddelelser om Grønland. 1915;**23**:1156-1212

[24] Jungersen HFE. Alcyonarian and Madreporarian corals in the Museum of Bergen, collected by Fram-expedition 1898-1900 and by the 'Michael Sars' 1900-1906. Bergen museums aarbok 1915-1916. Naturvidenskabelig Række. 1917;**6**:44

[25] Madsen FJ. Octocorallia: Stolonifera–Telestacea–Xeniidea–Alcyonacea–Gorgonacea. The Danish Ingolf-Expedition. 1944;**13**:1-65

[26] Carlgren O. Polyppdyr (Coelenterata) III. Koraldyr. Danmarks fauna 51. Copenhagen: GEC Gads; 1945

[27] Hecker B, Blechschmidt G, Gibson P. Final Report — Canyon Assessment Study in the Mid- and north Atlantic Area of the U.S. Outer Continental Shelf. (Contract No. BLM AA551-CT8-49) U.S. Washington DC: Department of Interior, Bureau of Land Management; 1980

[28] Zibrowius H. Les Scle'ractiniaires de la Me'diterrane'e et de l'Atlantique nord-oriental. Memoires de l'Institut oceanographique. 1980;**11**:226

[29] Genin A, Dayton PK, Lonsdale PF, Speiss FN. Corals on seamount peaks provide evidence of current acceleration over deep-sea topography. Nature. 1986;**322**:59-61

[30] Tendal OS. The North-Atlantic distribution of the octocoral *Paragorgia aborea* (L,1758) (Cnidaria,Anthozoa). Sarsia. 1992;**77**:213-217

[31] Cairns SD. Scleractinia of the temperate North Pacific. Smithsonian Contributions to Zoology. 1994;**557**:150

[32] Rogers AD. The biology of *Lophelia pertusa* (Linnaeus, 1758) and other deep-water reef forming corals and impacts from human activities. International Review of Hydrobiolgy. 1999;**844**:315-406

[33] Mortensen PB, Buhl-Mortensen L, Gordon DC Jr. Distribution of deep-water corals in Atlantic Canada. In: Proceedings of the 10th International Coral Reef Symposium. Tokyo, Japan: Japanese Coral Reef Society; 2006:1832-1848

[34] Carney RS. Zoonation of deep biota on continental margins. Oceanography and Marine Biology: An Annual Review. 2005;**43**:211-278

[35] Bonnin J, van Raaphorst W, Brummer G-J, van Haren H, Malschaert H, Intense mid-slope resuspension of particulate matter in the Faeroe–Shetland Channel: Short-term deployment of near-bottom sediment traps. Deep-Sea Research I. 2002;**49**:485-1505

[36] Beaulieu SE. Life on glass houses: Sponge stalk communities in the deep sea. Marine Biology. 2001;**138**:803-817

[37] Storm V. Oversigt over Throndheimsfjordens fauna (med et kort). Trondhjems Biologiske Station, Meddelelser fra stationsanleggets arbeidskomite. Trondheim: H Moe's Bog & Accidenstrykkeri; 1901:20

[38] Strømgren T. Vertical and horizontal distribution of *Lophelia pertusa* (Linné) in Trondheimsfjorden on the west coast of Norway. Det Kongelige Norske Videnskabers Selskabs Skrifter. 1971;**6**:1-9

[39] Fujita T, Ohta S. Photographic observations of the life-style of a deep-sea ophiurid *Asteronyx loveni* (Echinodermata). Deep Sea Research. 1988;**35**:2029-2044

[40] Buhl-Mortensen L, Mortensen PB. Symbiosis in deep-water corals. Symbiosis. 2004; **37**:33-61

[41] Baillon S, Hamel JFA. Diversity distribution and nature of faunal associations with deep-sea pennatulacean corals in the Northwest Atlantic. PLoS ONE. 2014;**9**(11):e111519. https://doi.org/10.1371/journal.pone.0111519

[42] Buhl-Mortensen L, Serigstad B, Buhl-Mortensen P, Olsen MN, Ostrowski M, Błażewicz-Paszkowycz M, Appoh E. Structure and megafaunal community of a large Lophelia reef on the Ivorian-Ghanaian margin (the Gulf of Guinea). Deep Sea Research II. 2017;**137**:148-156

[43] Vytopil E, Willis BL. Epifaunal community structure in *Acropora* spp. (Scleractinia) on the great barrier reef: Implications of coral morphology and habitat complexity. Coral Reefs. 2001;**20**:281-288

[44] Burkepile DE, Hay ME. Predator release of the gastropod *Cyphoma gibbosum* increases predation on gorgonian corals. Oecologica. 2007;**154**:167-173

[45] Garcia-Matucheski S, Muniain C. Predation by the nudibranch *Tritonia odhneri* (Opisthobranchia: Tritoniidae) on octocorals from the South Atlantic Ocean. Marine Biodiversity. 2011;**41**:287-297

[46] Wheeler AJ, Beyer A, Freiwald A, de Haas H, Huvenne VAI, Kozachenko M, Olu-Le Roy K, Opderbecke J. Morphology and environment of cold-water coral carbonate mounds on the NW European margin. International Journal of Earth Sciences. 2007;**96**:37-56

[47] Mortensen PB, Buhl-Mortensen L, Gebruk AV, Krylova EM. Occurrence of deep-water corals on the mid-Atlantic ridge based on MAR-ECO data. Deep-Sea Research II. 2008;**55**:142-152

[48] Buhl-Mortensen P, Gordon Jr. DC, Buhl-Mortensen L, Kulka DV. First Description of a Lophelia pertusa reef complex in the Atlantic Canada, Deep Sea Deep Sea Research I. 2017;**126**:21-30

[49] Kenchington E, Yashayaev I, Tendal OS, Jørgensbye H. Water mass characteristics and associated fauna of a recently discovered *Lophelia pertusa* (Scleractinia: Anthozoa) reef in Greenlandic waters. Polar Biology. 2017;**40**(2):321-337

[50] Mortensen PB, Buhl-Mortensen L, Dolan M, Dannheim J, Kröger K. Megafaunal diversity associated with marine landscapes of northern Norway: A preliminary assessment. Norwegian Journal of Geology. 2009;**89**:163-171

[51] Reynaud J-Y, Tessier B, Berne´ S, Chamley H, Debatist M. Tide and wave dynamics on a sand bank from the deep shelf of the Western Channel approaches. Marine Geology. 1999;**161**:339-359

[52] Mortensen PB, Hovland MT, Fosså JH, Furevik DM. Distribution, abundance and size of *Lophelia pertusa* Coral reefs in mid-Norway in relation to seabed characteristics. Journal of the Marine Biological Association of the UK. 2001;**81**:581-597

[53] Mortensen PB, Buhl-Mortensen L, Dolan M, Bellec V, Hassel A, Bogetveit FR Seascape description of an unusual coral reef area off Vesterålen, Northern Norway. 4th International Symposium on Deep-Sea Corals, Wellington, New Zealand; 1-5 Dec 2008. Poster

[54] Fosså JH, Lindberg B, Christensen O, Lundälv T, Svellingen I, Mortensen PB, Alvsvåg J. Mapping of Lophelia reefs in Norway: Experiences and survey methods. In: Freiwald A, Roberts JM, editors. Cold-Water Corals and Ecosystems. Berlin: Springer-Verlag; 2005. p. 359-391

[55] Mortensen PB, Hovland M, Brattegard T, Farestveit R. Deep water bioherms of the scleractinian coral *Lophelia pertusa* (L.) at 64° N on the Norwegian shelf: Structure and associated megafauna. Sarsia. 1995;**80**:145-158

[56] Mortensen PB, Fosså JH Species diversity and spatial distribution of invertebrates on *Lophelia* reefs in Norway. In: Proceedings of the 10th International Coral Reef. Tokyo, Japan: Japanese Coral Reef Society; 2006:1849-1868

[57] Reed JK, Gore RH, Scotto LE, Wilson KA. Community composition, structure, areal and trophic relationships of decapods associated with shallow- and deep-water *Oculina varicosa* Coral reefs: Studies on decapod Crustacea from the Indian River region of Florida, XXIV. Bulletin of Marine Science. 1982;**32**:761-786

[58] Reed JK, Mikkelsen PM. The molluscan community associated with the scleractinian coral, *Oculina varicosa*. Bulletin of Marine Science. 1987;**40**:99-131

[59] Jensen A, Frederiksen R. The fauna associated with the bank-forming deepwater coral *Lophelia pertusa* (Scleractinaria) on the Faroe shelf. Sarsia. 1992;**77**:53-69

[60] Mortensen PB. Aquarium observations on the deep-water coral *Lophelia pertusa* (L., 1758) (Scleractinia) and selected associated invertebrates. Ophelia. 2001;**54(2)**:83-104

[61] Fedotov DM. Einige Beobachtungen ueber die Biologie und Metamorphose von Gorgonocephalus. Zoologischer Anzeiger. 1924;**61**:303-311

[62] Mortensen T. Handbook of the Echinoderms of the British Isles. Edinburgh: Humphrey Milford Oxford University press; 1927. 471pp

[63] Hendler G. Echinodermata: Ophiuroidea. In: Giese AC, Pearse JS, Pearse VB, editors. Reproduction of Marine Invertebrates. Echinoderms and Lophophorates. Vol. 6. Pacific grove, CA: The Boxwood Press; 1991. p. 355-511

[64] Mortensen PB, Buhl-Mortensen L. Distribution of deep-water gorgonian corals in relation to benthic habitat features in the Northeast Channel (Atlantic Canada). Marine Biology. 2004;**144**:1223-1238

[65] Stone RP, Shotwell SK. State of the U.S. Deep Coral Ecosystems in the Alaska Region: Gulf of Alaska, Bering Sea and the Aleutian Islands. In: The State of Deep Coral Ecosystems of the United States. NOAA Technical Memorandum CRCP-3. Silver Spring MD. 2007:65-108

[66] Lumsden SE, Hourigan TF, Bruckner AW, Dorr G. The State of Deep Coral Ecosystems of the United States. Silver Spring MD: NOAA Technical Memorandum CRCP-3; 2007

[67] Wainwright SA, Dillon JR. On the orientation of sea fans (genus Gorgonia). Biological Bulletin. 1969;**136**:130-139

[68] Goh NKC, Ng PKL, Chou LM. Notes on the shallow water gorgonian associated fauna on coral-reefs in Singapore. Bulletin of Marine Science. 1999;**65**:259-282

[69] Emson RH, Woodley JD. Submersible and laboratory observations on *Asteroschema tenue:* A long-armed euryline brittle stare epizoic on gorgonians. Marine Biolology. 1987;**96**:31-45

[70] Buhl-Mortensen L, Olafsdottir SH, Buhl-Mortensen P, Burgos JM, Ragnarsson SA. Distribution of nine cold-water coral species (Scleractinia and Gorgonacea) in the cold temperate North Atlantic in light of bathymetry and hydrography. Hydrobiologia. 2015;**759**: 39-61

[71] De Clippele LH, Buhl-Mortensen P, Buhl-Mortensen L. Fauna associated with cold water gorgonians and sea pens. Continental Shelf Research. 2015;**105**:67-78

[72] Molodtsova TN. Black corals (Antipatharia: Anthozoa: Cnidaria) of the north-east Atlantic. In: Mironov AN, Gebruk AV, Southward AJ, editors. Biogeography of the North Atlantic Seamounts. Moscow: KMK Press; 2006. p. 141-151

[73] Braga-Henriques A, Porteiro FM, Ribeiro PA, de Matos V, Sampaio T, Ocaña O, Santos RS. Diversity, distribution and spatial structure of the cold-water coral fauna of the Azores (NE Atlantic). Biogeosciences 2013;**10**(6):4009-4036

[74] Bo MG, Bavestrello G, Canese S, Giusti M, Salvati E, Angiolillo M, Greco S. Characteristics of a black coral meadow in the twilight zone of the central Mediterranean Sea. Marine Ecology Progress Series. 2009;**397**:53-61

[75] Henry LA, Roberts JM. Developing an interim technical definition for coral gardens specific for UK waters and its subsequent application to verify suspected records. JNCC Report. 2014;**507**:38

[76] Buhl-Mortensen P, Buhl-Mortensen L. Diverse and vulnerable deep-water biotopes in the Hardangerfjord. Marine Biology Research. 2013;**10**:253-267. DOI: 10.1080/17451000.2013.810759

[77] Howell KL. A benthic classification system to aid in the implementation of marine protected area networks in the deep/high seas of the NE Atlantic. Biological Conservation. 2010;**143**:1041-1056

[78] Greathead C, Gonzalez-Irusta JM, Clarke J, Boulcott P, Blackadder L, Weetman A, Wright PJ. Environmental requirements for three seapen species: Relevance to distribution and conservation. ICES Journal of Marine Science. 2015;**72**:576-586

[79] Birkeland C. Interactions between a sea pen and seven of its predators. Ecological Monographs. 1974;**44**:211-232

[80] Morin JG. Probable functions of bioluminescence in the Pennatulacea (Cnidaria, Anthozoa) coelenterate ecology and behaviour. In: Coelenterate Ecology and Behavior. New York: Plenum Press. 1976:629-638

[81] Hughes DJ. Sea pens & burrowing megafauna (volume III). An overview of dynamics and sensitivity characteristics for conservation management of marine SACs. Scottish Association for Marine Science (UK Marine SACs Project). 1998:105

Chapter 3

Deep Gorgonians and Corals of the Mediterranean Sea

Michela Angiolillo and Simonepietro Canese

Additional information is available at the end of the chapter

http://dx.doi.org/10.5772/intechopen.69686

Abstract

Recent studies, carried out by means of innovative technological tools as remotely operated vehicles (ROVs), have highlighted the richness of the Mediterranean deep-sea environments, characterized by great diversity and abundance of organisms. In particular, corals, gorgonians, and sponges play the important ecological role of ecosystem engineers in deep marine environments, creating complex three-dimensional habitats enhancing high biodiversity and ecosystem functioning at every level. Coral forests and bathyal white coral communities, starting from depths of 50–70 m and below 300 m, respectively, represent the richest ecosystems known so far for the Mediterranean basin. The different assemblages show a strong heterogeneity, varying in terms of specific composition, abundance, size of colonies, and associated fauna, even on a small spatial scale. Unfortunately, the high commercial fishing effort of trawling and longline fleets mainly operating along this bathymetric range represents a major threat for these vulnerable marine ecosystems, particularly in consideration of their structuring organisms which are long-lived species with slow growth rates and recovery ability. Further knowledge on deep coral assemblages is urgently needed to implement effective management and proper conservation measures. This approach is now an international priority that proceeds together with the inclusion of the structuring species in numerous directives.

Keywords: corals, deep-sea benthic communities, animal forest, Mediterranean Sea, fishing impact, ROV-Imaging

1. Introduction

The Mediterranean basin, considered a biodiversity *hotspot* [1], has been widely investigated for centuries. Nevertheless, for obvious logistic reasons, most of the investigations have been carried out within the depth range of traditional scuba diving (~40 m depth) [2–5]. So, knowledge about Mediterranean coral and gorgonian assemblages in coastal areas and photic zone had significantly increased in recent decades [3]. Conversely, knowledge about the deep realm and gorgonian assemblages located on the continental shelf and slope still remains largely unknown [1, 6–7]. Occurrence, in the Mediterranean Sea, of coral ecosystems dwelling at depth greater than 50 m depth was already reported in the 18th century. Their presence were mostly observed due to specimens by-cached by fishermen or withdrawn by blind destructive techniques, such as grabs and dredgers [8–9] for taxonomical studies [10]. In the Mediterranean, the scientific curiosity for these ecosystems arose for the first time about 10 years ago, after the discovery, in the Ionian Sea of living white corals reefs [11–12] between 500 and 600 m depth, hosting a very rich associated fauna [13, 14]. Successively, rich coral biocoenoses, the so-called coral forest [*sensu* 15], made of arborescent gorgonians and antipatharians, were reported also at bathymetric range starting within the circalittoral zone (50 m depth) and ending at the limit of the continental shelf, around 200 m depth, both along coastal areas and seamounts [16–28].

The most recent innovative technological development and increased availability of video-equipped towed gears, remotely operated vehicles (ROVs), and manned submersibles, coupled with multibeam echo-sounder (MBES), have significantly increased accessibility to deeper areas, allowing for controlled sampling and quantitative study of deep bottoms communities [4, 5, 16–31]. The use of ROV technology has revolutionized the knowledge of such environments, allowing scientists to obtain, through many operating hours of direct observation in not destructive way, new ecological data such as habitat preferences, bathymetric distribution, and species association. Moreover, the acquisition of small fragmented samples allowed scientist to obtain biological data on reproduction, fecundity, recruitment, genetics, and growth.

With increasing depth and as a result of light attenuation, benthic sessile organisms replace algal and seagrass "forests" and become progressively the most important contributors to the three-dimensional complexity of deeper ecosystems [32]. In particular, large benthic cnidarians can play an important ecological role in deep realm. From a structural point of view, they act as ecosystem engineers [*sensu* 33] forming complex structures that provide a suitable habitat, a colonizable substrate [2, 14, 18, 25], a refuge for numerous species, and a nursery area for fish [15, 19, 31]. From a functional point of view, corals promote a significant flow of matter and energy from the pelagic to the benthic system by capturing plankton and suspended particulate organic matter [2, 3, 25]. Current flow, food availability, and sediment re-suspension vary widely within the complex structures formed by the colonies, and this heterogeneity increases the abundance and functional diversity of the associated fauna [20, 25–32], sustaining high biodiversity levels in both epibenthic and proximal interstitial surroundings [7, 13, 23, 34]. In the Mediterranean Sea, the recent discovery of the two major coral ecosystems, mesophotic coral forests [15] (between 50 and 300 m depth) and bathyal white coral mounds (from 300 m depth), is progressively increasing awareness of their ecological relevance underlying their paramount ecological role. Unfortunately, these deepwater ecosystems are highly vulnerable to a wide spectrum of direct or indirect anthropogenic impacts [35]. In particular, high commercial fishing effort of trawling and longline fleets mainly operating along this bathymetric range represents a major threat for deepwater ecosystems, whose structuring organisms are long-lived species with slow growth rates and recovery ability [36–38]. However, there is still a great lack of information on communities' structure, extension and distribution, environmental constraints, and adaptive responses to stress of these habitats, mainly due to the technical difficulties associated with deep exploration.

The present chapter intends to focus its attention on these Mediterranean deep-sea coral environments, whose studies were carried out with the aim to give a biocenotic characterization, to describe the biodiversity and the ecological role of coral assemblages, and to evaluate the anthropic impact on the structuring species.

2. Materials and methods

Since the first years of the 2000s, an increasing number of expeditions were carried out along the western Mediterranean coasts with the aim of studying biodiversity and ecology of the benthic communities and of evaluating the anthropic impact on the structuring species in the depth range of 70–500 m through ROV-Imaging and taxonomic sampling. Investigations and samplings focused on areas mainly located along the Italian, France, and Spanish coast (**Figure 1**). The surveys were carried out on board the oceanographic vessels, mainly instrumented with multibeam echo-sounder (MBES) and the remotely operated vehicle (ROV)

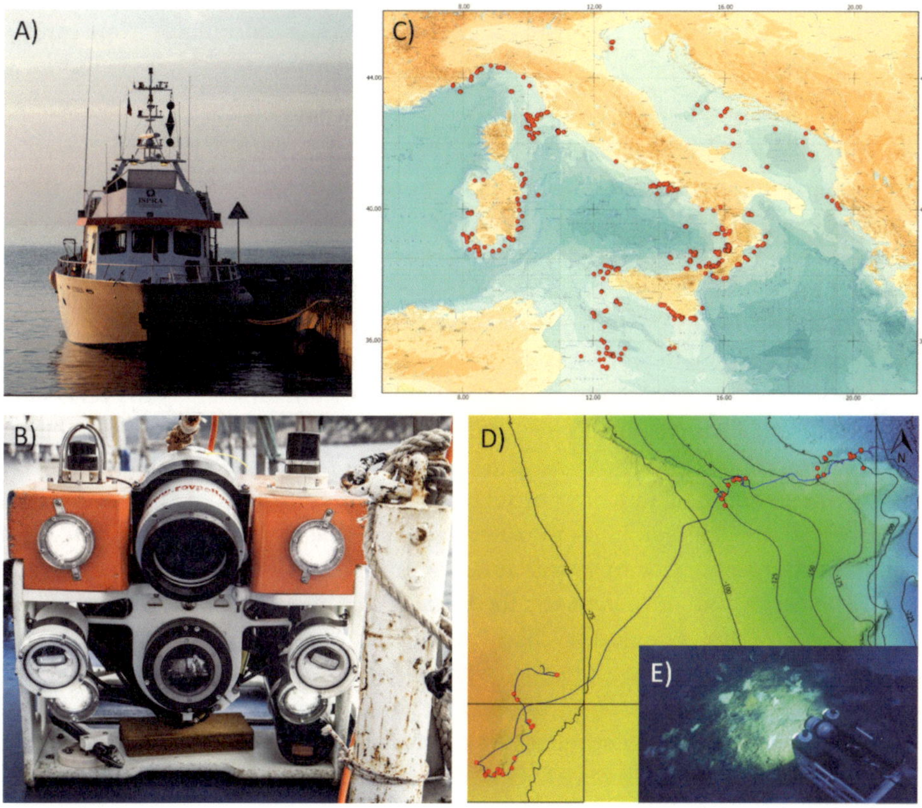

Figure 1. (A) The Italian R/V Astrea of ISPRA used for the surveys. (B) The Italian ROV "Pollux III" equipped for scientific purpose. (C) Study areas: example of location of ROV dives (dots) along the Italian seas. (D) example of morpho-bathymetric maps of a study areas and ROV transect (line). (E) The depth environment with the ROV in exploration phase.

equipped for scientific purpose (**Figure 1**). MBES provided high detailed morpho-bathymetric three-dimensional maps in order to localize rocky outcrops then explored by means of the ROV. The MBES is a device used to determine the depth of water and the nature of the seabed. It works by transmitting a broad acoustic fan shaped pulse from a transducer every seconds. The data processing of returning soundwaves is used to generate accurate high resolution three-dimensional maps of sea bottom. The ROV is an underwater unmanned vehicle that is able to navigate at a speed of approximately 2–3 knots equipped with video camera and connected to the surface through an umbilical cable (a long electrical cable) that carries the power source and control signals deriving from the vessel and transmits the acquired video and sensor data. The surface unit is composed of a console for the remote controls, a unit for the storage of data, and monitors to observe in real time the acquired images. The winch with the umbilical cable completes the set of equipment to guarantee the full operability of the system. The ROV has a depth sensor, a compass, and an Ultra Short Baseline underwater acoustic tracking position system, providing detailed records of the tracks along the seabed and a navigation Sonar. Moreover, some ROVs are furnished with specific scientific tools such as a digital high definition photo camera, underwater strobes, HD video camera, ultra-led power system providing light for the cameras, and laser beams placed at fixed distance and used as a metric scale of the images and the visual field. In addition, manipulator arms can be used to collect any biological samples.

The ROV was generally conducted ~1.5 m above the seabed, at constant speed (approximately 0.5 knots). Explorative or vertical transects, from the shallow to depth, were carried out, and geographical positions and depth were registered from the beginning to the end of each transect, every 1 second. Along each transect, videos were constantly recorded and HD photos were systematically acquired in order to better identify and quantify the target species, the marine benthic litter, or misunderstanding items. ROV-Imaging analysis of video transects was carried out to examine diversity, spatial distribution, abundance, demography, and vulnerability of the gorgonian and coral species detected in order to obtain information about ecology of a single species or a benthic communities characterization. So data on occurrence, occupancy (frequency of coral patches m^{-1}), density (number of colonies m^{-2}), population size-frequency distribution by means of measure of the main morphometric parameters (colony height, basal diameter, and branching order), and community composition were extracted and analyzed. Usually, some large anthozoans were used as surrogate descriptors of mega-benthic marine biodiversity due to their richness, abundance, identification easiness, and ecological role. The correspondence between the visual assignment of an individual to a certain coral species and its actual taxonomic classification was checked through the collection of at least one sample per species. The collection of photographic material and high resolution videos helped to define the species composition of these populations. Moreover, environmental variables, such as substrate exposure, substrate slope, sediment cover, co-occurrence with other species, or debris presence, were collected to habitat characterization and benthic community's assessment.

To assess some anthropic impacts, such as the presence of marine debris, observed items were usually divided into categories: fishing gears, plastics, recreational litter, and so on. The presence of debris was evaluated both by occurrence (frequency of debris types), relative abundance (debris items 100 m^{-2}), or taking into consideration the percentage of frames showing lost debris. To evaluate the impact of debris on the benthic fauna, different types of impact were identified (covering, abrasion, hanging, lying), and every damaged or entangled colony was annotated.

3. Animal forest

In the past, deep-sea ecosystems were among the least studied and explored marine regions of the world [4, 5, 35], and until a few years ago, a common assumption was to consider deep-sea areas to be lacking in terms of biodiversity. The exploration carried out with ROV leading to the discover that deep-sea ecosystems are characterized by a great heterogeneity of assemblages and dominated by sessile suspension feeder organisms (such as sponges, corals, gorgonians and bivalves) that vary in dimension, richness, and abundance, representing important natural heritages in the world [31, 35]. In the Mediterranean, cnidarian-rich deep-sea habitats, dominated by antipatharians, gorgonians, and scleractinians, as major habitat forming taxa, for the typical bush or arborescent development similar to the trees in the terrestrial forest, finally build up the so-called animal forest [15]. Comparable to earth forests in terms of stability, complexity, biodiversity, role, and longevity, these communities are widely distributed along rocky areas of Mediterranean Sea and represent unique ecosystems, very oasis of biodiversity [15, 20, 24–30, 32, 39].

Terrestrial forests supply food, protection, and support to a great variety of organisms, which may establish occasional or highly specialized relationships with both the dead and living portions of the trees [24, 40]. Similarly, oceans host astonishing examples of forests in the deep-sea, entirely structured by colonial animals, which represent the most frequently observed coral taxa in these ecosystems [18, 24, 25, 31].

Thanks to their flexible organic skeletons, which offer them a weak resistance to the current, gorgonians and black corals can tolerate strong currents, thus enabling them to obtain a greater quantity of food and favoring larval dispersal [24]. The local turbulent conditions that develop among the branches encourage the persistence of food in suspension and thus attracting numerous organisms both epibiontic, such as molluscs, platyhelminthes, hydroids, and bryozoans, and vagile organisms, such as fish, crustaceans, and echinoderms. So, they represent an attractive pole for a very rich associated fauna of small invertebrate, increase the possibility of new ecological niches, and also play key site and species-specific roles on the early-stage recruitment of other epibenthic assemblages [31]. Moreover, they provide refuges for numerous species and host nursery areas for several commercial fish, favoring the development of plentiful high-quality fish. As a consequence, the richness and biodiversity of these environments are increased. Hence, the important ecological role of these large anthozoans: major elements in the formation of forests and hosts of a rich associated fauna into and around their wide branches. Moreover, like the oldest terrestrial trees, also some of these coral species may live for thousands of years [24]. The most complex forests are the result of a long history of growth and structure. The oldest, for example, are formed by tall and branched corals such as the 1000-year-old smooth black coral, *Leiopathes glaberrima*, able to reach 2 m height [24]. These deep benthic communities show their best form on the rocky bottoms due to habitat heterogeneity. The different assemblages vary strongly not only based on the specific composition, but also in terms of abundance, size of the colonies, and associated fauna, even on a small spatial scale. Their distribution is strongly influenced by multiple factors, as their limited larval dispersal, slow growth rates, low ability of recovery, late maturity age, availability of nutrients, heterogeneity of the habitat, rate of sedimentation, colonies vulnerability to impacts and habitat destruction, harvesting by fishery operation, or other threats are determined by human activity. In fact, these animal forests, of long-living, slow-growing organisms, are fragile ecosystems. They are able to adapt to environmental standard changes and to overcome occasional perturbations, but those which are more important and recurring, such as professional and recreational fishing, trawling, hypertrophication, and pollution, can have devastating effects when they all play in synergy, making recovery impossible.

For their importance from the ecological point of view, from their capacity to contribute to the creation of habitat, from the *facies* of deep coral to white coral reefs of the Mediterranean bathyal zone, certain species of anthozoan can be considered emblematic of the state of health of the benthic population of Mediterranean seas, heavily influenced by multiple human activities.

3.1. Gorgonians

Gorgonians are one of the most important and diverse bioengineering organisms in the Mediterranean Sea, forming dense assemblages that extended over vast areas. The majority of them have a fan-like structure that can be as wide as 2 m, but they can also be found in the arborescent or finely branched form, bush-like, or also devoid of branches. They are typical of the hard seabeds, but they are also able to tolerate high levels of sedimentation or anchor themselves on soft bottoms. They have a wide bathymetric and geographical distribution. According to the species, they can be predominant in deep areas, forming different association and assemblages. In the Mediterranean basin, the most abundant and frequent deep gorgonian species are *Eunicella singularis* (Esper, 1791), *Eunicella cavolini* (Koch, 1887), *Paramuricea clavata* (Risso, 1826), *Paramuricea macrospina* (Koch, 1882), *Corallium rubrum* (Linnaeus, 1758), *Acanthogorgia hirsuta* (Gray, 1857), *Callogorgia verticillata* (Pallas, 1766), *Ellisella paraplexuroides* (Stiasny, 1936), *Viminella flagellum* (Johnson, 1863), *Swiftia pallida* (Madsen, 1970), *Villogorgia bebrycoides* (Koch, 1887), *Bebryce mollis* (Philippi, 1842), and *Muriceides lepida* (Carpine & Grasshoff, 1975). These gorgonians can form dense monospecific *facies* that gradually shift to other *facies* or more complex assemblages, made by sponges, antipatharians, scleractinians, alcyonacean, and associated fauna [15, 18, 20–26, 28, 32, 39, 41]. Depth clearly segregates gorgonians and the assemblage composition varied strongly from sites to another. However, a general pattern of high gorgonian diversity is observed (**Figure 2**), since these species have been recently found with increasing frequency on rocky substrates at similar depths in different areas of the Mediterranean Sea [7, 16–28, 32, 37–39, 41–44, 46–52].

Overall, spatial structure of these gorgonian distribution can be generalized: i) coastal species extending their distribution to deep coralligenous banks (*E. singularis, E. cavolini,* and *P. clavata; E. paraplexuroides; C. rubrum*); ii) dominance of one single species on the continental shelf (*P. macrospina, E. cavolini, C. rubrum*); iii) concentration of several gorgonian species on the

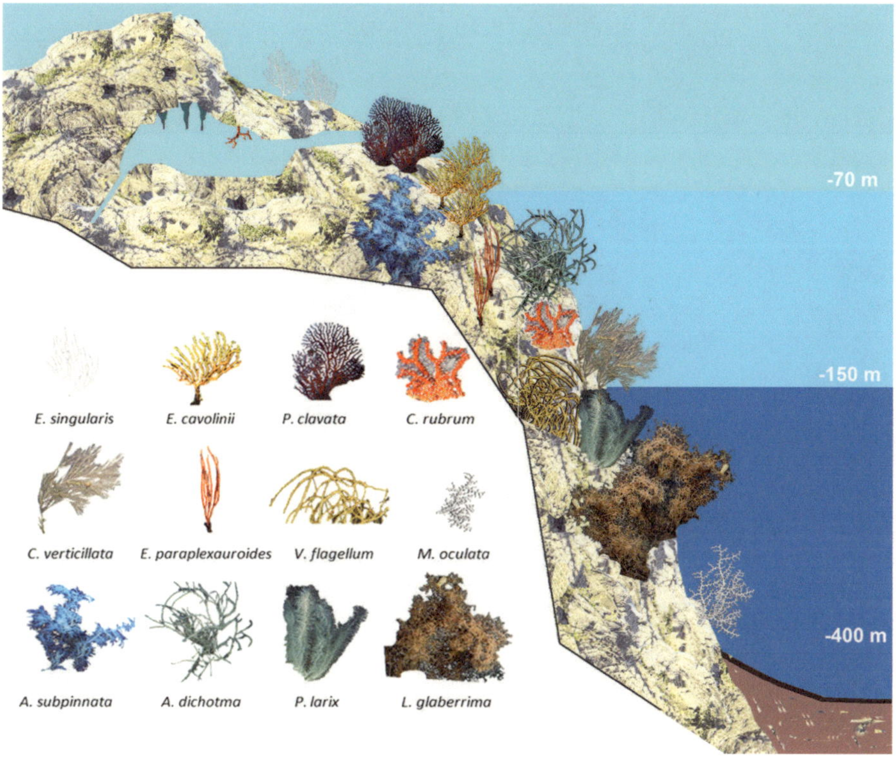

Figure 2. Zonation of the megabenthic assemblages. The figure presents the distribution of the most conspicuous and abundant components of the megafauna dwelling at about 50–500 m depth.

shelf edge (*E. cavolini, P. macrospina, S. pallida, A. hirsuta, V. flagellum, B. mollis* and *C. verticillata,* and *C. rubrum*); iv) species extending their distribution deeper into the upper slope (*B. mollis, C. verticillata, V. flagellum, V. bebrycoides,* and *S. pallida*).

E. singularis dominated assemblages located on the continental shelf on rocky and horizontal substrate (**Figure 3 (A)**) and in the western side of basin (along the Spanish coast), in particular water transparency condition, it can extend its distribution to deeper water [26, 42]. In deeper environments, populations of *P. clavata* were found in low-density patches on coralligenous banks at 70–90 m depth [26, 30, 39, 42] and few colonies on the shelf edge and slope at 100–120 m depth [20, 37, 39]. Colonies, generally, achieved larger sizes than in shallower environments [3], as a possible consequence of the higher environmental stability of deeper areas. The common Mediterranean coastal species *E. cavolini* showed a wide distribution: it was found on sloping and vertical rocky on deep coralligenous banks on the continental shelf and on the shelf edge (45–150 m depth), where it can form dense *facies*. Commercially harvested since ancient times, the red coral *C. rubrum* is emblematic species occurring on rocky bottoms over a wide bathymetric range (15–800 m) [15, 43–45]. Recent deep investigation led to discovery of new thriving deep population [45] (**Figure 3 (B)**). The brick-red candlestick colonies of *E. paraplexuroides* were rare and discontinuously distributed in the Mediterranean. They were very shallow (12–35 m) and abundant in Chafarinas Islands and around Alboran

Figure 3. Coral assemblages and impacts of fishing litter on the rocky bottom of the Tyrrhenian Sea. (A) Rocky elevation covered by *E. singularis facies*, 50 m. (B) Example of coral forest, composed of *C. rubrum, C. verticillata, E. cavolini,* and *B. mollis*, 140 m. (C) Mixed assemblage of *V. falgellum, E. cavolini,* and juvenile colony of *C. verticillata*, 180 m. (D) Multi-species meadows of *E. cavolini, V. bebrycoides,* and the new species of Mediterranean soft coral *Chironephthya Mediterranean n. sp.*, recently described, 120 m. (E) Arborescent colony of *A. subpinnata*, 97 m. (F) Alive white corals *M. oculata*, associated with the crustacean *Paromola cuvieri*, which often carries a sponge on its exoskeleton, 470 m. (G) The precious red coral colonies entangled and abraded by an old line snagged on the rocks, 114 m. (H) Lines, colonized by alcyonarians and hydroids, are fully incorporated in the yellow scleractinia *D. cornigera*, 80 m. (I) Example of the so-called ghost fishing: *Munida rugosa* caught by an abandoned or lost net, 150 m.

sea [46–48], whereas only isolated deep colonies (70–200 m) were found in the southern western basin [49–50]. Populations of the whip-like gorgonian *V. flagellum* were known to have a patchy distribution, alternating dense meadows with sparse colonies, in vertical and subhorizontal rocky bottoms [10, 50]. Recent records [17, 32, 49–51] have showed wider distribution of this species in the deep water (100–250 m) of the western basin, where it is one of the most abundant and frequent [39] (**Figure 3 (C)**). *A. hirsuta* is a species showing restricted distribution, occurring in few locations of the shelf edge at 150–200 m depth [39]. This species usually occurred on lightly silted rocky substrates and it often observed together with other anthozoan species [10, 52], suggesting a certain adaptability to various environmental conditions. *P. macrospina* dwelled on rocky substrates and never as a dominant species [39], but it was also observed that it associated with horizontal Maërl beds on the continental shelf where it can raise very high abundances [39]. Moreover, *P. macrospina* can also grow as a fouling and epibiontic organism, suggesting a fast growth rate [19]. Populations of *C. verticillata* formed dense *facies* [20, 23, 27, 32] mainly in areas characterized by high sedimentation rates and the low values of coral abundance, providing a general "forest-like" aspect to the coral communities [32]. The small size gorgonian *B. mollys, S. pallida,* and *V. bebrycoides* were observed to extend their distribution under 200 m depth [39] from the shelf edge to the upper slope. Together with sponges, they can create a sort of "underbrush," living at the base of bigger species. So, a rich community of filtrating organisms is supported (**Figure 3 (D)**), composed principally of corals and erect sponges of large dimensions, thus representing an important passage in transferring energy between the benthic and the pelagic zone. On the scarce rocky substrates occurring in these deeper environments, gorgonian abundance tends to decrease with increasing depth, and antipatharians become the dominant arborescent species [24, 53].

At these depth (100–250 m) are also recorded some rare species like the stylasterid *Errina aspera* in the Messina Strait [28], or the parasitic zoanthid *Isozoanthus primnoidus* (Carreiro-Silva, 2011) that when mechanically stimulated, produced an intense green-blue bioluminescence that disappeared after some seconds [23]. Moreover, *facies* of the scleractinians as *Dendrophyllia ramea* in Cyprus and Sicily and *D. cornigera* were recorded often in the Western Mediterranean [23, 41]. Finally, the soft sediments of these deeper environments can be colonized by the bamboo coral *Isidella elongata* [24, 36], which can form extended assemblages in bathyal muds [8] or in natural protected enclaves [24].

Distribution of coral communities may be determined by the combination of both biological and environmental factors that can synergistically affect spawn, larval development and settling, growth, and death rates of individuals and geomorphological characteristics of settlement substrates that affect sediment accumulation rates, bottom currents, and the rates of food supply [39]. Overall, these factors shape Mediterranean coral communities, which show very high variability within the smallest spatial scale [32].

3.2. Black corals

Black corals are spread throughout all the oceans of the world. They are found especially in tropical and subtropical areas, where they can colonize very shallow waters, and in temperate and polar regions, where they dwell in deep water. The black corals have a black skeleton, from which they derive their common name, while the living organisms have a whitish color and heavily branched tufts. In the past, it was believed that they had magical and curative properties—in fact, the scientific name, Antipatharia, refers to the Latin term which indicated a talisman against witchcraft. Some species was used for jewelry trade [54–55]. Black corals seem to be keystone species, essential for the maintenance of high biodiversity levels. Six species are described for the Mediterranean Sea: *Antipathes dichotoma* (Pallas, 1766), *Antipathes fragilis* (Gravier, 1918), *Parantipathes larix* (Esper, 1790), *Leiopathes glaberrima* (Esper, 1792), *Antipathella subpinnata* (Ellis and Solander, 1786), and *Antipathella wollastoni* (Johnson, 1899) [56]. In this basin, black corals are usually restricted to deep habitats, probably to avoid strong seasonal fluctuations of temperature [57]. The shallowest species were found never shallower than 50 m in depth, while the majority occurs between 100 and 300 m depth only on rocky slopes and isolated shoals [56] (**Figure 2**). These organisms carry out an important ecological role in the deep sea. Considered to be among the rarest coral species of the Mediterranean

Sea [56], instead, new findings indicate that black corals are among the most conspicuous and widely distributed components of the Mediterranean deep-circalittoral coral communities, where they can form huge meadows [7, 18–19, 22–24, 57]. They may reach impressive abundances and sizes, forming important *facies* in the deep-sea realm [22]. The large, white, branched *A. subpinnata* is a common component of the lower fringe of the circalittoral (60–150 m) and is the most widespread [57]. It is able to form dense meadows (**Figure 3 (E)**), counting more than 30,000 colonies [18], starting at 60 m, and is also occasionally recorded at up to 500 m associated with white coral mounds [14]. The congeneric *A. wollastoni* forms conspicuous populations in both shallow and deep Atlantic waters, but has also recently been recorded in the Mediterranean basin in the vicinity of the Gibraltar Strait [22, 58]. *A. dichotoma* is characterized by tall arborescent colonies with loose, long, flexible branches and large polyps [19, 22]. Sparse colonies were found at 100 m and deeper in several Mediterranean areas in benthic assemblage characterized by several coral species [14, 16–17, 41, 59]. *A. fragilis* is a doubtful species. It has never been found again with certainty, and since the type specimen is lost, there is no possibility to verify, at present, its taxonomic status [22]. *L. glaberrima* is a tall arborescent, bright orange or white, black coral species, among the most common black corals of the Mediterranean basin [22]. Occasionally found at 100 m, it forms dense forests only from 200 m along the rocky bench terraces [23–24, 52, 60] or among white coral reefs [14, 61]. It is a frequent bycatch of long-line fishermen or trawlers [55]. *P. larix* is a monopodial or sparsely branched species showing a characteristic bottle-brush pinnulation pattern, up to 2 m tall [22]. This Atlantic-Mediterranean species living on rocky bottoms both along the continental shelf and in very deep waters (up to 2300 m) [54, 62] is usually very sparsely distributed [22, 62]. Exceptional forests dominated by dense monospecific populations of *P. larix* were recorded only off the Island of Montecristo (Tuscan Archipelago, Tyrrhenian Sea) [22]. Similar to other filtering colonial organisms, black corals tend to settle in areas of moderate to strong current, maximizing food capture and larval dispersion. The black coral population is inhabited by a huge variety of sessile, encrusting, and vagile organisms, searching for a refuge or a source of food, supporting the hypothesis that the existence of the coral canopy is able to enhance biodiversity at every level [4, 24, 56].

It is intriguing how these species, known for centuries and representing a conspicuous structure-forming component of relatively deep coral ecosystems in a well-known basin such as the Mediterranean Sea, have avoided attention for such a long time. In comparison, recent explorations are progressively unveiling the characteristics of Mediterranean black coral forests [18–19, 55].

3.3. White corals

The so-called white coral bathyal community or cold-water coral (CWC) [8] consists of the scleractinians species as *Lophelia pertusa* (Linnaeus, 1758), *Madrepora oculata* (Linnaeus, 1758), and *Desmophyllum dianthus* (Esper, 1794). These taxa, known to be well alive in the Atlantic Ocean at present, were considered almost eradicated from the Mediterranean basin from the last deglaciation age [65]. In the Mediterranean Sea, the last ROV exploration and fishery operations [12, 64] step-by-step led to discovery of many new still-alive CWC sites [65]. Live stocks were and are unevenly distributed in the Mediterranean Sea under 300 m depth (**Figure 2**), with six cold-water coral provinces identified thus far: the Southwestern Adriatic CWC (Bari Canyon), the Northern Ionian CWC (Santa Maria di Leuca), the Strait of Sicily CWC (South Malta), the Nora canyon, offshore the southern Sardinia [63], the Alboran Sea CWC (Melilla), and the Catalan-Provençal-Ligurian canyons CWC, to which to add more spotty records of live CWC [65]. The distribution of living CWC in the Mediterranean does not overlap everywhere the occurrence of subfossil counterparts, and this is particularly evident in the eastern basin [63]. These areas include both highly-structured ecosystems covering substantial surfaces, and patchy or spot records [27, 29, 60, 64–69]. The structural complexity offered by CWC habitats allows for the development of highly diverse associated communities that usually result in considerably higher biodiversity than the surrounding environment [30]. Antipatharians, gorgoniaceans, the scleractinians *Dendrophyllia cornigera* and sponges such as *Pachastrella compressa* and *Poecillastra monilifera* are also characteristic components of CWC communities elsewhere [21, 63, 67, 69]. The living coral colonies offer shelter to many invertebrates, such as the decapod crustaceans *Munida* cf. *tenuimana* and *Anamathia rissoana*. The vagrant macrofauna within or around coral grounds includes also a few echinoderms, such as the echinoids *Cidaris* sp. and *Echinus* sp., the sea star *Peltaster placenta*, and the spider crab *Paromola cuvieri* (**Figure 3F**). CWC bioconstructions are found and work as nurseries for several deepwater species of commercial interest (i.e., gadids or lophids) and are probably important for the maintenance of superficial stocks [5, 9, 10]. Demersal fish such as blackbelly rosefish *Helicolenus dactylopterus*, the Atlantic wreckfish *Polyprion americanus*, the large lender rockfish *Scorpaena elongata*, and the Blackspot seabream *Pagellus bogaraveo* are the principal species of fishes associated with CWC. The rare scale-rayed wrasse *Acantholabrus palloni* and the rare Ophidiidae *Benthocometes robustus* and several shark species such as *Hexanchus griseus* are also observed [63]. The dead coral portions are characterized by iron oxide and are fouled by hydroids, zoanthids, bryozoans, and brachiopods [63]. The solitary coral *D. dianthus* is almost ubiquitous preferentially settling live and dead colonies. The bivalve oyster *N. zibrowii* serves as substrate to living *D. dianthus* and *M. oculata*, but one living specimen was observed settling a dead *M. oculata* frame [63]. As it is the common rule in CWC grounds, the highest diversity is encountered in the dead part of the coral reef and within the interspersed soft sediment. Recent research has shown that canyons may be home to lush CWC communities [32, 63, 70–72]. This holds true for the Mediterranean Sea as well where important CWC presence has been documented at a number of canyons [27, 29, 44, 63, 65, 67, 73]. Because of their general patchiness inside canyon systems and intrinsic exposure to a number of direct and indirect dangers, canyon coral habitats are classified as vulnerable marine ecosystems (VME) [27] and obviously deserve protection [73–75].

4. Impacts and threats

In recent years, a marked increase of the human activities impact on deep-sea habitats has been observed [75, 76]. Dumping, increasing littering, oil spilling, mineral extraction, on-bottom framework works (e.g., pipes, cables), and fishing activity [34] are affecting the health status of such bottom communities. In addition, indirectly, global warming trends affect vital hydrographical attributes and progress in ocean acidification [34]. However, major direct threats are connected to fishing practices, which include (accidental) trawling, long lining, and crab trapping [77]. The effects of fishing activities on the benthic biocoenoses represent a worldwide problem particularly relevant in the Mediterranean Sea, a site characterized by intense historical fishing traditions. The majority of the studies concerning fishing impact have been addressed on the soft bottom assemblages subjected to trawling activities [33, 78], while very few information is available on the response of hard bottom communities, with the exception of some priority habitats such as seamounts [79] and white coral forests [29, 61, 73]. The recent ROV investigations in the western Mediterranean basin have revealed the heavy impact of fishing activities on the benthic communities as an omnipresent feature in all the surveyed localities [20, 23–24, 27, 29, 37–38, 80]. In the Mediterranean Sea, the rocky bottoms between 50 and 700 m depth, being generally avoided by trawlers, represent important fishing grounds for artisanal and recreational fleets. The rocky areas populated by coral forests attract a rich associated fauna, including fish species of commercial interest, which therefore increase the importance of these habitats for fishermen [25, 37]. Coral bycatch data suggest a dramatic effect of certain gears, estimated in various dozens of eradicated colonies per fishermen per year [73]. Moreover, due to complex structure of rocky bottom, some gears (e.g., trammel nets, gill nets, and long lines) can become entangled in the surrounding rocks, breaking and covering all the organisms that elevate on the substrate (e.g. corals, sponges, bryozoans) and accumulating in great quantity and at any depth (**Figure 3 (G)**). The gears also produce a continuous mechanical scouring on the sea bottom, determining sediment re-suspension. As a result, progressive and extended habitat degradation is observed [81, 82, 33] with substantial modifications of the structure and functioning of deep ecosystems through a shift in species composition toward opportunistic species with a faster growth rates and adapted to live on more silted or unstable habitats [37]. The abrasive action due to entangled gears on the colonies causes a continuous removal of their tissues (**Figure 3 (H)**), making them more vulnerable to parasite or bacterial infections as well as the epibiont agglomerates covering [37–38], which in the long term can also cause the colony mortality [83]. The introduction of very strong fishing gears, made of non-biodegradable synthetic materials, contributes to the phenomenon of "ghost fishing": the abandon nets, still in their working position, may continue to catch for a long span of time a large spectrum of organisms [82, 84] (**Figure 3 (I)**). Due to the extremely slow degradation of nylon, lost gears may persist for a centuries [85, 86] accumulating on the sea bottom, altering the surrounding habitat, and covering wide portions of the settled communities impeding the re-colonization for large anthozoans [87]. In the sea bottom, also other kind of litters, such as tyres, cans, glass bottles, and bigger objects (washing machines, bins, etc.), were observed. This debris can be adapted as shelters and colonized by encrusting or sessile organisms, vagile fauna, formed by echinoderms, fish, crabs, and other crustaceans [86, 88]. Although such artificial substrata enhance biodiversity, they are an alteration of the natural environment and of the communities which might settle down [89].

Generally, the most impacted species are those with a medium-large colony size, an arborescent morphology, and a flexible skeleton, that easily remain entangled. Coral skeletal characteristics, such as stiffness, flexibility, or fragility, are known to play an important role in the resistance of friction, which explains the different responses of coral to mechanical impacts [27, 37–38, 52]. Some species are often removed, while other are frequently crushed or only scrape the soft tissue of the branches [37–38, 52]. It is likely that such colonies can be completely eradicated from some heavily fished sites where coral forests have almost disappeared. Finally, in some highly impacted sites, the high percentage of damaged colonies that are partially or entirely covered by epibionts may suggest a general state of stress of the community [52]. The continuous pressures of human activity on such particular animal forests make them extremely vulnerable and are one of the major causes of degradation of the marine environment [82, 86, 88].

Different regional fishing traditions, accessibility of the fishing grounds, distance from the coast and the exploited depth range, hence efforts, as well as the typology of employed gears, the species composition of the community and, therefore, the response of the community to the impact and its recovery ability can play a synergistic role in defining different impact among the investigated areas [37]. In the Mediterranean basin, with impressive peaks along some areas [27, 37–38], the deep shoals are progressively depleting with dramatic consequences that have a negative effect also on the productivity of artisanal fishing. These data suggest that it is highly probable that pristine coral forests no longer exist also on rocky sea bottoms, especially in traditionally exploited fishing grounds [37].

5. Conclusion

The Mediterranean Sea offers an example on how advancement in marine technology and the recent increasing effort to investigate the deep-sea frontier have led to a revolution in the knowledge about deep realm [63]. The deep-sea benthos of the Mediterranean basin did target preferentially soft bottom (mobile) substrates, easier to sample by trawls, dredges, grabs, and corers. On the contrary, hard substrates have been understandably little touched by the biological exploration especially when associated with canyons and seamounts because of the intrinsic risk to imperil sampling gears [65, 89]. ROV exploration has opened new perspectives in the study of the deep benthic fauna, providing qualitative and quantitative description of benthic assemblages and progressively unveiling the characteristics of Mediterranean coral forests [22, 18–19, 55]. The high-density deep coral assemblages may provide an approximate idea of how Mediterranean continental shelves and upper slopes stood before decades of commercial fishing. These coral communities include the most common species which suffer by-catch of bottom trawling, trammel nets, and longline fishing which can have far-reaching and long-lasting negative effects [24]. Effects of fishing activities on the seabed and benthic communities have been indeed compared to those of forest clear-cutting, with an immediate reduction of the structural diversity and following alteration of biogeochemical cycles, species recovery, and settling rates [24]. Animal forest has now been internationally recognized as unique habitats characterized by numerous structuring species, generating complex and fragile ecosystems, which act as important oases of biodiversity in the deep realm. For these reasons, the international scientific community has recently proposed the inclusion of some of these species in many protection lists, recognizing their ecological value and their vulnerability to human activities: the Convention of Biological Diversity [17]; "Coral Garden" habitats were added to the OSPAR (Oslo and Paris Conventions for the Protection of the Marine Environment of the North-East Atlantic) "List of threatened and/or declining species and habitats" in 2007 [17]. Moreover, due to the fact that they are considered sensitive habitats, they may be identified as vulnerable marine ecosystems (VMEs). Recently, the Food and Agriculture Organization (FAO) [40] recommended the establishment of protected areas where such VMEs are known to be or likely to occur in order to put into action an ecosystem-based fishery management of deep-sea ecosystems [27]. Despite the European Commission has now proposed regulations against the use of trawling nets on important ecosystems as coralligenous, seamounts, and white coral mounds [90], in the Mediterranean waters, the interdiction of these deep-sea coral sanctuaries located on trawling routes or within traditional artisanal fishing grounds would raise numerous socio-economic problems [24].

Future research efforts should be paid to better understand the factors driving deep coral biodiversity in the Mediterranean Sea in a more "holistic" key, focused on understanding how, at different spatial scales, the disappearance of animal forests could represent a point of no-return whose consequences are still not clear to the scientific community [15, 24, 37].

Acknowledgements

The authors like to thank the crew of the research vessel R/V Astrea for the great availability and the constant help and support in all conditions and all researchers, whose precious collaboration and the work done together over the last years allowed the realization of this work.

Author details

Michela Angiolillo* and Simonepietro Canese

*Address all correspondence to: michela.angiolillo@isprambiente.it

Istituto Superiore per la Protezione e Ricerca Ambientale (ISPRA), Rome, Italy

References

[1] Bianchi CN, Morri C. Marine biodiversity of the Mediterranean Sea: Situation, problems and prospects for future research. Marine Pollution Bulletin. 2000;**40**:367-376

[2] Gili JM, Coma R. Benthic suspension feeders: Their paramount role in littoral marine food webs. Trends in Ecology & Evolution. 1998;**13**:316-321. DOI: 10.1016/s0169-5347(98)01365-2

[3] Linares C, Coma R, Garrabou J, Diaz D, Zabala M. Size distribution, density and disturbance in two Mediterranean gorgonians: *Paramuricea clavata* and *Eunicella singularis*. Journal of Applied Ecology. 2008;**45**:688-699

[4] Menza C, Kendall M, Hile S. The deeper we go the less we know. Revista de Biologia Tropical. 2008;**56**:11-24

[5] Danovaro R, Company JB, Corinaldesi C, D'Onghia G, Galil B, et al. Deep-sea biodiversity in the Mediterranean sea: The known, the unknown, and the unknowable. PLoS One. 2010;**5**:e11832

[6] Coll M, Piroddi C, Steenbeek J, Kaschner K, Lasram FBR. The biodiversity of the Mediterranean sea: Estimates, patterns, and threats. PLoS One. 2010;**5**:e11842

[7] Bo M, Bertolino M, Borghini M, Castellano M, Covazzi Harriague A, Di Camillo CG, Gasparini GP, Misic C, Povero P, Pusceddu A, Schroeder K, Bavestrello G. Characteristics of the mesophotic megabenthic assemblages of the Vercelli seamount (North Tyrrhenian Sea). PLoS One 2011;**6**:e16357

[8] Pérès JM, Picard J. Nouveau manuel de bionomie benthique de la Méditerranée. Recueil des Travaux de la Station marine d'Endoume 31. Endoume: Station Marine d'Endoume; 1964. 137 p

[9] Relini G, Peirano A, Tunesi L. Osservazioni sulle comunità dei fondi strascicabili del Mar Ligure Centro-Orientale. Bollettino dei Musei e Degli Istituti Biologici dell'Università di Genova. 1986;**52**:139-161

[10] Carpine C, Grasshoff M. Les gorgonaires de la Méditerranée. Bulletin de l'Institut océanographique de Monaco. 1975;**71**:1-140

[11] Tursi A, Mastrototaro F, Matarrese A, Maiorano P, D'onghia G. Biodiversity of the white coral reefs in the Ionian Sea (Central Mediterranean). Chemistry and Ecology. 2004;**20**:107-116

[12] Taviani M, Remia A, Corselli C, Freiwald A, Malinverno E, Mastrototaro F, Savini A, Tursi A. First geo-marine survey of living cold-water *Lophelia* reefs in the Ionian Sea (Mediterranean basin). Facies. 2005;**50:**409-417

[13] Carlier A, Le Guilloux E, Olu K, Sarrazin J, Mastrototaro F, Taviani M, Clavier J. Trophic relationships in a deep Mediterranean cold-water coral bank (Santa Maria di Leuca, Ionian Sea). Marine Ecology Progress Series. 2009;**397**:125-137

[14] Mastrototaro F, D'Onghia G, Corriero G, Matarrese A, Maiorano P, Panetta P, Gherardi M, Longo C, Rosso A, Sciuto F, Sanfilippo R, Gravili C, Boero F, Taviani M, Tursi A. Biodiversity of the white coral and sponge community off Cape Santa Maria di Leuca (Mediterranean Sea): An update. Deep-Sea Research Part II. 2010;**57**:412-430. DOI 10.1016/j.dsr2.2009.08.021

[15] Rossi S. The destruction of the 'animal forests' in the oceans: Towards an oversimplification of the benthic ecosystems. Ocean & Coastal Management. 2013;**84**:77-85

[16] Aguiliar R, Pastor X, de la Torriente A, García S. Deep-sea coralligenous beds observed with ROV on four seamounts in the Western Mediterranean. In: Pergent-Martini C, Brichet M, editors. UNEP-MAP-RAC/SPA: Proceedings of the 1st Mediterranean Symposium on the Conservation of the Coralligenous and Others Calcareous Bio-Concretions; 15-16 January 2009; Tabarka. Tunis: UNEP-MAP-RAC/SPA; 2009. p. 147-149

[17] Aguiliar R, Marín P. Mediterranean Deep-Sea Corals: Reasons for Protection Under the Barcelona Convention [Internet]. 2013. Available from: http://oceanaorg/sites/default/files/euo/OCEANA_Brief_Deep-sea_Corals.pdf

[18] Bo M, Bavestrello G, Canese S, Giusti M, Salvati E, Angiolillo M, Greco S. Characteristics of a black coral meadow in the twilight zone of the central Mediterranean Sea. Marine Ecology Progress Series. 2009;**397**:3-61

[19] Bo M, Bavestrello G, Canese S, Giusti M, Angiolillo M, Cerrano C, Salvati E, Greco S. Coral assemblage off the Calabrian Coast (South Italy) with new observations on living colonies of *Antipathes dichotoma*. Italian Journal of Zoology. 2011;**78**:231-242

[20] Bo M, Canese S, Spaggiari C, Pusceddu A, Bertolino M, Angiolillo M, Giusti M, Loreto MF, Salvati E, Greco S, Bavestrello G. Deep coral oases in the South Tyrrhenian Sea. PLoS One 2012;**7**:e49870

[21] Bo M, Bertolino M, Bavestrello G, Canese S, Giusti M, Angiolillo M, Pansini M, Taviani M. Role of deep sponge grounds in the Mediterranean Sea: A case study in southern Italy. Hydrobiologia. 2012;**687**:163-177

[22] Bo M, Canese S, Bavestrello G. Discovering Mediterranean black coral forests: *Parantipathes larix* (Anthozoa: Hexacorallia) in the Tuscan Archipelago, Italy. Italian Journal of Zoology. 2014;**81**:112-125. DOI: 10.1080/11250003.2013.859750

[23] Bo M, Cerrano C, Canese S, Salvati E, Angiolillo M, Santangelo G, Bavestrello G. The coral assemblages of an off-shore deep Mediterranean rocky bank (NW Sicily, Italy). Marine Ecology. 2014;**35**:332-342

[24] Bo M, Bavestrello G, Angiolillo M, Calcagnile L, Canese S, Cannas R, Cau Al, D'Elia M, D'Oriano F, Follesa MC, Quarta G, Cau A. Persistence of pristine deep-sea coral garden in the Mediterranean Sea (SW Sardinia). PLoS One. 2015;**10**:e0119393

[25] Cerrano C, Danovaro R, Gambi C, Pusceddu A, Riva A, Schiaparelli S. Gold coral (*Savalia savaglia*) and gorgonian forests enhance benthic biodiversity and ecosystem functioning in the mesophotic zone. Biodiversity and Conservation. 2010;**19**:153-167

[26] Gori A, Rossi S, Berganzo E, Pretus JL, Dale MRT, Gili JM. Spatial distribution patterns of the gorgonians *Eunicella singularis, Paramuricea clavata* and *Leptogorgia sarmentosa* (Cap of Creus, Northwestern Mediterranean Sea). Marine Biology. 2011;**158**:143-158

[27] Fabri MC, Pedel L, Beuck L, Galgani F, Hebbeln D, Freiwald A. Megafauna of vulnerable marine ecosystems in French Mediterranean submarine canyons: Spatial distribution and anthropogenic impacts. Deep-Sea Research Part II. 2014;**104**:184-207

[28] Salvati E, Angiolillo M, Bo M, Bavestrello G, Giusti M, Cardinali A, Puce S, Spaggiari C, Greco S, Canese S. The population of *Errina aspera* (Hydrozoa: Stylasteridae) of the Messina Strait (Mediterranean Sea). Journal of the Marine Biological Association of the United Kingdom. 2010;**90**:1331-1336

[29] Orejas C, Gori A, Lo Iacono C, Puig P, Gili JM, Dal, MRT. Cold-water corals in the Cap de Creus canyon, northwestern Mediterranean: Spatial distribution, density and anthropogenic impact. Marine Ecology Progress Series. 2009;**397**:37-51

[30] Rossi S, Tsounis G, Orejas C, Padron T, Gili JM, Bramanti L, Teixido N, Gutt J. Survey of deepdwelling red coral (*Corallium rubrum*) populations at Cap de Creus (NW Mediterranean). Marine Biology. 2008;**154**:533-545

[31] Buhl-Mortensen L, Vanreusel A, Gooday AJ, Levin LA, Priede IG, Buhl-Mortensen P, Gheerardyn H, King NJ, Raes M. Biological structures as a source of habitat heterogeneity and biodiversity on the deep ocean margins. Marine Ecology. 2010;**31**:21-50

[32] Cau A, Follesa MC, Moccia D, Alvito A, Bo M, Angiolillo M, Canese S, Paliaga EM, Orrù PE, Sacco F, Cannas R. Deepwater corals biodiversity along roche du large ecosystems with different habitat complexity along the south Sardinia continental margin (CW Mediterranean Sea). Marine Biology. 2015;**162**:1865-1878

[33] Jones CG, Lawton JH, Shachak M. Organisms as ecosystem engineers. Oikos. 1994; **29**:373-386

[34] Buhl-Mortensen L, Mortensen PB. Distribution and diversity of species associated with deep-sea gorgonian corals off Atlantic Canada. In: Freiwald A, Roberts JM, editors. Cold-water Corals and Ecosystems, Proceedings of the Second International Symposium on Deep Sea Corals. Erlangen: Springer; 2005. p. 849-879

[35] Ramirez-Llodra E, Tyler PA, Baker MC, Bergstad OA, Clark MR, Escobar E, Levin LA, Menot L, Rowden AA, Smith CR, Van Dover CL. Man and the last great wilderness: Human impact on the deep sea. PLoS One. 2011;**6**:e22588

[36] Maynou F, Cartes JE. Effects of trawling on fish and invertebrates from deep-sea coral facies of *Isidella elongata* in the western Mediterranean. Journal of the Marine Biological Association of the United Kingdom. 2011;**92**:1501-1507

[37] Bo M, Bava S, Canese S, Angiolillo M, Cattaneo-Vietti R, Bavestrello G. Fishing impact on deep Mediterranean rocky habitats as revealed by ROV investigation. Biological Conservation. 2014;**171**:167-176

[38] Angiolillo M, di Lorenzo B, Farcomeni A, Bo M, Bavestrello G, Santangelo G, Cau A, Mastascusa V, Cau A, Sacco F, Canese S, Distribution and assessment of marine debris in the deep Tyrrhenian Sea (NW Mediterranean Sea, Italy). Marine Pollution Bulletin. 2015;**92**:149-159

[39] Grinyó J, Gori A, Ambroso S, Purroy A, Calatayud C, Dominguez-Carrió C, Coppari M, Lo Iacono C, López-González PJ, Gili J-M. Diversity, distribution and population size structure of deep Mediterranean gorgonian assemblages (Menorca Channel, Western Mediterranean Sea). Progress In Oceanography. 2016;**145**:42-56

[40] Food and Agriculture Organization (FAO). Global forest resources assessment 2005. Progress towards sustainable forest management. Rome: FAO Forestry Paper. 2006;**147**:350

[41] Fourt M, Goujard A, Harmelin Jg, Vacelet J, Verlaque M. the scientific team of the MedSeaCan and CorSeaCan cruises. French Mediterranean submarine canyons and deep rocky banks: A regional view for adapted conservation measures. In: Proceedings of the Symposia on the conservation of the Mediterranean Marine Key Habitats; 27-31 October 2014; Portorož. Tunis: UNEP/MAP–RAC/SPA. 2014. pp. 33-38

[42] Gori A, Rossi S, Linares C, Berganzo E, Orejas C, Dale MRT, Gili JM. Size and spatial structure in deep versus shallow populations of the Mediterranean gorgonian *Eunicella singularis* (Cap de Creus, northwestern Mediterranean Sea). Marine Biology. 2011;**158**:1721-1732

[43] Angiolillo M, Gori A, Canese S, Bo M, Priori C, Bavestrello G, Salvati E, Erra F, Greenacre M, Santangelo G. Population structure of a long lived, overharvested octocoral: A ROV survey on deep-dwelling red coral populations in W Mediterranean. Marine Ecology. 2016;**37**:294-310. DOI: 10.1111/maec.12274

[44] Freiwald A, Beuck L, Rüggeberg A, Taviani M, Hebbeln D, R/V METEOR Cruise M70-1 Participants. The white coral community in the central Mediterranean Sea revealed by ROV surveys. Oceanography. 2009;**22**:58-74

[45] Cau A, Bramanti L, Cannas R, Follesa MC, Angiolillo M, Canese S, Bo M, Cuccu D, Guizien K. Habitat constraints and self-thinning shape Mediterranean red coral deep population structure: Implications for conservation practice. Scientific Reports 2016;**6**:23322

[46] Arroyo Tenorio MC, Domenech AB, Lampreave DM, López-González PJ. *Ellisella paraplexauroides* Stiasny, 1936. In: Barea-Azcón JM, Ballesteros-Duperón E. Moreno D, editors. Libro rojo de los invertebrados de Andalucía. 4 Tomos. Sevilla: Consejería de Medio Ambiente, Junta de Andalucía; 2008. pp. 239-242

[47] Maldonado M, López-Acosta M, Sánchez-Tocino L, Sitjà C. The rare, giant gorgonian Ellisella paraplexauroides: Demographics and conservation concerns. Marine Ecology Progress Series. 2013;**479**:127-141

[48] Angiolillo M, Bo M, Bavestrello G, Giusti M, Salvati E, Canese S, Record of *Ellisella paraplexauroides* (Anthozoa: Alcyonacea: Ellisellidae) in Italian waters (Mediterranean Sea). Marine Biodiversity Records. 2012;**5**:e4

[49] Angiolillo M, Bavestrello G, Bo M, Cau Al, Cau A, Giusti M, Salvati E, Tunesi L, Canese S. Distribution of deep-dwelling *Viminella flagellum* in the Italian western Mediterranean by means of multi-year ROV surveys. In: Proceedings of the Symposia on the Conservation of the Mediterranean Marine Key Habitats; 27-31 October 2014; Portorož. Tunis: UNEP/MAP–RAC/SPA; 2014. pp. 65-66

[50] Giusti M, Bo M, Bavestrello G, Angiolillo M, Salvati E, Canese S. Record of *Viminella flagellum* (Alcyonacea: Ellisellidae) in Italian waters (Mediterranean Sea). Marine Biodiversity Records. 2012;**5**:e34

[51] Giusti M, Bo M, Angiolillo M, Cannas R, Cau A, Follesa MC, Canese S. Habitat preference of *Viminella flagellum* (Alcyonacea: Ellisellidae) in relation to bathymetric variables in the Mediterranean Sea. Continental Shelf Research. 2017;**138**:41-50

[52] Bo M, Angiolillo M, Bava S, Betti F, Canese S, Cattaneo-Vietti R, Cau A, Sandulli R, Santangelo G, Tunesi L, Bavestrello G. Fishing impact on Italian coral gardens and management of Vulnerable Marine Ecosystems. In: Proceedings of the Symposia on the conservation of the Mediterranean Marine Key Habitats; 27-31 October 2014; Portorož. Tunis: UNEP/MAP–RAC/SPA; 2014. pp. 21-26

[53] Deidun A, Andaloro F, Bavestrello G, Canese S, Consoli P, Micallef A, Romeo T, Bo M. First characterization of a *Leiopathes glaberrima* (Cnidaria: Anthozoa: Antipatharia) forest in Maltese exploited fishing grounds. Italian Journal of Zoology. 2014;**82**:1-10

[54] Opresko DM. Three new species of *Leiopathes* (Cnidaria: Anthozoa: Antipatharia) from Southern Australia. Records of the South Australian Museum. 1998;**31**:99-111

[55] Deidun A, Tsounis G, Balzan F, Micallef A. Records of black coral (Antipatharia) and red coral (*Corallium rubrum*) fishing activities in the Maltese Islands. Marine Biodiversity Records. 2010;**3**:e90

[56] Opresko DM, Försterra G. Orden Antipatharia (corales negros o espinosos). In: Hofrichter R, editor. El Mar Mediterraneo: Fauna, Flora, Ecologia. Barcelona: Omega 2; 2004. pp. 506-509

[57] Bo M, Tazioli S, Spanò N, Bavestrello G. *Antipathella subpinnata* (Antipatharia, Myriopathidae) in Italian seas. Italian Journal of Zoology. 2008;**75**:185-195

[58] Ocaña O, Opresko DM, Brito A. First record of the black coral *Antipathella wollastoni* (Anthozoa: Antipatharia) outside of Macaronesian Waters. Revista de la Academia Canaria de Ciencias. 2007;**18**:125-138

[59] Vertino A, Savini A, Rosso A, Di Geronimo I, Mastrototaro F, Sanfilippo R, Gay G, Etiope G. Benthic habitat characterization and distribution from two representative sites of the deep-water SML Coral Mound Province Mediterranean). Deep-Sea Research Part II. 2010;**57**:380-396. DOI: 10.1016/j.dsr2.2009.08.023

[60] Mytilineou C, Smith CJ, Anastasopoulou A, Papadopoulou KN, Christidis G, Bekas P, Kavadas S, Dokos J. New cold-water coral occurrences in the Eastern Ionian Sea: Results from experimental long line fishing. Deep-Sea Research Part II. 2014;**99**:146-157

[61] D'Onghia G, Maiorano P, Sion L, Giove A, Capezzuto F, et al. Effects of deep-water coral banks on the abundance and size structure of the megafauna in the Mediterranean Sea. Deep-Sea Research Part II, in press. 2010;**57**:397-411

[62] Fabri MC, Pedel L, Habitats particuliers du bathyal et de l'abyssal/SRM MO. 2012. Available from: http://archimer.ifremer.fr/doc/00230/34118/32581.pdf

[63] Taviani M, Angeletti L, Canese S, Cannas R, Cardone F, Cau A, Cau A, Follesa MC, Marchese F, Montagna P, Tessarolo C. The "Sardinian cold-water coral province" in the context of the Mediterranean coral ecosystems. Deep-Sea Research Part II, in press. DOI: 10.1016/j.dsr2.2015.12.008

[64] Schembri PJ, Dimech M, Camilleri M, Page R. Living deep-water *Lophelia* and *Madrepora* corals in Maltese waters (Strait of Sicily, Mediterranean Sea). Cahiers de Biologie Marine. 2007;**48**:77-83

[65] Taviani M, Angeletti L, Antolini B, Ceregato A, Froglia C, López Correa M, Montagna P, Remia A, Trincardi F, Vertino A. Geo-biology of Mediterranean Deep-Water coral ecosystems. Marine Geology. 2011;**DTA/06**:705-719

[66] Angeletti L, Taviani M, Canese S, Foglini F, Mastrototaro F, Argnani A, Trincardi F, Bakran-Petricioli T, Ceregato A, Chimienti G, Mačić V, Poliseno A. New deep-water cnidarian sites in the southern Adriatic Sea. Mediterranean Marine Science. 2014;**15**:263-273

[67] Gori A, Orejas C, Madurell T, Bramanti L, Martins M, Quintanilla E, Marti-Puig P, Lo Iacono C, Puig P, Requena S, Greenacre M, Gili JM. Bathymetrical distribution and size structure of cold-water coral populations in the Cap de Creus and Lacaze-Duthiers canyons (northwestern Mediterranean). Biogeosciences. 2013;**10**:2049-2060

[68] Taviani M, Vertino A, López Correa M, Savini A, De Mol B, Remia A, Montagna P, Angeletti L, Zibrowius H, Alves T, Salomidi M, Ritt B, Henry P. Pleistocene to recent scleractinian deep-water corals and coral facies in the Eastern Mediterranean. Facies. 2011;**57**:579-603

[69] Calcinai B, Moratti V, Martinelli M, Bavestrello G, Taviani M. Uncommon sponges associated with deep coral bank and maerl habitats in the Strait of Sicily (Mediterranean Sea). Italian Journal of Zoology. 2013;**80**:412-423

[70] Stiles ML, Ylitalo-Ward H, Faure P, Hirshfield MF. There's no place like home: Deep seafloor ecosystems of New England and Mid-Atlantic. OCEANA; 2007:38

[71] Vetter EW, Smith CR, De Leo FC. Hawaiian hotspots: Enhanced megafaunal abundances and diversity in submarine canyons on the oceanic islands of Hawaii. Marine Ecology. 2010;**31**:183-199

[72] Würtz M. Mediterranean Submarine Canyons: Ecology and Governance. Gland, Switzerland; Málaga, Spain: IUCN; 2012. p. 216

[73] D'Onghia G, Maiorano P, Carlucci R, Capezzuto F, Carluccio A, Tursi A, Sion L. Comparing deep-sea fish fauna between coral and non-coral "megahabitats" in the Santa Maria di Leuca cold-water coral province. PLoS One. 2012;**7**:e44509

[74] WWF/IUCN. The Mediterranean Deep-Sea Ecosystems: An Overview of their Diversity, Structure, Functioning and Anthropogenic Impacts, with a Proposal for Conservation. Malaga: IUCN; 2004. p. 64

[75] Davies AJ, Roberts JM, Hall-Spencer J. Preserving deep-sea natural heritage: emerging issues in offshore conservation and management. Biological Conservation. 2007; **138**:299-312

[76] Jones DOB, Wigham BD, Hudson IR, Bett BJ. Anthropogenic disturbance of deep-sea megabenthic assemblages: A study with remotely operated vehicles in the Faroe-Shetland Channel, NE Atlantic. Marine Biology. 2007;**151**:1731-1741

[77] Pusceddu A, Bianchelli S, Martín J, Puig P, Palanques A, Masqué P, Danovaro R. Chronic and intensive bottom trawling impairs deep-sea biodiversity and ecosystem functioning. Proceedings of the National Academy of Sciences of the United States of America. 2014;**111**:8861-8866

[78] Smith CJ, Papadopoulou KN, Diliberto S. Impact of otter trawling on an eastern Mediterranean commercial trawl fishing ground. ICES Journal of Marine Science Journal Du Conseil. 2000;**57**:1340-1351

[79] Freiwald A, Boetius A, Bohrmann G. Deep water ecosystems of the Eastern Mediterranean 2006 – Cruise No. M70 – September 24 – December 8, 2006 – La Valletta (Malta) – Heraklion (Greece). METEOR-Berichte, M70. Bremen: DFG Senatskommission für Ozeanographie; 2011. p. 146

[80] Pham CK, Ramirez-Llodra E, Alt CHS, Amaro T, Bergmann M, et al. Marine litter distribution and density in European seas, from the shelves to deep Basins. PLoS ONE. 2014;**9**:e95839

[81] Fosså JH, Mortensen PB, Furevik DM. The deep-water coral *Lophelia pertusa* in Norwegian waters: Distribution and fishery impacts. Hydrobiologia. 2002;**471**:1-12

[82] Brown J, Macfadyen G. Ghost fishing in European waters: Impacts and management responses. Marine Policy. 2007;**31**:488-504

[83] Bavestrello G, Cerrano C, Zanzi D, Cattaneo-Vietti R. Damage by fishing activities to the Gorgonian coral *Paramuricea clavata* in the Ligurian Sea. Aquatic Conservation Marine and Freshwater Ecosystems. 1997;**7**:253-262

[84] Matsuoka T, Nakashima T, Nagasawa N. A review of ghost fishing: Scientific approaches to evaluation and solutions. Fisheries Science. 2010;**71**:691-702

[85] Thompson RC, Olsen Y, Mitchell RP, Davis A, Rowland SJ, John AWG, McGonigle D, Russell AE. Lost at sea: Where is all the plastic? Science. 2004;**304**:838

[86] Watters DL, Yoklavich MM, Love MS, Schrodeder DM. Assessing marine debris in deep seafloor habitats off California. Marine Pollution Bulletin. 2010;**60**:131-138

[87] Saldanha HJ, Sancho G, Santos MN, Puente E, Gaspar MB, Bilbao A, Monteiro CC, Gomez E, Arregi L. The use of biofouling for ageing lost nets: a case study. Fisheries Research. 2003;**64**:141-150

[88] Mordecai G, Tyler PA, Masson DG, Huvenne VAI. Litter in submarine canyons off the west coast of Portugal. Deep-Sea Research Part II. 2011;**58**:2489-2496

[89] Katsanevakis S, Verriopoulos G, Nicolaidou A, Thessalou-Legaki M. Effect of marine litter on the benthic megafauna of coastal soft bottoms: A manipulative field experiment. Marine Pollution Bulletin. 2007;**54**:771-778

[90] European Commission. Proposal for a Council Regulation concerning management measures for the sustainable exploitation of fishery resources in the Mediterranean Sea and amending Regulations (EC) No. 2847/93 and (EC) No.973/2001 (COM(2003) 589 final-2003/0229 (CNS)). Brussels: Official Journal of the European Union; 2003 Available from: http://eur-lex.europa.eu/legal-content/EN/TXT/?uri=CELEX%3A52004AE0320

Chapter 4

Precious Coral

Luwei Fan

Additional information is available at the end of the chapter

http://dx.doi.org/10.5772/intechopen.73149

Abstract

Precious coral, with attractive pink-to-red color, has been used for ornamental purposes for several thousand years. According to the related science, attractive red precious corals material, are defined to the class Anthozoa, subclass Octocorallia, order Alcyonacea, suborder Scleraxonia, and family Coralliidae in zootaxy. About 30 species are discovered in Coralliidae family compared to the huge Cnidaria phylum. *Corallium rubrum, Corallium japonicum, and Corallium elatius* are the three main species in Coralliidae family used for jewelry material in the gem market. The purpose of this chapter is to show the nature of animals in Coralliidae family, analyze the nondestructive test methods to identify the natural species from the imitations, and discuss the origin of color and the interactions in between the organic matrix and mineral. The chapter was organized in six parts. The first part reviews the history of precious coral used for different purposes by humans and then describes exact affiliation of precious coral on zoology and taxonomy. The second part deals with the biology and formation of precious coral. The paragraph also presents the information about Coralliidae colonies' sexual maturity, life span, growth rate, and mortality. The trade market and conservation are also summarized in this part. The gemological properties of *C. rubrum, C. japonicum, and C. elatius*, the main species in precious coral market, are introduced in the third part. As a consequence, an effective and nondestructive identification method to distinguish natural precious corals from their imitations was stated with Raman spectra as demonstrated in the fourth part. In the fifth part, the origin of precious coral color based on the results of Raman scattering measurements and PeakFit analysis is demonstrated. Three different excitation wavelengths (785, 633, and 514 nm) were used for the same samples at the same points. The result shows that all of the samples are colored by a mixture of pigments. Different colors are explained by different mixtures, not by a single pigment. Organic composition, even present in a small amount, plays an important role in the color of precious corals. The sixth part concludes the text by presenting what we have learned from the experimental data of microscope, SEM, TEM, and EBSD. The spatial relationship between the organic matrix and mineral components is determined by SEM observation on decalcification-treated samples. By integration of the results from nanometer to centimeter-scale detections, a hierarchical structure of precious coral is revealed. In the skeleton of precious coral, building blocks are arranged into several hierarchical levels of oriented modules. The modules in each hierarchical level assemble into larger unit that comprises the next higher level of the hierarchy. Precious coral, as a member of biomineral family, assembled skeleton as a delicate arrangement of a hierarchy of crystals with well-defined orientations under the control of organic matrix. Organic matrix works in both color pigment and architecture field for the precious coral.

Keywords: precious coral, organic matrix, oriented module

1. Introduction

Precious coral, as an organic gem material with attractive pink-to-red color, has been used for ornamental purposes for several thousand years. The earliest history of this material used for decoration could be traced back to 8000 BC for the red coral amulets uncovered in Neolithic graves in Switzerland. It was also reported the discovery of fine coral jewelry in Sumeria and Egypt around 3000 BC [1]. Coincidently, precious coral was valued highly in Asian Culture throughout history.

In China, precious coral was recognized treasure of the ocean since Three Kingdoms Times around 2000 years ago. It appeared constantly in palace tribute, imperial trappings, Buddha beads, medicinal material during the complete China's history after then. The first description about precious coral was the geographic monograph of the South China Sea with the book name "*Fu Nan Zhuan*". The morphological characteristics of coral were demonstrated in that ancient writing (**Figure 1**). Later, precious coral was also mentioned as medicinal material in literatures of both Li Jing (Tang Dynasty, 659 AD) and Li Shizhen (Ming Dynasty, 1578 AD). Due to its distinctive tree-like appearance and hard texture, precious coral was mistaken for plant or mineral for its early history. The misunderstanding about the nature of precious coral was not by chance, even distinguished naturalist Pliny the Elder named it *Zoophyte*, which habituated with the features of both animal and plant.

Being benefited from the developments of biology and taxonomy, the biological behavior and habits on coral's growth, reproduction, nutrition, and habitat are much well known nowadays than before. According to the related science, attractive red precious corals, which are commonly used as jewelry material, are defined to the class Anthozoa, subclass Octocorallia, order Alcyonacea, suborder Scleraxonia, and family Coralliidae in zootaxy [2].

Since fine specimens of red coral, which is also called gem-class coral or precious coral (**Figure 2**), are the most desirable yet among the least available [3], series of enhancement and treatment technologies, such as dyeing, bleaching, and polymer impregnation, are applied to precious coral's imitations to get higher value. Since a significant amount of imitations share the similar color distribution and structural characteristics with the natural precious corals, the

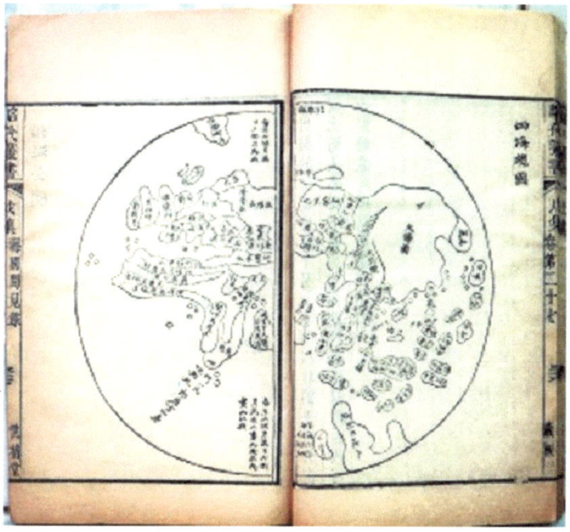

Figure 1. *"Fu Nan Zhuan"* in Three Kingdoms Times described the morphological characteristics of coral for the first time.

traditional identification methods are facing the difficulties to distinguish the precious corals from the others. The present chapter looks the current statues of precious coral from zoological taxonomy, gemological properties, and adopted methods to explore the essence of precious coral's attractive color and to determine the spatial relations between the organic matter and mineral component in order to provide information for identification, appraisal, and biomineral research.

Figure 2. Fine precious coral carving designed as religious blessing pattern.

2. Biology and formation

2.1. General biological background of precious coral

The term coral, as it is commonly intended by divers and enthusiasts, leads itself to numerous misunderstandings in strictly zoological terms. This happens mainly because many books on these extraordinary organisms broadly use the word coral to describe all creatures with a hard skeleton. This definition may thus be used to group organism very different one from the others. The term precious coral ought to be used only for the Coralliidae family in zootaxy. About 30 species are discovered in Coralliidae family compared to the huge Cnidaria phylum, which is containing over 10,000 species of animals.

Despite the obvious differences in form, shape, and dimension between one species and another, animals in Coralliidae family share the same basic structural distinction. They keep the sedentary living habits of organisms, mobile only at the larva stage. Every animal can be divided according to the infinite patterns of symmetry along the radius of incredible regular circle. The animal responsible for skeletal formation is precious coral's primary unit-polyp, which is a sort of sac adhering at the base to a rigid substratum and having an aperture facing upwards, surrounded by a variable number of tentacles (**Figure 3**). The number of these (six, eight, or multiples thereof) allows the Anthozoa group to be divided into Octocorallia or Hexacorallia. The Coralliidae belongs to Alcyoniina suborder, Alcyonacea order, and Octocorallia subclass accordingly.

Precious corals are colonial animals, which consist of thousands of small polyps. They reproduce sexually by releasing eggs and sperm polyps of gametes of the same species simultaneously over a period of one to several nights around a full moon. The polyps remain connected and continue to grow and reproduce on their own. These colonies can live for several centuries, during which their continuous calcification creates layered skeletal archives that used to be the material for the jewelry, medicine, and so on. The formation process of the precious coral's skeleton inspired the understanding on bio-mineralization and environmental impact on biological growth. Coralliidae corals grow under strict habitat requirements, which include deep water, rocky bottom, typically aggregate on banks, seamounts, under ledges, and

Figure 3. The arrangement of the polyps of coral colonies.

Figure 4. Coralliidae corals inhabit deep water, rocky bottom habitats.

in and around caves (**Figure 4**); generally, where there are strong bottom currents. The suitable growth temperature for precious corals is from 8 to 20°C.

Coralliidae colonies reach sexual maturity until 10–12 years old after then adhere to the bottom. Compared with other invertebrate animals, Coralliidae colonies live a long life, grow at slow rate, and die at low mortality. S. Giacomelli, F. Cicognae, and G. Bavestrello studied the biology of Coralliidae colonies from 1965 to 1966. The results indicate that the colonies can attain height of only 1 cm during 1 year. It is impossible to artificially breed the Coralliidae colonies until nowadays [4].

2.2. Trade market

The traditional trade market of precious coral was concentrated in Italy at the very beginning, which was supposed to be the earliest and biggest market once upon a time. There was significant trade in precious coral between the Mediterranean and India. Italy, as an irreplaceable trade market, stood head above the other trade regions could be boiled down to its unique integrated functions of precious corals' producing region, design center, and commercial market. A creative drum-like precious coral bead was first designed in Italy and became widely spread to Tibet and Japan along the Silk Road. The trade market on precious coral's season began and lasted for several centuries.

In 1847, precious coral was explored in Sea of Japan infused the fresh blood to the market. As a result, the output production of precious coral went straightly upwards while the fine designed jewelry was more popular than ever before. The history occurred similarly in 1923, a new fishery of precious coral was discovered in Taiwan that year. After the development over decades, Taiwan caught up from behind to be the world's biggest producing region for precious coral in 1964. It was reported that the output of precious coral in Taiwan accounted for 80% compared with the total production of the world in 1984. Over 90% precious corals produced in Taiwan were exported to Japan and Italy during that time.

The red coral species produced in Italy grew in relatively shallow region of the sea. The diameter of the Italian precious coral was small with a diameter less than 10 cm as the consequence of its habitat, which provided easier conditions for collecting. It explained that the common coral jewelry style in Italy was delicate, exquisite, and always set in gold or silver. Precious coral species grew in Taiwan sea were in the depth of over 200 m and could reach the size of several meters. Thus, the jewelry designs of precious coral vary from sculpture, carving, snuff bottle, headwear to common jewelry. Since precious coral is a non-renewable resource, many regulations or bans on protection are launched by governments all over the world.

2.3. Conservation

Since the species of Coralliidae family could not be cultivated, the limited resources raised public concerns on their protection. Recently, on April 8, 2008, China, which now has domestic laws to protect these species, requested that CITES include four *Corallium* species (*C. elatius, C. japonicum, C. konojoi,* and *C. secundum*) under Appendix III [5].

Meanwhile, the rising price of precious coral resulted in a variety of imitations is flooding the market. Consequently, the research on the composition and structural properties of precious coral, which will promote the development of identification of precious corals from their imitations, is more desired for both the researchers and public than ever before.

3. Basic gemological properties of main commercial species in Coralliidae family

C. rubrum, C. japonicum, and *C. elatius* are the three main species in Coralliidae family used for jewelry material in the gem market. Since they are belonging to the same zoological family, they have some components or structural characteristics in common and also some distinctive features of each own (**Table 1**).

3.1. Chemical composition and mineral components

As being determined by EPMA, LA-ICP-MS, FTIR, and XRD, the samples from Coralliidae family show similar principle chemical composition of $CaCO_3$ and major mineral component of high-Mg calcite (**Figures 5** and **6**). Beyond that, the samples also contain a small amount of organic matters, which play a magic role in the construction of precious coral skeleton (will be explained in Sections 4 and 5). One interesting fact is the theory several decades ago indicated that the color of precious coral was caused by metal ions such as Cu and Fe, of which they absorbed from the seawater, is negated by the test results of EPMA with small to non-detective amount of these elements (**Table 2**).

3.2. Optical properties

3.2.1. Color

Most precious corals show attractive even pink-to-red color, while some species have white spots, white core, or entire white on the body. To conclude, three main precious coral species'

http://dx.doi.org/10.5772/intechopen.73149

Trade name	Biological term	Producing area
Aka Coral, Aka Red Coral	*C. japonicum*	Taiwan, Japan
Momo Coral, Formosa Coral, Pink Coral	*C. elatius*	Taiwan, Japan
Midway Coral	*C. secundum*	Midway
Deep Sea Coral	*Corallium* sp.	Midway
Sardinia Coral, Red Coral	*C. rubrum*	Southern Europe, North Africa
White Coral	*C. konojoi*	Pacific Ocean

Table 1. Main precious coral species' trade names and biological term.

Figure 5. *C. japonicum* always has glass bright luster.

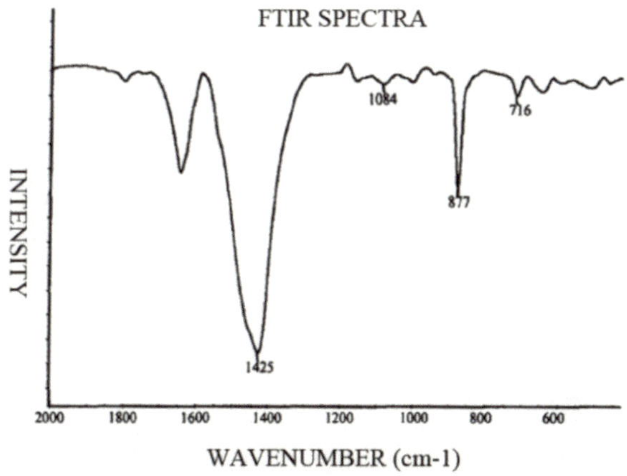

Figure 6. The FTIR spectra of precious samples identifies the biogenic calcium carbonate phase of the skeleton ($CaCO_3$), with peaks positioned at 1084, 817, and 716 cm^{-1}.

	Composition elements (ω_B%)								
	Na	Ca	Mn	Mg	Ba	Fe	Si	Cu	Sr
Max	0.503	38.775	0.015	4.413	0.161	0.381	7.652	0.0752	1.063
Min	0.227	28.022	—	0.108	—	—	—	—	0.192

Table 2. The value of principle composition elements in precious corals.

features of color are described in **Table 3** and **Figures 7–9**. The features of color are the important information to distinguish the different species of precious coral when they are kept of original dendritic shape.

3.2.2. Luster and transparency

Precious corals show waxy to glass luster when they are polished. Among the three species of precious corals, *C. japonicum* obtained the best transparency of subtranslucent, while *C. elatius* and *C. rubrum* are opaque.

3.2.3. Refractive index

The refractive indexes of precious corals are ranged from 1.49 to 1.65.

Species	Color features
C. japonicum	Orange-to-red color has white core certainly
C. elatius	Pink-to-red color, some specimens are white entirely, white core appears in most specimens
C. rubrum	Deep red color does not have white spot or core

Table 3. Precious coral species' color features.

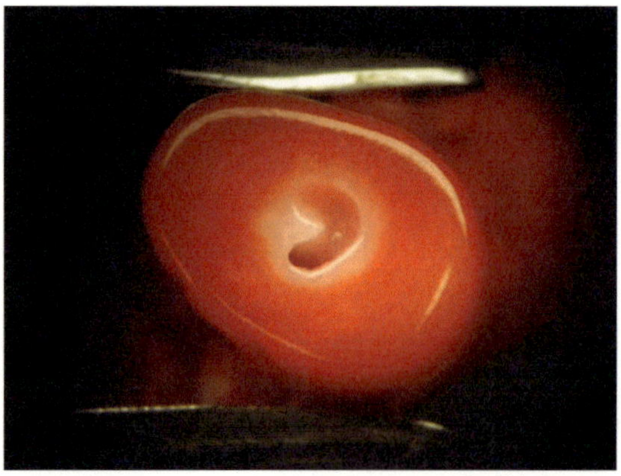

Figure 7. White core of *C. japonicum*.

http://dx.doi.org/10.5772/intechopen.73149

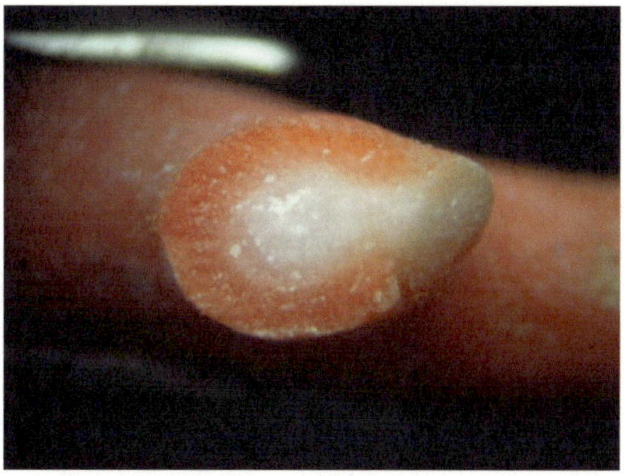

Figure 8. White core of *C. elatius*.

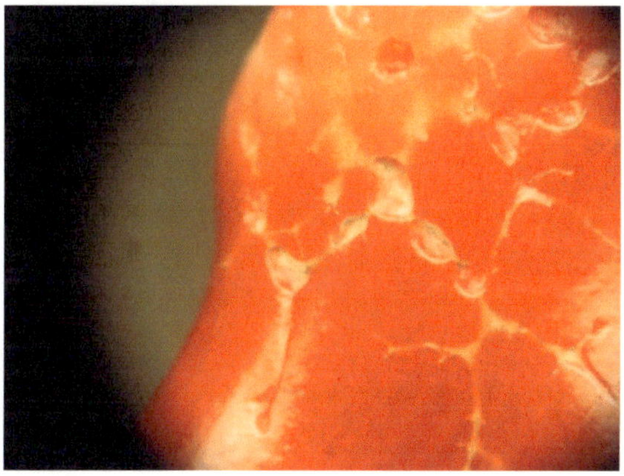

Figure 9. Worm cavities distributed on *C. rubrum*.

3.3. Mechanical properties

3.3.1. Hardness

Moh's hardness of precious corals is 3–4. However, brittleness of precious coral species is determined by environmental factors like ocean depth they grow at. Some data indicate that the precious corals, which living in deep sea, are easier to get fragile.

3.3.2. Relative density

As tested by hydrostatic weighing method, the results show the slight difference among the three species of precious coral. The relative density of *C. japonicum* is various from 2.55 to 2.65, while *C. elatius* with the relative density of 2.68–2.70, and 2.65–2.70 for *C. rubrum*.

3.4. Morphology features

The structures of most precious corals are typically consisting of two patterns. The first is ribbed or striated pattern that extends roughly parallel to the length of the coral branch. The other is a concentric, scalloped structure. Paralleled grooves are typically appear on the surface of *C. elatius* and *C. rubrum* (**Figures 10** and **11**), while the surfaces are relatively smooth on *C. japonicum* (**Figure 12**). In addition, natural dotting the coral surface, which may be described as pits and pockmarks, is only observed on *C. rubrum*. However, the paralleled stripes exist in the internal vertical section of all three species, no matter how different they look like on the surface (**Figures 13** and **14**).

These two patterns are easily to be understood by recognizing the formation of polyps in precious coral. Actually, the longitudinal section of precious coral is made in correspondence with the oral disc at the bottom.

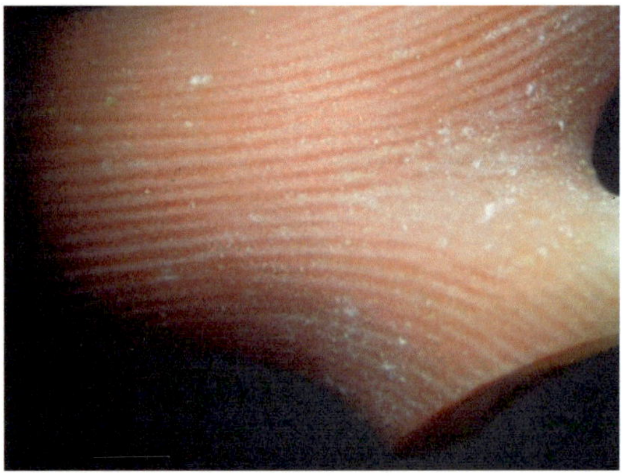

Figure 10. Paralleled grooves appear on the surface of *C. elatius*.

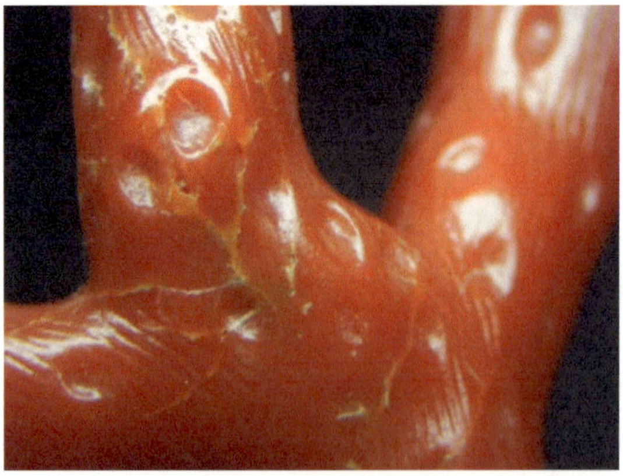

Figure 11. Paralleled grooves and pit marks appear on the surface of *C. rubrum*.

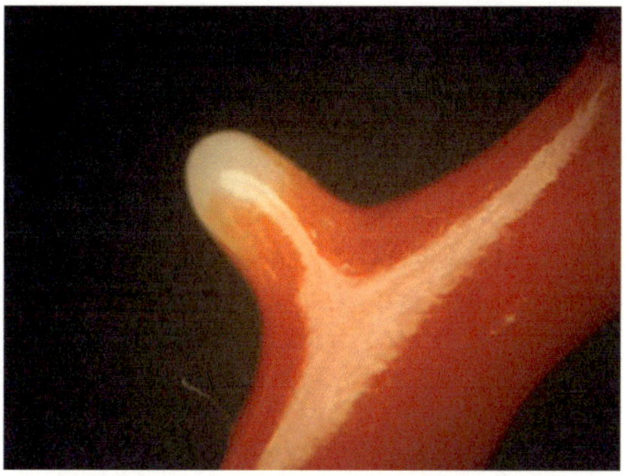

Figure 12. Relatively smooth surface of *C. japonicum*.

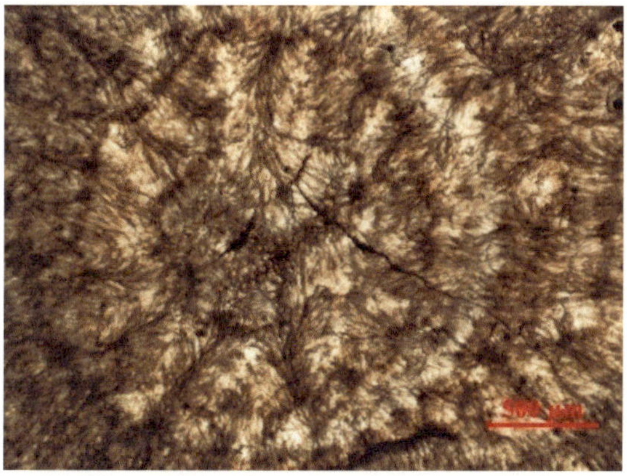

Figure 13. Micrograph of the cross section of precious coral.

Figure 14. Micrograph of the vertical section of precious coral.

4. General identification

Precious coral with its unique appearance and distinctive structure is easy to distinguish them from the common imitations such as dyed bone artifacts, dyed shell, dyed marble, conch pearl, Gilson coral, red glass, red plastic, and dyed wood. The identification features are listed in **Table 4**.

Among all these imitations, dyed corals are the most complicated. Some of the dyed corals are sharing the similar features on cross and vertical sections. Fan [6] acquired four specimens covering *C. japonicum*, *C. elatius*, *C. rubrum*, and *Isis hippuris*, which are the common species coral in trade market.

I. hippuris, also known as the name "bamboo coral", is a white coral belonging to family Isididae, phylum Cnidaria, class Anthozoa, and subclass Octocarallia (**Figures 15** and **16**). The chemical composition and mineral component are keeping the same as the precious coral in Coralliidae family. It is also hard to tell the distinction from the texture and structure. As a result, dyed *I. hippuris* is always selected to be the material to imitate the precious coral (**Figures 17** and **18**). All the specimens are tested by laser Raman spectroscopy. The result

Properties varieties	Color	Transparency	Luster	Refractive index	Relative density	Moh's hardness	Fracture	Others
Precious coral	White, pink-to-red	Opaque to semitransparent	Oily luster	1.48–1.65	2.70 (±0.05)	3–4	Even fracture	Color distribution is naturally uneven; concentric annulus on cross section; parallel stripes on vertical section acid blistering
Gilson coral	Red	Opaque	Waxy luster	1.48–1.65	2.44	3.5–4	Even fracture	No chromatic difference, fine grained texture; acid blistering
Dyed bone	Red	Opaque	Waxy luster	1.54	1.70–1.95	2.5	Splintery fracture	Uneven color on the surface and inside; round hole structure
Dyed marble	Red	Opaque	Glass luster	1.48–1.65	2.70 (±0.05)	3	Uneven fracture	Fine grained texture; acid blistering
Red plastic	Red	Transparent to opaque	Waxy luster	1.49–1.67	1.4	<3	Even fracture	Hot needle test (acrid odor); bubble inclusion
Red glass	Red	Transparent to opaque	Glass luster	1.635	3.69	5.5	Conchoidal fracture	Bubble inclusion
Dyed coral	Red	Transparent to opaque	Waxy	1.48–1.65	2.70 (±0.05)	3–4	Even fracture	Acetone reaction (color developing effect)
Conch pearl	Light salmon	Opaque	Waxy	1.486–1.658	2.85	3.5	Splintery fracture	Flame structure; acid blistering

Table 4. Identification features for precious coral and its imitations.

http://dx.doi.org/10.5772/intechopen.73149

Figure 15. Raw *I. hippuris*.

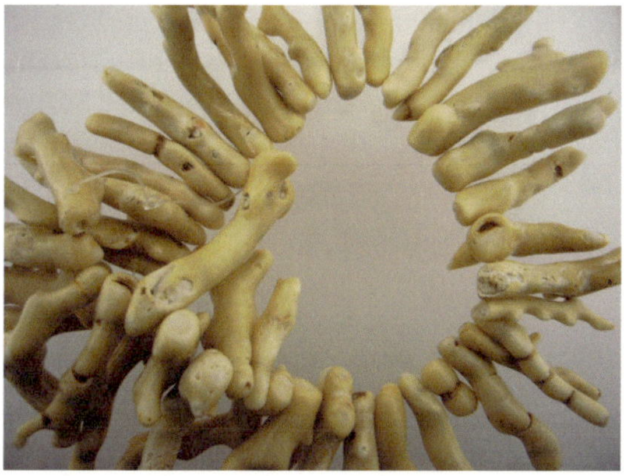

Figure 16. Polished *I. hippuris*.

Figure 17. Dyed and polished *I. hippuris* (claret-colored).

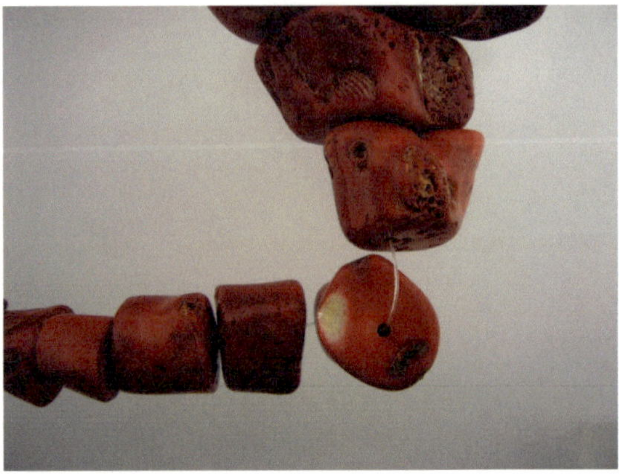

Figure 18. Dyed and polished *I. hippuris* (scarlet color).

shows that the white part of the precious coral and *I. hippuris* has the same and distinctive spectrum of calcite. The interesting thing is that the red part of all precious corals shows a suit of peaks at 1016, 1128, 1296, 1518, 2147, 2250, 2633, 3032, 3361, and 3470 cm^{-1}, while the dyed red *I. hippuris* has no featured peaks within this range. The study indicates that laser Raman spectroscopy is the useful, rapid, and nondestructive identification methods to distinguish precious coral from dyed *I. hippuris*.

5. Determination on the origin of precious coral's color

In the end of last century, carotenoids are determined to be responsible for a broad range of colors of plants and animals depending on the complex form and its incorporation into the host [7]. Rolandi et al. [8] attributed that various carotenoids are responsible for yellow, orange, brown, and blue-to-violet hues in coral skeletons. Kaczorowska et al. [9] opined that color of the red coral is caused by partial degradation and "leakage" of organic matter including beta-carotene from plant and algae material.

Raman effect arose from the inelastic collision between light and the molecule. The number, intensity, and shape of the Raman spectrum are correlative with molecular vibration or base group vibration. Raman scattering effect is a unique tool for the in situ study of biomineral [10]. Some previous studies using Raman spectroscopy have suggested that this method is useful to detect pigments in red corals. Understanding the nature of the pigments in the natural red corals can help to separate them from their equivalents. The present study shows that precious corals may contain more than one pigment.

This study was carried out on various coral samples of *C. japonicum*, *C. elatius*, and *C. rubrum*. Samples of *C. japonicum* and *C. elatius* are selected the ones with white portions. Samples were tested by 514, 633, and 785 nm laser of Renishaw inVia micro-Raman spectrometer at normal temperature and pressure. The power on the sample was 2 mW, whereas the acquisition time

http://dx.doi.org/10.5772/intechopen.73149

was 20 s and the slits were set at 40 μm. Raman tests were carried on at red color zone and white color zone of the specimens for the detection of chromatic composition. PeakFit 4.12 software was adopted to overlap the peaks at approximately 1500 cm^{-1} for finer analysis.

Not surprisingly, all the specimens present the diagnostic peaks of calcite at 1085 cm^{-1} among 100–1800 cm^{-1} range. In **Figure 19**, Raman spectra of a white portion in *C. japonicum* and *C. elatius* using three different excitation wavelengths are shown. Raman peaks are due to calcite at 1085 cm^{-1} (υ1, symmetric mode of carbonate), 712 cm^{-1} (in-plane bending mode of the carbonate), and 282 cm^{-1} (crystal lattice vibration of calcite).

In **Figure 20**, Raman spectra of four different precious coral samples, using the 514 nm excitation, show two intense additional bands in the region 1000–1800 cm^{-1} compared to the spectra of white core in samples. In all the samples tested in this study, these peaks can be found. It is clear that these peaks, which are characteristics of polyenic chains, are responsible for the pigmentation in the samples.

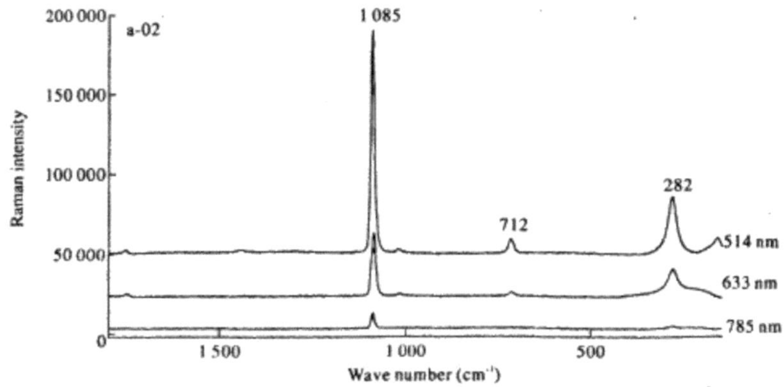

Figure 19. Raman spectra of the white core in the sample, using three different excitation wavelengths.

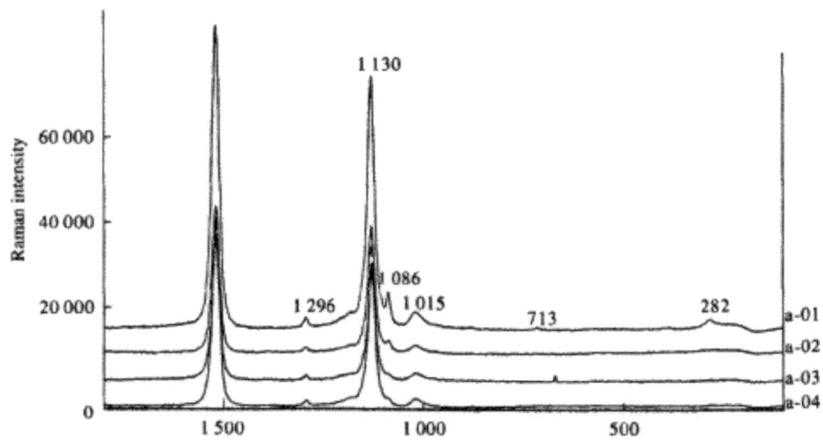

Figure 20. Raman spectra of the red parts in the samples, using 514 nm excitation wavelength.

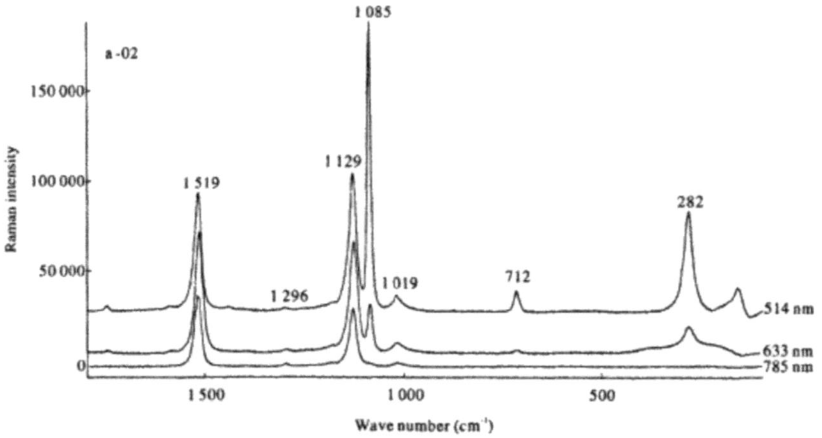

Figure 21. Raman spectra of the red parts in the samples, using three excitation wavelengths.

Figure 18 shows the Raman spectra of samples presented in **Figure 21**, which is measured by three different excitation wavelengths. In these figures, changes in shape and relative intensities occur at the same point by changing the excitation wavelength. Region 1500 cm^{-1} is sensitive to the length of the polyenic chain.

All the samples show the two major Raman peaks at 1130 and 1519 cm^{-1}. These peaks were not detected in the white core of the precious coral samples. C=C (υ1 near 1500 cm^{-1}) and C—C (υ2 near 1130 cm^{-1}) stretching vibration observed only in red coral samples suggests that these compounds play the main role with regard to the color. The series of bands observed by decomposition can be assigned to a series of polyenic molecules lacking methyl groups. Different colors of the precious corals are because of different mixtures of pigments in varying relative proportions. These kinds of pigments are not found free in nature but rather are complexed with some forms of a protein. A $CaCO_3$ — pigment complex has been proposed for the calcareous skeletons of corals and some shells. In the absence of an adequate protein complex, the pigments in the precious corals could possibly be complexed with carbonates. All of the samples are colored by a mixture of pigments. Different colors are explained by different mixtures, not by a single pigment. As we mentioned in Section 3.1, organic composition, even is a small amount, plays an important role on the color of precious corals.

6. Determination on organic matrix and biogenic calcite in precious corals

In the process of determining the origin of precious corals' color, we are very clear about the existence of organic matter and its role. In spite of detailed studies, the internal structure of the axial skeleton of precious corals is not understood. In particular, the spatial relation between the organic matrix and biogenic calcite remains in great part unexplored. Curiously, the answers for how is the spatial distribution of the organic matter, the interface between the organic matter and biogenic calcite, the effect of the organic matter produced during the mineral construction of precious corals are still need to be studied.

http://dx.doi.org/10.5772/intechopen.73149

Different techniques, such as SEM and HR TEM, will be applied to study how each hierarchical layer of precious corals assembles into larger units. EBSD and TEM studies will show the degree of crystallographic misorientation between the building blocks. The approach of in situ analysis will be carried out by means of sputtering ion, decalcification, and structural analysis techniques. The method is to observe the spatial distribution of organic framework and organic/inorganic interface spatial relationship in multilevel. The study is to understand a multiscale physico-chemistry characterization of mineral part of precious coral and its three-dimensional architecture.

6.1. Organic matrix

Since the amount of organic matter is very slight in precious coral, an inclined decalcification experiment was conducted in the study. First, fine polished vertical and cross sections of precious coral samples were fixed on an inclined specimen holder. Then EDTA (Ethylenediaminetetraacetic acid disodium salt) solution in certain concentration was dropping from dropper to etch the surface of the samples gradually. Several hours later, a transition surface including fine polished part, semi-etched part, and fully decalcified part was formed. Spicules and organic matrix of precious coral were observed by SEM.

We observed several types of spicules, such as cross spicule (**Figure 22**), eight-axial spicule (**Figure 23**), six-axial spicule (**Figure 24**), and double eggplants spicule (**Figure 25**). Organic matrix could be easily found around the surface of the spicules.

The organic matrix net is distributed in layers to form a three-dimensional cavity (**Figure 26**), which provides the growth space for the spicule. Most of spicules were wrapped by organic matrix (**Figure 27**). But, we also observed some flaw parts of the samples (**Figures 28** and **29**) lacked of organic matter leaded a sprawling growth habit, which indicate the growth of precious coral were under the control of organic matter.

Figure 22. Cross spicule.

Figure 23. Eight-axial spicule.

Figure 24. Six-axial spicule.

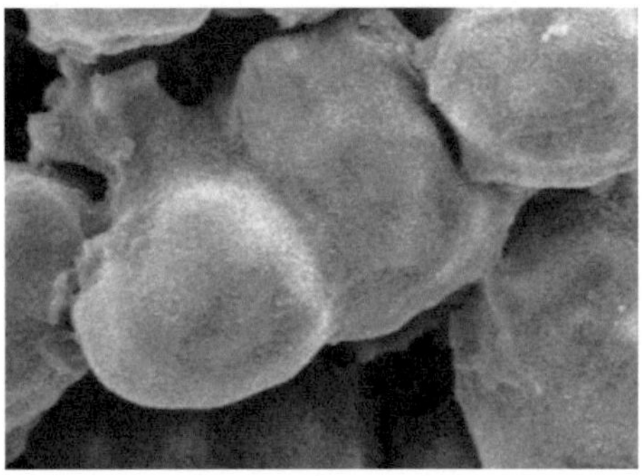

Figure 25. Double eggplants spicule.

http://dx.doi.org/10.5772/intechopen.73149

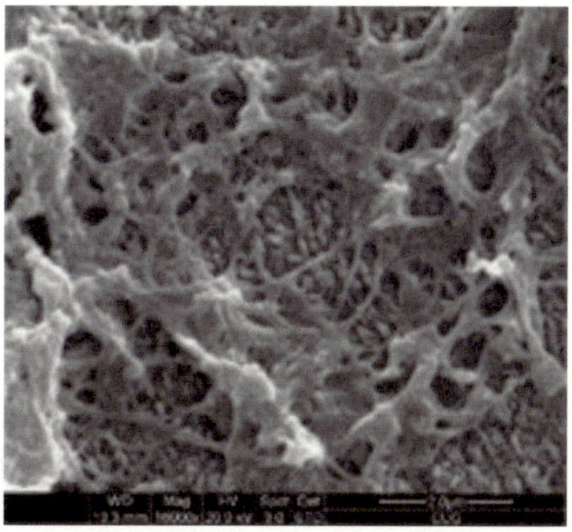

Figure 26. The organic matrix net is distributed in layers to form a three-dimensional cavity.

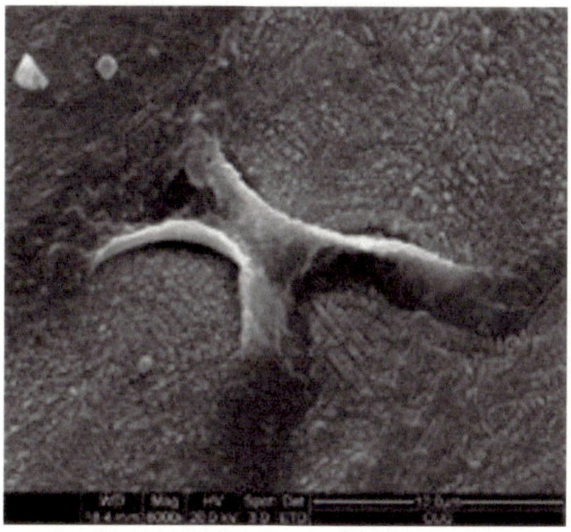

Figure 27. Spicules were wrapped by organic matrix.

Figure 28. Sample with flaws.

Figure 29. Radial growth of calcite crystal without the control of organic matter.

6.2. Multilevel modular mesocrystalline organization in precious coral

Biominerals often display morphological, chemical, and crystallographic patterns at length scales ranging from the nanoscale to the macroscale (**Figure 30**).

In the skeleton of precious coral, we observed that building blocks were arranged into several hierarchical levels of oriented modules. The modules in each hierarchical level assemble into larger unit that comprises the next higher level of the hierarchy. The EBSD and TEM studies show the degree of crystallographic misorientation between the building blocks, which decreases with decreasing module size (**Figure 31**). The crystal units of axial skeleton are columnar with two perpendicular directions. Throughout the region, the orientation of c axis is consistent with its long axis of each unit. The orientation of a axis and b axis of each unit is relatively complex with four kinds of orientation. One orientation is parallel to the plane. The other orientation is perpendicular to the plane. The remained orientation crosses the plane

Figure 30. Precious coral skeleton formed by nanoscale to the macroscale calcite crystals.

http://dx.doi.org/10.5772/intechopen.73149

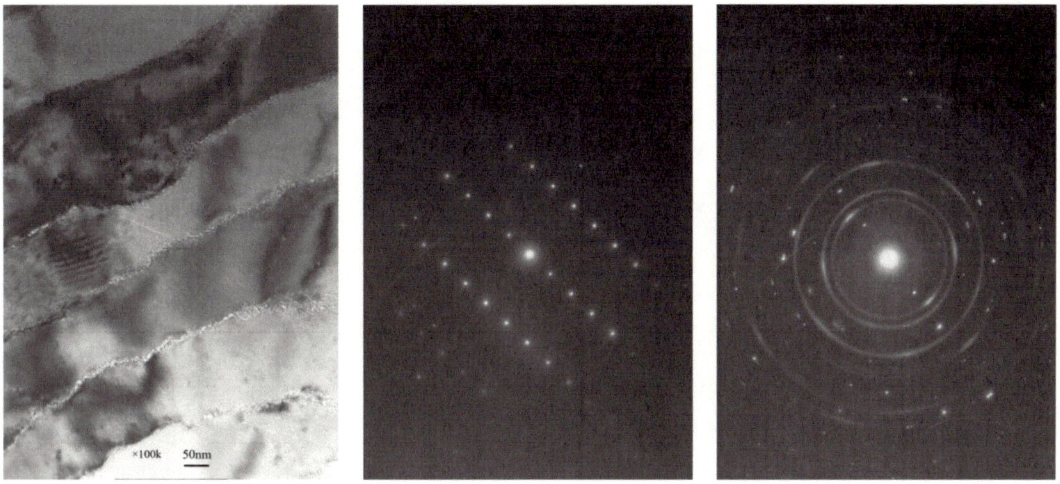

Figure 31. TEM test shows the crystallographic orientations of precious coral sample.

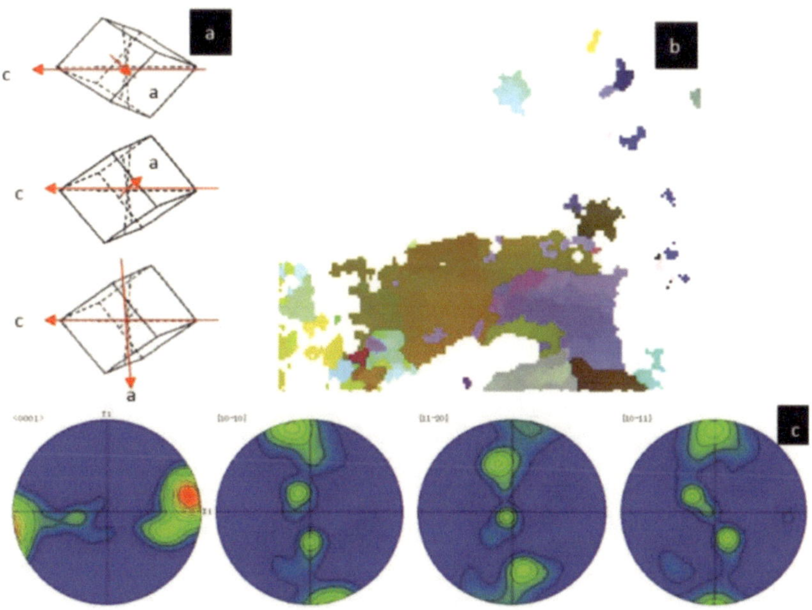

Figure 32. The misorientation angle of different units of precious coral is between 33 and 48°. The misorientation angle is nearly 5° in the area of similar orientation by EBSD test.

with different skew angles. The misorientation angle of different units is between 33 and 48°. The misorientation angle is nearly 5° in the area of similar orientation (**Figure 32**).

That is to say, precious coral, as a member of biomineral family, assembled skeleton as a delicate arrangement of a hierarchy of crystals with well-defined orientations under the control of organic matrix.

Author details

Luwei Fan

Address all correspondence to: luwei.fan@gmail.com

China University of Geosciences, Beijing, China

References

[1] Smith CP, Mcclure SF, Eaton-Magaña S, Kondo DM. Pink-to-red coral: A guide to determining origin of color. Gems & Gemology. 2007;**43**(1):4-15. DOI: Doi.org/10.5741/gems.43.1.4

[2] Zou Renlin, Gan Zijun, Chen Shaomou, Ke Peihui, Qiu Dezhong. Red Coral. 1st ed. Beijing: Science Press; 1993. 4 p. ISBN: 7-03-003972-6/Q.481

[3] Karampelas S, Fritsch E, Rondeau B, Andouche A, Metivier B. Identification of the endangered pink-to-red stylaster corals by Raman spectroscopy. Gems & Gemology. 2009;**45**(1):48. DOI: Doi.org/10.5741/gems.45.1.48

[4] Cognetti G. FAO Congress on *Corallium rubrum*. Marine Pollution Bulletin. 1989;**20**(2):95. DOI: Doi.org/10.1016/0025-326x(89)90237-3

[5] Red Coral (*Corallium rubrum*). 2017. Available from: http://www.arkive.org/red-coral/-corallium-rubrum/ [Accessed: Nov 15, 2017]

[6] Fan L, Lv L, Wang Y, Yan K. Laser Raman spectroscopic chareacteristics of gem-quality red coral. Journal of Gems & Gemology. 2007;**9**(3):1-3. DOI: 10.3969/j.issn.1008-214X.2007.03.001

[7] Karampelas S, Fritsch E, Sklavounos S, et al. Determination by Raman scattering of the nature of pigment in cultured freshwater pearls from mollusk *Hyriopsis cumingi*. Journal of Raman Spectroscopy. 2007;**38**:217-230. DOI: Doi.org/10.1002/jrs.1626

[8] Rolandi V, Brajkovic A, Adamo I, Bocchio R, Landonio M. Gem corals: Classification and spectroscopic features. Australian Gemmologist. 2005;**22**(7):285-297

[9] Kaczorowska B, Hacura A, Kupka T, Wrzalik R, Talik E, Pasterny G. Spectroscopic characterization of natural corals. Analytical and Bioanalytical Chemistry. 2003;**377**:1032-1037. DOI: Doi.org/10.1007/s00216-003-2153-1

[10] Fan L, Yang M. In situ resonance Raman spectra of organic pigments in momo coral. Journal of Earth Science. 2008;**19**(2):146-151. DOI: Doi.org/10.1016/s1002-0705(08)60033-3

Section 3

Corals in Changing Environments

Chapter 5

Coral Reef Bleaching: An Ecological and Biological Overview

Norma Olguín-López, Carolina Gutiérrez-Chávez,
Víctor Hugo Hérnández-Elizárraga,
César Ibarra-Alvarado and Alejandra Rojas-Molina

Additional information is available at the end of the chapter

http://dx.doi.org/10.5772/intechopen.69685

Abstract

Coral reefs are central to the biology of our planet, but in the past few decades, they have suffered a severe decline due to a variety of natural and anthropogenic disturbances. On a worldwide scale, the main disturbance is bleaching, which can be defined as the loss of endosymbiotic dinoflagellates and/or of their photosynthetic pigments from their cnidarian host; with that, the normal pigmentation of the tissue of cnidarians is generally lost and the white calcium carbonate skeleton becomes visible through the transparent tissue of the host. Coral bleaching can be triggered by multiple factors, but most of the bleaching observed in the field is a result of elevated sea surface temperature. It has been widely documented that bleaching is deleterious to coral reefs, significantly altering the biological and ecological processes that maintain reef communities; yet populations resistant to climate change have recently been identified, and it has been reported that acclimatization occurs in less than two years. The aim of this review is to provide up-to-date information regarding cnidarian-dinoflagellate symbiosis; causes of coral bleaching; mechanisms underlying this phenomenon; consequences of bleaching; and the survival mechanisms by which coral reefs face this threat.

Keywords: cnidarians, coral bleaching, global warming, marine ecosystem, symbiotic dinoflagellates, *Symbiodinium*

1. Introduction

Despite occupying only 0.1% of the ocean, coral reefs play a critical role in marine ecology, and in human sustainability, they are invaluable from a variety of perspectives. They are home to more than a quarter of all the species that inhabit the ocean, provide coastal protection, and support more than 10 millions of people living on tropical coasts [1, 2]. Economic goods and ecosystem services of coral reefs are worth more than US $20 trillion annually [3].

Coral reefs are central to the biology of our planet; in terms of biodiversity, they are the most rich, complex, and productive marine ecosystem on Earth. It has been estimated that over 90% of the species inhabiting coral reefs have yet to be described [4]. Therefore, it is not surprising that the organisms that constitute these ecosystems produce a great variety of molecules with unique structural characteristics that exhibit numerous biological activities [5], and are considered a rich source of novel bioactive agents with great pharmaceutical and biotechnological potential [6].

Unfortunately, coral reefs are extremely susceptible to the stress related to greenhouse gas emissions, particularly ocean warming and acidification [7–10], which provoke disturbances that can seriously affect and break down the homeostatic capacity of coral reefs to overcome stressors [11]. One of these disturbances is the event called "bleaching," in which the tissues of corals and hydrocorals lose their photosynthetic endosymbiotic zooxanthellae (dinoflagellate algae of the genus *Symbiodinium*) or their pigments, which exposes the white exoskeleton of calcium carbonate [12–14].

Numerous studies have shown that bleaching is harmful to coral reefs, since it significantly alters the biological and ecological processes that maintain equilibrium in the reef communities. Bleaching episodes have resulted in massive damage to coral reefs around the world, with serious effects on the maintenance of biodiversity in the marine tropics. Bleaching is also responsible for other declines in reef health, such as the increase in coral diseases, declines in reef calcification, the breakdown of reef framework by bioeroders, and the loss of critical habitat for associated reef organisms [15–18]. Climate models predict that, if CO_2emissions continue to rise at the current rate, bleaching events will increase in frequency and severity, threatening the survival of coral reefs. Actually, it has been projected that 90% of coral reefs around the world will be at risk if bleaching events occur annually [15, 16, 19].

The US National Oceanic and Atmospheric Administration (NOAA) coral reef watch, the Global Coral Reef Monitoring Network (GCRMN), as well as other environmental agencies around the world have been monitoring mass coral bleaching events for more than a decade, in order to understand conditions that cause bleaching and to develop measures to rehabilitate reefs [1–20]. On the other hand, academic researchers have addressed the study of this phenomenon employing different approaches, including genomics, transcriptomics, and proteomics to better comprehend the molecular mechanisms that provoke it, its consequences, and the potential adaptive response of cnidarians to this condition [21–29]. Undoubtedly, bleaching represents an enormous threat to the survival of coral reefs, as bleached organisms display an increased susceptibility to pathogens, a decreased resistance to predators and bioeroders, declines in reef calcification, depressed growth and reproduction rates, and a lessening of ability to repair damages [15–18]. In this context, the purpose of this review is to provide updated information regarding cnidarian-dinoflagellate symbiosis; causes and consequences of coral bleaching; the molecular processes underlying this phenomenon; and the survival mechanisms by which coral reefs face this threat.

1.1. Search strategy

The literature consulted corresponds to the main reviews and articles explaining the phenomenon of bleaching, which contributes to the understanding of the relationship between the causes, mechanisms, and consequences of bleaching.

2. Cnidarian-dinoflagellate symbiosis

An important feature of coral reef ecosystems is that most of the 798 reef-forming species have developed a mutualistic symbiosis with unicellular dinoflagellate algae of the genus *Symbiodinium*, commonly referred to as zooxanthellae (**Figure 1**). This symbiosis is essential in the formation of large and important structures in coral reefs [30]. The genus *Symbiodinium* encompasses nine major clades (A to I), most of which were identified based on the ribosomal DNA small subunit [31]. These tiny organisms (8–10 μm in diameter) live within cnidarian cells, inside a host-derived vacuole (symbiosome) located within the gastrodermal cell layer. Under normal conditions, the population density of symbionts ranges from 0.5 to 5 × 10^6 cells per cm^2 of coral surface, although the limits of this range may vary [30, 32]. Regulation of symbiont density involves different mechanisms, such as limiting the nutrients delivered from corals to symbionts, digesting photosynthetic algae, expulsing excess symbionts, rearranging excess symbionts into new coral cells, and producing growth inhibition factors [30, 33–35].

Symbiodinium algae carry out photosynthesis, but instead of retaining the sugars and amino acids that result from this activity for their own growth and reproduction, they deliver more than 95% of their photosynthetic production to their host. In return, *Symbiodinium* has direct access to the waste products of coral metabolism, such as carbon dioxide, which is used in photosynthesis. The recycling of nutrients between coral host and its endosymbionts is extremely efficient and allows them to live in nutrient-poor waters [30, 35, 36]. Endosymbionts translocate molecular oxygen to their hosts and most of their photosynthetically-fixed carbon in the form of glycerol, glucose, amino acids, and lipids. Moreover, the oxygen produced during photosynthesis helps maintain the high levels of ATP required for the calcification process [30]. In exchange, cnidarians afford inorganic nitrogen, phosphorus, and carbon, as well as a lighted environment that provides refuge from herbivores (**Figure 1**) [36, 37]. Signaling molecules regulating host-symbiont interaction have not been fully characterized. It has been suggested that translocation of photosynthetic products may be controlled by host release factors (HRFs) [30, 36], some of which are of proteinaceous nature (approximately 10 KDa) [38], although free amino acids and mycosporine-like amino acids have been found to induce the release of photosynthetically fixed carbon [39, 40].

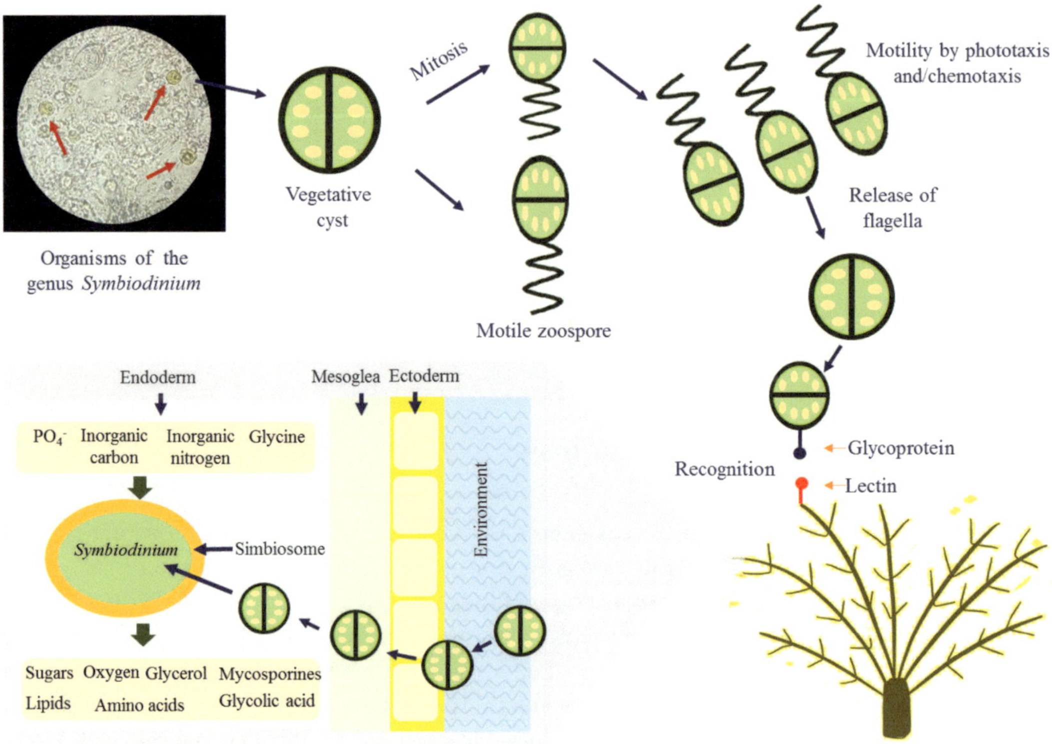

Figure 1. Cnidarian-*Symbiodinium* symbiosis.

Another important characteristic of reef-forming cnidarians is their capacity to produce calcified skeletons. Calcification is a crucial process for building the largest biological structures in the world [38, 41]. Despite its importance, the mechanism of calcification, or, more broadly, biomineralization, is little known. The coral exoskeleton is made of aragonite, a crystalline form of calcium carbonate ($CaCO_3$), and it is completely covered by a thin single layer of epithelial cells called calicoblastic epithelium (calicodermis). These cells play an important role in coral calcification, controlling the ionic composition of the medium in which calcification occurs (extracellular calcifying medium, ECM) [38, 41, 42].

Two hypotheses have been proposed to explain the stimulation of calcification by the presence of dinoflagellate symbionts [38]. One of them considers that absorption of CO_2 and release of O_2 by the symbionts [43] indirectly alter the pH and/or modify inorganic calcium concentration inside the gastrovascular cavity of the cnidarian, which leads to calcification by the precipitation of $CaCO_3$ [44]. The other hypothesis states that symbiotic algae produce organic molecules, such as nitrogenous compounds, glucose, and glycerol, that carry out the following functions: provide energy for calcification, modify the buffering capacity of the ECM, or are precursors for organic matrix synthesis [38, 45].

Colonies of reef-building cnidarians exhibit a great variety of colors due to the different types and concentrations of pigments conferred by the presence of the millions of symbionts found per square centimeter in their tissues [46]. Symbionts can be acquired by either vertical or horizontal transmission. Vertical transmission occurs when symbionts are transferred from parents to offspring through direct inheritance. On the other hand, horizontally transmitted symbionts are acquired from the environment, such is the case of 80% of scleractinian symbiotic corals, whose larvae acquire their symbionts during a nutritional process [47–50]. The process of the establishment and maintenance of symbiosis comprises six phases: (i) initial contact; (ii) immersion of the symbiont; (iii) dynamic intracellular sorting of the symbionts; (iv) proliferation of symbionts in gastrodermal cells; (v) stabilization dynamics; and (vi) symbiosis dysfunction and breakdown [38]. According to some authors, the recognition of symbionts works in the same way as the recognition of pathogens, i.e., it involves pattern recognition receptors (PRRs) that are able to recognize and bind to specific conserved components of the cell walls of the algae (carbohydrates, lipids, and proteins) [51–53].

During their life cycle, *Symbiodinium* cells interchange between a vegetative cyst, which is the dominant form in the endosymbiotic state, and a motile zoospore that possesses thecal plates and two flagella, one transverse and one longitudinal [54]. The current hypothesis about the establishment of symbiosis between host cnidarians and *Symbiodinium* states that in the first phase, lectins, secreted by the host cell, induce symbionts to progress to cyst stage [30, 53]. Subsequently, lectins on the host cell surface attach to glycoproteins present on the surface of non-motile *Symbiodinium* cysts, which are subsequently phagocytized and carried into an early endocytic compartment by Rab5 proteins. Afterward, cells that were successfully recognized end up in a symbiosome, whereas damaged symbionts are digested by fusion with lysosomes after transiting through the late endocytic compartment. This traffic takes place through Rab7 and Rab11 proteins [30]. The relationship between cnidarians and symbiotic algae is regulated by environmental and physiological conditions of the host [30, 55].

It is clearly evident that establishment and maintenance of cnidarian-*Symbiodinium* symbiosis are critical to preserve homeostasis in coral reef ecosystems. On one side, algal symbionts obtain from their hosts protection and inorganic compounds, which are essential for their metabolism. On the other hand, cnidarians receive from their symbionts a great percentage of their energy demand and a balanced pH and precursor molecules needed for the calcification process. Undoubtedly, disruption of this symbiotic relationship can trigger numerous adverse effects, not only for the reef-forming organisms, but also for the great variety of organisms that depend on coral reefs.

3. Coral bleaching: causes and global episodes

Bleaching has been described as a visual effect of the stress that occurs when the symbiosis between reef-forming cnidarians and their symbiotic algae breaks down (**Figure 2**). It comprises the loss of pigmentation in coral reefs due to decreased *Symbiodinium* population, reduction in the concentration of their photosynthetic pigments, or both [12, 17, 56, 57]. Up to now, at least four general cellular mechanisms of algal loss have been proposed: *in situ*

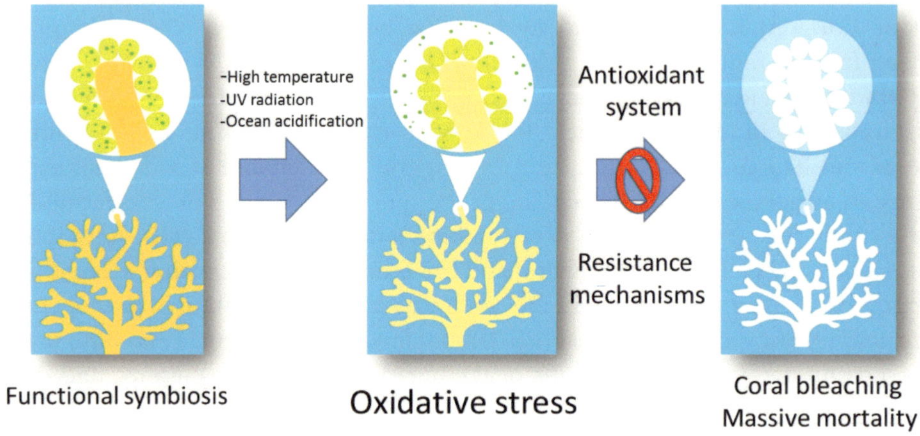

Figure 2. Coral reef bleaching. Modified from Baird et al. (2008) and http://oceanservice.noaa.gov/facts/coral_bleach.html.

degradation, symbiont (intact or degraded) expulsion, host-cell detachment, and host-cell death [58, 59].

It has been widely documented that several factors can induce coral bleaching, including reduced salinity [60]; amplified or reduced solar irradiation [61–64]; elevated [62, 65] or low water temperature [66, 67]; and the presence of chemical contaminants in the sea water, such as herbicides, pesticides, and cyanide [68]. However, field and laboratory investigations have indicated that elevated temperature is the principal cause of bleaching. Some of these studies are listed below.

(a) Hoegh-Guldberg and Smith (1989) studied the conditions required to achieve experimental bleaching of *Stylophora pistillata* and *Seriatopora hystrix*; they found that exposing fragments of these species to 32°C for 7 h caused a rate of expulsion of zooxanthellae 1000 times greater than in the control, which was kept at 27°C [62].

(b) Jokiel and Coles (1990) concluded that bleaching can be induced by short-term exposure (i.e., 1–2 days) to temperature increases of 3–4°C above normal summer temperature, or by long-term exposure (i.e., several weeks) to temperature increases of 1–2°C [66].

(c) Lesser et al. (1990) showed that increases in temperature significantly reduced the total number of zooxanthellae per polyp of the class Anthozoa. At the same time, temperature, solar radiation, and ultraviolet radiation independently increased the activities of superoxide dismutase (SOD), catalase, peroxidase, and ascorbate in the symbionts of *Palythoa caribaeorum* [69].

(d) Warner et al. (1999) sampled bleached colonies of *Montastraea faveolata* and *Montastraea franksi* in Florida. The bleaching was caused by a rise in temperature to over 30°C for 6 weeks during the summer of 1997. The analysis of these samples indicated damage in Photosystem II and a decrease in the expression of the D1 protein [70].

(e) Eakin et al. (2005) reported that the most serious bleaching event that affected Caribbean and Atlantic coral reefs happened in 2005, the warmest year ever documented in the Northern Hemisphere. Whitening of coral reefs provoked mortality of living reef-forming organisms by up to 40% [71].

Among the other environmental factors that can cause bleaching, besides thermal stress, solar radiation [17] has the most significant influence. This stressor can act independently of, or synergistically with, elevated sea water temperature [14, 72]. Fitt and Warner (1995) found a substantial reduction in symbiont photosynthesis in the coral *Montastraea annularis* after exposure to ultraviolet and blue light [73], whereas Gleason and Wellington (1993) demonstrated that bleaching occurred more readily when corals were exposed to high energy, short wavelength solar radiation [61]. Other studies have confirmed that exposure to high temperature and ultraviolet light exacerbates the stress on the symbiont population, contributing to bleaching [74, 75]. Along with increased temperature and high irradiance, elevated atmospheric CO_2 concentration is also affecting coral reefs by lowering ocean pH levels, which decreases calcification, increasing the rates of erosion that exceed the capacity for accretion of the reefs in tropical and subtropical zones [76–78].

Bleaching of coral reefs was first recorded in the 1870s [79, 80], and since 1980, coral reef bleaching and mortality episodes linked to elevated temperatures have been monitored. Ever since 1979, nine major coral bleaching episodes have affected coral reefs around the world, with an alarming rise in scale and frequency. At first, the phenomenon was regarded as linked to El Niño Southern Oscillation (ENSO) events [77, 81], which caused a high mortality in colonies of *Millepora platyphylla* in the Eastern Pacific coral reefs. However, since the early 1990s, it is known that bleaching is correlated in near real time with anomalously high satellite-derived sea surface temperature (SST) [81]. This relationship was clearly observed in the Caribbean basin during the 1980s and 1990s, when annual coral bleaching increased logarithmically with SST anomalies [82]. A minimal rise of only 0.1°C in regional SSTs caused a 35% increase in the extent and intensity of bleaching, while when SST was increased by 0.2°C and above, massive bleaching events took place [15].

The GCRMN (with support from more than 30 countries, the IOC-UNESCO, UNEP, IUCN, and the World Bank) has provided annual reports on the state of coral reefs. The report from 1997 to 1998 indicated the occurrence of an unprecedented global episode of mass coral bleaching. Before 1998, GCRMN surveys reported a 9.5% loss of corals in six regions. During 1998, one of the hottest years on record, these regions lost an average of 17.7% of their live reef-building corals [83]. The 1998 record was surpassed in 2005 and again in 2010 [71, 84]. However, 2014 [71, 85] and 2015 were considered the hottest years ever recorded, and 2016 continued this tendency, which has led to the longest and most damaging global coral bleaching event officially documented [86].

A summary of the major bleaching events reported since the 1980 is shown in **Table 1**. Unfortunately, anthropogenic activities that affect environment have provoked that coral bleaching events increase in frequency and severity, which represents a very serious threat to coral reefs worldwide. Reef-forming organisms have survived to mass extinction events in the past; however, it is unknown if they will be able to adapt fast enough to cope with rapid environmental change.

Date	Bleaching event
79/80	Great Barrier Reef
82/83	Eastern Pacific, Indonesia, Tokelau, Panama, Galapagos, Moorea, S-Japan
84	Moorea
86/87	Great Barrier Reef, Moorea, Caribbean Galapagos
88	Andaman Sea
89	Jamaica
90	Jamaica
91	Moorea, Andaman Sea
92/93	Great Barrier Reef, Galapagos
94	Pacific, E-Africa, Great Barrier Reef, Moorea
96	Arabian Gulf, Hawaii
97/98	**Worldwide**
00	South Africa, Easter Island
02	Great Barrier Reef, Arabian Gulf, Hawaii
05	Eastern Caribbean, Southern Africa
06	Great Barrier Reef
07	Iran
08/09	Queensland Australia
10	**Worldwide**
11	Western Australia Coast, Ningaloo Reef
15/16	**Worldwide**

Table 1. Documented bleaching events.

4. Molecular mechanisms underlying bleaching

It has been shown that, under elevated sea water temperatures and high solar radiation, *Symbiodinium* photosynthesis leads to intense oxidative stress in the two partners of the mutualistic symbiosis between reef-forming cnidarians and dinoflagellate endosymbionts [35, 87]. Oxidative stress involves the production and accumulation of reactive oxygen species (ROS), which can damage lipids, proteins, carbohydrates, and DNA [7, 52]. ROS play a key role in signal transduction of cell damage mediators and in processes such as apoptosis, autophagy, and necrosis [35, 88, 89]. In fact, numerous evidences indicate that the first step of bleaching involves inhibition of photosystem II (PSII) activity, a phenomenon referred to as photoinhibition, which often results in the overproduction of ROS, especially H_2O_2 [90, 91]. These cell damage response pathways are important during bleaching and depend on stress intensity and duration [17, 35]. Moreover, it has been proposed that bleaching itself is a control mechanism to minimize the harmful effects caused by the metabolic imbalance in cnidarian-algal symbiosis [56].

Photosynthetic damage in *Symbiodinium* spp. has been proven using non-invasive chlorophyll fluorescence techniques, which demonstrated a significant loss of PSII photochemical efficiency during bleaching [70, 92, 93]. Elevated temperatures and high radiation cause photoinhibition and damage to the chloroplasts and the photosynthetic apparatus of the symbionts, through at least three pathways:

- Damage to the D1 protein, a core component of PSII, which is involved in the photolysis of water. This hypothesis proposes that, under stress conditions, the CO_2 fixation is limited and the electrons of PSI are transferred to O_2. This results in the formation of superoxide, which is quickly converted into H_2O_2 by superoxide dismutase (SOD). H_2O_2 hinders the repair of PSII by impairing the synthesis of the precursor D1 protein (pre-D1) [91]. During bleaching, the rate of photoinhibition and damage of D1 protein exceeds the rate of the PSII repair cycle [87].
- Inactivation of ribulose-1,5-bisphosphate carboxylase oxygenase (Rubisco), one of the key enzymes of the Calvin-Benzon cycle [94]. This mechanism was proposed when Bhagooli (2013) found that inhibition of the Calvin-Benson cycle by glycolaldehyde induced photoinhibition and coral bleaching, even at optimal temperatures [95].
- Injury of the thylakoidal membranes by ROS, which elicits an energy decoupling of the electron transport in both PSI and PSII, resulting in diminished ATP and NADP production [52]. The excess electrons reduce oxygen instead of NADPH with the subsequent generation of superoxide ion, which is reduced by SOD to H_2O_2. This last molecule reacts with ferrous ion and yields, the even more reactive, hydroxyl radical. Furthermore, excess electrons can react with photosynthetic pigments and molecular oxygen to produce atomic highly reactive oxygen. All the above-mentioned ROS spread to the host tissues triggering innate immunity, via the transcription factor NF-kB, leading to apoptosis [52]. Activation of NF-κB also induces iNOS expression, increasing the levels of nitric oxide, which reacts with superoxide ion to form peroxynitrite (ONOO-). This anion is highly reactive and damages the mitochondrial membrane, which releases pro-apoptotic molecules, such as AIF, promoting apoptosis (**Figure 3**) [96].

Another hypothesis to explain the molecular events that lead to bleaching proposes disruption to the carbon-concentrating mechanisms of the coral host. According to this hypothesis, bleaching is initiated by the inability of the coral to efficiently supply its symbionts with CO_2, mainly in periods of high solar radiation, when the algal CO_2 demand is very high. The lack of CO_2 needed for "dark reactions" reduces consumption of ATP and NADPH, blocking electron transport components. Sustained concentration of light excitation energy into the over-reduced electron transport chain triggers photoinhibition, damage to the photosynthetic components (mainly PSII), and ROS generation [97]. Excessive ROS production stimulates antioxidant defenses in the cnidarians and their symbionts, and it is also related to the exit of photosynthetic algae [98]. Moreover, bleaching can be caused by damage to "dark reactions" in the absence of thermal stress [99].

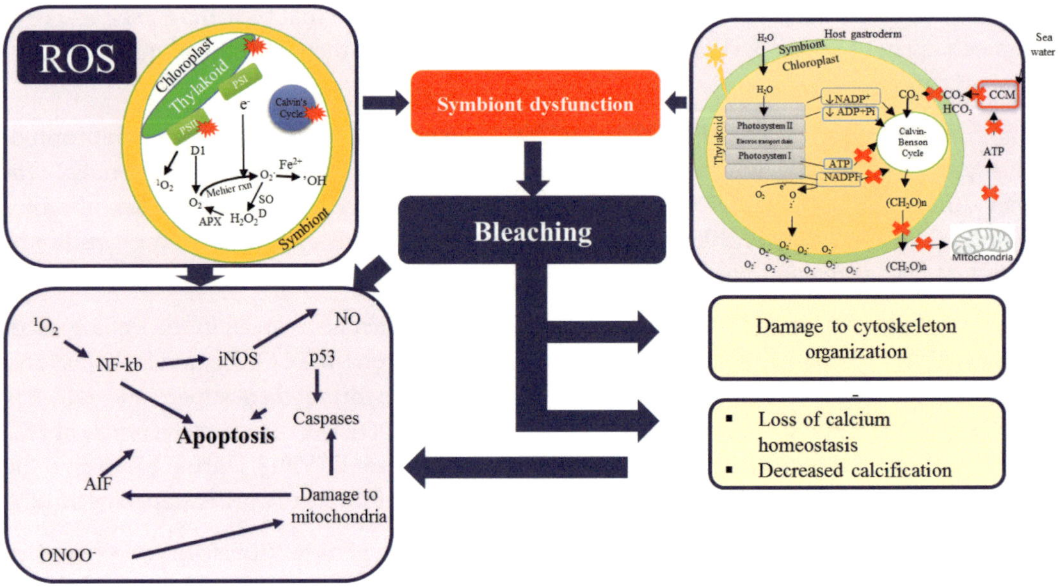

Figure 3. Molecular mechanisms of coral bleaching. Modified from Weis [52] and Wooldridge [98].

Recent investigations have examined the bleaching phenomenon employing genomic and transcriptomic approaches in order to measure changes in the expression of genes and transcripts during thermal stress and bleaching in different cnidarian species [100], including *Acropora nana* [21], *Stylophora pistillata* [22], *Acropora millepora* [101], *Acropora palmate* [24], *Aiptasia pallida* [25], *Orbicella faveolata* [23], and *Acropora hyacinthus* [27]. The results from those studies revealed that differential expression patterns occur between normal and bleached specimens, providing evidence that several important cell processes are affected by bleaching, such as stress response, Ca^{2+} homeostasis, cytoskeleton organization, cell transport, cell proliferation, apoptosis, calcification, protein expression, immune response, and metabolism, among others [21, 23–25, 27, 29, 101].

On the other hand, proteomic approaches have been applied to assess the effect of post-translational environmental stress on marine organisms [29. A recent study carried out on *Acropora palmata* showed that bleaching induced a differential protein expression response in this cnidarian. Thirty-eight key proteins were differentially expressed, primarily transcription factors involved in heat stress/UV responses, immunity, apoptosis, biomineralization, the cytoskeletal organization, and endo-exophagocytosis [24].

It is evident that up to now, the studies of climate-related stress on coral reefs have indicated that bleaching does not result from a single process, but rather from a complex interaction of environmental and genetic factors, operating at several levels within the reef-forming holobionts [99, 102, 103], which highlights the need for further detailed investigations directed toward a better understanding of the mechanisms underlying bleaching.

5. Mechanisms by which cnidarians face bleaching

Coral bleaching events have increased in frequency and intensity. Actually, some studies project that if this trend continues, tropical coral reefs might disappear this century [104, 105]. Mass coral bleaching and mortality events that have been registered worldwide over the past three decades have raised serious concerns about the future of coral reef ecosystems [77, 106]. Ecological extinction of corals reefs in some regions has been predicted to occur within the next 20–50 years, if reef-forming cnidarians are unable to adapt sufficiently rapidly to keep pace with global warming [107].

The ability to recover from a bleaching event has been associated with the energy reserves and heterotrophic feeding capacity of the cnidarian host [16, 108, 109]. *Symbiodinium* can provide up to 100% of a healthy coral's daily fixed carbon requirements; however, following bleaching, recovering corals may heavily rely on alternate sources of fixed carbon, which is acquired via catabolism of energy reserves and/or by increased heterotrophy [16, 110, 111].

Some evidences suggest that cnidarians are able to deal with thermal stress, through a series of adaptive processes (e.g., acclimatization, genetic adaptation, and symbiont shuffling), which may help reduce the harmful consequences and mortality provoked by bleaching [112, 113].

Acclimatization or acclimation is a type of phenotypic plasticity in which organisms, such as corals and their symbionts, can optimize their physiological performance in response to environmental changes [114]. The capacity for acclimation and adaptation of cnidarians and *Symbiodinium* clades is currently poorly known. Differentially expressed genes in *Acropora hyacinthus*, under physiological and stress conditions, suggested that the acclimatization that occurs less than 2 years after exposure to thermal stress is the same as that might have been expected from natural selection over many generations. Incorporating these adaptive capacities to ecosystem models will likely reduce the predictions of the disappearance of coral reef ecosystems [115].

Some studies have determined the existence of coral species that have survived high temperatures, indicating that they have higher thermal tolerance thresholds than others. Resistant species belong to the genera *Cyphastrea, Goniopora, Galaxea,* and *Pavona,* whereas species of the genera *Stylophora, Pocillopora,* and *Acropora* are more vulnerable to thermal stress [59]. In the case of *Symbiodinium* algae, some clades have been recognized as being more resistant than others. A study carried out on the major reef-building species in the Caribbean showed that *Symbiodinium* algae of the clade D display a significant thermal tolerance [116]. This was confirmed in a further study on *Orbicella faveolata* that demonstrated that under nonstressful conditions, thermally sensitive *Symbiodinium* of the clades B17 and C7 are prevalent, whereas after a bleaching event, clades D1 and A3 proliferated and repopulated cnidarian tissues previously evacuated by clades B17 and C7 [117].

Various mechanisms have been proposed to explain adaptive capacity of cnidarian species to face heat and radiation stress [7]. One of them involves photoprotection provided by pigments within the host tissues. Evidences supporting this proposal were obtained from a study of the coral *Montipora monasteriata*, which was selected since it occurs in multiple color morphs (tan, blue, brown, green, and red). That study demonstrated that two of the non-fluorescent host pigments (CP-420 and CP-580) are up-regulated in response to elevated irradiance. This behavior appeared to favor retention of antennal chlorophyll by endosymbionts and therefore, photosynthetic capacity. It was found that up-regulated host pigments can facilitate the establishment of new or restructured dinoflagellate chloroplasts by modifying the internal light field. Particularly, high concentrations of CP-420 provided photoprotection for *Symbiodinium* by intercepting photons destined for photochemical quenching by dinoflagellate algae [118].

Another adaptive process entails non-photochemical quenching (NPQ) that dissipates excess light as heat. Reef corals produce fluorescent pigments belonging to a family related to green fluorescent proteins, which through the absorption, scattering, and dissipation of high-energy solar radiation by fluorescence reduce photoinhibition and the severity of bleaching [113]. Some of these sunscreen compounds are mycosporine-like amino acids, such as shinorine. This pigment showed no detectable fluorescence when excited at a frequency of 312–348 nm, which corresponds to the band of its mean-maximum absorption. Furthermore, electron paramagnetic resonance spectroscopy (EPR) revealed that purified shinorine (50 μM) produced no detectable radicals when irradiated with 305–700 nm. The lack of free radical formation by UV irradiation and the lack of fluorescence are consistent with the high efficiency dissipation of thermally absorbed UV energy [113, 119].

A well-known biochemical adaptation to thermal stress is the heat-shock response (HSR), which comprises the induced expression of a conserved set of molecular chaperones, known as heat-shock proteins. These molecules are critical for protein homeostasis, defense mechanisms, the refolding of denatured proteins, and the breakdown and replacement of non-replaceable proteins [120].

An additional adaptive response implies increased expression of enzymes that sequester oxygen radicals. The use of recent molecular techniques has improved our understanding of the magnitude of the transcriptional response of corals to various stressors, including high temperatures. An analysis carried out on *Acropora millepora* indicated that thermally stressed samples experienced a significant overexpression of four genes associated to cnidarian oxidative stress responses (HSP70, MnSOD, ferritin, and Zn^{2+}-metalloprotease) [121].

A caspase-mediated apoptotic cascade, caused by reactive oxygen species mainly generated by the algal symbionts, is a well-known factor that triggers bleaching and death of the host cnidarian. It has been shown that under high temperatures, some corals are able to naturally suppress caspase activity and significantly reduce caspase concentration as a mechanism to avoid colony death from apoptosis. It has been hypothesized that variability in response to thermal stress is determined by a four-element, combinatorial genetic matrix intrinsic to the specific symbiotic association [109].

On the other hand, symbiont shuffling and/or switching are considered other important mechanisms for explaining the way coral reef communities can counter environmental stress conditions. In some cases, the distribution of certain clades of *Symbiodinium* can substantially change after bleaching, via shuffling (a shift in symbiont dominance) or by symbiont switching (algae are acquired from exogenous sources). A case study was performed on *Stylophora pistillata* that belonged to a coral reef that had suffered bleaching in March 2006. The objective of the study was to determine the percentages of different *Symbiodinium* clades populating the host after bleaching, and after the coral recovered from the event. Samples were collected in November 2006 and July 2007, and the results indicated the initial presence of clades C79, C78, C8/a, and C35/a, the last two in greater proportion. However, once the coral recovered from bleaching and repopulated, the proportions of C79, C35/a, and C78 decreased, while the population of C8/a increased considerably. Furthermore, other more resistant *Symbiodinium* clades appeared [122].

It is worth mentioning that studies directed toward understanding the causes and effects of bleaching, as well as the tolerance mechanisms that counteract this phenomenon have focused on Anthozoa species, showing that bleaching affects various processes that are essential to the survival of cnidarians. However, an integrative analysis is still needed to understand the molecular mechanisms underlying the different responses (adaptation or death) of cnidarians to coral bleaching.

6. Consequences of bleaching in coral reefs ecology

Undoubtedly, the bleaching phenomenon has seriously affected corals reefs. Although it has been observed that some reef-building cnidarians that suffer the loss of their symbiotic algae can, in some cases, survive and recover [90], bleaching can also lead to massive death [79, 104, 123], inflicting an enormous damage to the ecological balance of entire reef communities.

Coral reef monitoring programs exist in all regions of the world, recording reef features such as coral cover, fish biomass, and macroalgal cover, among others. However, at present the way of interpreting coral reef monitoring data has not been globally standardized, which hampers decision making directed toward protecting and rehabilitating coral reefs [124].

Several aspects regarding coral resistance and resilience are still unclear. For example, it is uncertain which *Symbiodinium* clades remain fairly viable following a stressful event and if they are able to significantly contribute to the recovery of bleached corals [30]. In this regard, some researchers reported that a majority of the expelled symbionts remained healthy, whereas others claimed that the survival of *Symbiodinium* did not last; some *Symbiodinium* cells, released into the water column after a thermal stress, had a drastically reduced photosynthetic activity after 5 days, suggesting that their survival could be compromised. Therefore, their contribution to the free-living stocks may be limited [125].

At present, coral resilience ability is the object of various research projects, since it has been widely documented that many corals that apparently recover from a bleaching event display a decrease in growth rate and calcification [114, 126], and are more susceptible to disease [127, 128], which is not surprising, since their supply of energy available for fundamental processes is diminished.

Gene expression biomarkers are emerging as powerful diagnosis tools for identifying and characterizing coral stress. Among the most plausible candidates are genes related to expression of heat-shock proteins, immune and oxidative stress responses, some metabolic processes, and structural genes. Other genes, such as hsp16, Cacna1, MnSOD, SLC26, and Nf-kB, are showing excellent potential as reliable indicators of thermal stress in corals [107].

The impact on the reduction of the biodiversity of corals has not been fully envisaged. However, it is predicted that if reef-forming cnidarians do not evolve rapidly toward thermal tolerance, they will no longer dominate the reefs in the coming years [129, 130]. A possible scenario, according to some ecology specialists, is that if global warming continues at the present rate, a phase shift in coral reefs to an alternate state dominated by a different kind of organisms might occur [131, 132]. This alternate state might represent a substantial degradation of the coral reefs. Furthermore, degraded states can also be resilient to change, which complicates their possible reversal [11, 105].

7. Concluding remarks

Mass coral bleaching and mortality events that have been registered worldwide over the past three decades have raised serious concerns about the future of coral reef ecosystems. In fact, ecological extinction of corals reefs in some regions has been predicted to occur within the next 20–50 years, if reef-forming cnidarians are unable to adapt sufficiently rapidly to keep pace with global warming [104, 117, 133].

It is clear that the survival of coral reefs strongly depends on the functionality of the symbiosis between cnidarians and *Symbiodinium* algae, and up to now, we are just beginning to understand the molecular and cellular mechanisms underlying this relationship. Omics approaches have enabled a better comprehension about the way cnidarian-algal symbiosis functions and how holobiont physiology is modified by bleaching. Studies of the effects of global warming and coral bleaching have indicated that this event is the result of a complex interaction of environmental and genetic factors. Investigations of coral bleaching have also provided important insight into the mechanisms responsible for coral resistance to thermal stress.

It is evident that the key to understanding the future of coral reefs requires an insightful comprehension of the molecular and physiological mechanisms that promote thermal tolerance in the cnidarians and their symbionts, and to identify the genetic characteristics responsible for the variety of responses that occur in a coral bleaching event. Therefore, it is very important to continue studies in this regard to better understand cnidarian-*Symbiodinium* symbiosis, causes and effects of bleaching, the survival mechanisms of hosts and symbionts, as well as their ecological importance. Surely, the results derived from these studies will be useful to design strategies and policies to restore coral reefs and to promote their conservation.

Acknowledgements

The authors thank support from Universidad Autónoma de Querétaro. Norma Olguín, Carolina Gutiérrez, and Víctor Hugo Hérnandez acknowledge Consejo Nacional de Ciencia y Tecnología (CONACYT, México) for their graduate school scholarships.

Appendix

Acronyms	Definition
IOC	Intergovernmental Oceanographic Commission
UNESCO	United Nations Educational, Scientific and Cultural Organization
UNEP	United Nations Environment Programme
IUCN	International Union for Conservation of Nature
HRFs	Host release factors
ECM	Extracellular calcifying medium
PRRs	Pattern recognition receptors
UV	Ultraviolet radiation
SST	Sea surface temperature
ENSO	El Nino Southern Oscillation
GCRMN	The Global Coral Reef Monitoring Network
PSII	Photosystem II
ROS	Reactive oxygen species
RuBisCO	Ribulose-1,5-bisphosphate carboxylase oxygenase enzyme
SOD	Superoxide dismutase enzyme
PSI	Photosystem I
AIF	*Apoptosis-inducing factor*
NPQ	Non-photochemical quenching
EPR	Paramagnetic resonance spectroscopy
HSR	Heat-shock response
APX	Catalase
NADPH	Nicotinamide adenine dinucleotide phosphate
NF-KB	Nuclear factor kappa-light-chain-enhancer of activated B cells
p53	p53-gen

Author details

Norma Olguín-López, Carolina Gutiérrez-Chávez, Víctor Hugo Hérnández-Elizárraga, César Ibarra-Alvarado and Alejandra Rojas-Molina*

*Address all correspondence to: rojasa@uaq.mx

School of Chemistry, Autonomous University of Querétaro, Santiago de Querétaro, México

References

[1] Anthony KR. Ecosystem Damages from Loss of Coral Reefs Under Climate Change. Annu Rev Environ Resour. 2016;**41**:59-81

[2] Williams SM, Sánchez-Godínez C, Newman SP, Cortés J. Ecological assessments of the coral reef communities in the Eastern Caribbean and the effects of herbivory in influencing coral juvenile density and algal cover. Mar Ecol. 2017;**38**:e12395

[3] Roth MS. The engine of the reef: photobiology of the coral-algal symbiosis. Front Microbiol. 2014;**5**:422

[4] Rogers CS, Miller J. Measuring, Interpreting, and Responding to Changes in Coral Reefs: A Challenge for Biologists, Geologists, and Managers. Coral Reefs at the Crossroads. Springer Netherlands. 2016;277-92

[5] Jain R, Sonawane S, Mandrekar N. Marine organisms: Potential source for drug discovery. Curr Sci-Bangalore-. 2008;**94**(3):292

[6] Cragg GM, Newman DJ. Natural products: a continuing source of novel drug leads. Biochim Biophys Acta. 2013;**1830**(6):3670-95

[7] Baird AH, Bhagooli R, Ralph PJ, Takahashi S. Coral bleaching: the role of the host. Trends Ecol Evol. 2009;**24**(1):16-20

[8] Lesser MP. Coral reef bleaching and global climate change: Can corals survive the next century? Proc Natl Acad Sci. 2007;**104**(13):5259-5260

[9] Lough JM. Coral reefs: Turning back time. Nature. 2016;**531**(7594):314-315

[10] Mumby PJ, Van Woesik R. Consequences of ecological, evolutionary and biogeochemical uncertainty for coral reef responses to climatic stress. Curr Biol. 2014;**24**(10):R413-R423

[11] Mora C, Graham NAJ, Nyström M. Ecological limitations to the resilience of coral reefs. Coral Reefs. 2016;**35**(4):1271-80

[12] Lesser MP. Oxidative stress in marine environments: biochemistry and physiological ecology. Annu Rev Physiol. 2006;**68**:253-278

[13] Ainsworth TD, Hoegh-Guldberg O, Heron SF, Skirving WJ, Leggat W. Early cellular changes are indicators of pre-bleaching thermal stress in the coral host. J Exp Mar Biol Ecol. 2008;**364**(2):63-71

[14] Brown BE, Dunne RP. Coral Bleaching: The Roles of Sea Temperature and Solar Radiation. ResearchGate. 2015;**99**(C10):266-83

[15] Baker AC, Glynn PW, Riegl B. Climate change and coral reef bleaching: An ecological assessment of long-term impacts, recovery trends and future outlook. Estuar Coast Shelf Sci. 2008;**80**(4):435-471

[16] Grottoli AG, Warner ME, Levas SJ, Aschaffenburg MD, Schoepf V, McGinley M, et al. The cumulative impact of annual coral bleaching can turn some coral species winners into losers. Glob Change Biol. 2014;**20**(12):3823-3833

[17] Lesser MP. Coral bleaching: causes and mechanisms. In: Coral reefs: an ecosystem in transition. Springer Netherlands. 2011:405-419

[18] Okazaki RR, Towle EK, van Hooidonk R, Mor C, Winter RN, Piggot AM, et al. Species-specific responses to climate change and community composition determine future calcification rates of Florida Keys reefs. Glob Change Biol. 2016;**23**(3):1023-1035

[19] van Hooidonk R, Maynard JA, Manzello D, Planes S. Opposite latitudinal gradients in projected ocean acidification and bleaching impacts on coral reefs. Glob Change Biol. 2014;**20**(1):103-12

[20] Jackson J, Donovan M, Cramer K, Lam V. Status and trends of Caribbean coral reefs: 1970-2012 Global Coral Reef Monitoring Network, International Union for the Conservation of Nature Global Marine and Polar Program, Washington, DC

[21] Bay RA, Palumbi SR. Rapid acclimation ability mediated by transcriptome changes in reef-building corals. Genome Biol Evol. 2015;**7**(6):1602-1612

[22] Maor-Landaw K, Levy O. Gene expression profiles during short-term heat stress; branching vs. massive Scleractinian corals of the Red Sea. Peer J. 2016;**4**:e1814

[23] Pinzón JH, Kamel B, Burge CA, Harvell CD, Medina M, Weil E, et al. Whole transcriptome analysis reveals changes in expression of immune-related genes during and after bleaching in a reef-building coral. R Soc Open Sci. 2015;**2**(4):140214

[24] Ricaurte M, Schizas NV, Ciborowski P, Boukli NM. Proteomic analysis of bleached and unbleached *Acropora palmata*, a threatened coral species of the Caribbean. Mar Pollut Bull. 2016;**107**(1):224-232

[25] Sloan R, Sawyer S. The role of cellular signaling during bleaching in the sea anemone, *Aiptasia pallida*. Proc W Va Acad Sci. 2016;**88**:1

[26] DeSalvo MK, Estrada A, Sunagawa S, Medina M. Transcriptomic responses to darkness stress point to common coral bleaching mechanisms. Coral Reefs. 2012;**31**(1):215-228

[27] Seneca FO, Palumbi SR. The role of transcriptome resilience in resistance of corals to bleaching. Mol Ecol. 2015;**24**(7):1467-1484

[28] Anderson DA, Walz ME, Weil E, Tonellato P, Smith MC. RNA-Seq of the Caribbean reef-building coral *Orbicella faveolata* (Scleractinia-Merulinidae) under bleaching and disease stress expands models of coral innate immunity. Peer J. 2016;**4**:e1616

[29] Tomanek L. Environmental Proteomics: Changes in the Proteome of Marine Organisms in Response to Environmental Stress, Pollutants, Infection, Symbiosis, and Development. Annu Rev Mar Sci. 2011;**3**(1):373-99

[30] Fransolet D, Roberty S, Plumier JC. Establishment of endosymbiosis: The case of cnidarians and Symbiodinium. J Exp Mar Biol Ecol. 2012;**420**:1-7

[31] Baker AC, Correa AMS, Cunning R. Diversity, Distribution and Stability of Symbiodinium in Reef Corals of the Eastern Tropical Pacific. Coral Reefs of the Eastern Tropical Pacific. Springer Netherlands; 2017;405-20

[32] Yellowlees D, Rees TAV, Leggat W. Metabolic interactions between algal symbionts and invertebrate hosts. Plant Cell Environ. 2008;**31**(5):679-694

[33] Baghdasarian G, Muscatine L. Preferential expulsion of dividing algal cells as a mechanism for regulating algal-cnidarian symbiosis. Biol Bull. 2000;**199**(3):278-86

[34] Dimond J, Carrington E. Symbiosis regulation in a facultatively symbiotic temperate coral: zooxanthellae division and expulsion. Coral Reefs. 2008;**27**(3):601-604

[35] Muller-Parker G, D'Elia CF, Cook CB. Interactions between corals and their symbiotic algae. In: Coral Reefs in the Anthropocene. Springer Netherlands. 2015:99-116

[36] Venn AA, Loram JE, Douglas AE. Photosynthetic symbioses in animals. J Exp Bot. 2008;**59**(5):1069-1080

[37] Yellowlees D, Rees TAV, Leggat W. Metabolic interactions between algal symbionts and invertebrate hosts. Plant Cell Environ. 2008;**31**(5):679-694

[38] Davy SK, Allemand D, Weis VM. Cell biology of cnidarian-dinoflagellate symbiosis. Microbiol Mol Biol Rev. 2012;**76**(2):229-261

[39] Biel KY, Gates RD, Muscatine L. Effects of free amino acids on the photosynthetic carbon metabolism of symbiotic dinoflagellates. Russ J Plant Physiol. 2007;**54**(2):171-83

[40] Cook CB, Davy SK. Are free amino acids responsible for the `host factor' effects on symbiotic zooxanthellae in extracts of host tissue? Hydrobiologia. 2001;**461**(1-3):71-8

[41] Osinga R, Schutter M, Griffioen B, Wijffels RH, Verreth JAJ, Shafir S, et al. The Biology and Economics of Coral Growth. Mar Biotechnol N Y N. 2011;**13**(4):658-71

[42] Muscatine L, Tambutte E, Allemand D. Morphology of coral desmocytes, cells that anchor the calicoblastic epithelium to the skeleton. Coral Reefs. 1997;**16**(4):205-213

[43] Furla P, Bénazet-Tambutté S, Jaubert J, Allemand D. Functional polarity of the tentacle of the sea anemone *Anemonia viridis*: role in inorganic carbon acquisition. Am J Physiol. 1998;**274**(2 Pt 2):R303-310

[44] Venn A, Tambutté E, Holcomb M, Allemand D, Tambutté S. Live Tissue Imaging Shows Reef Corals Elevate pH under Their Calcifying Tissue Relative to Seawater. PLoS ONE. 2011;**6**(5):e20013

[45] Colombo-Pallotta MF, Rodríguez-Román A, Iglesias-Prieto R. Calcification in bleached and unbleached *Montastraea faveolata*: evaluating the role of oxygen and glycerol. Coral Reefs. 2010;**29**(4):899-907

[46] Oswald F, Schmitt F, Leutenegger A, Ivanchenko S, D'Angelo C, Salih A, et al. Contributions of host and symbiont pigments to the coloration of reef corals. FEBS J. 2007;**274**(4):1102-9

[47] Fabricius KE, Mieog JC, Colin PL, Idip D, H Van Oppen M. Identity and diversity of coral endosymbionts (zooxanthellae) from three Palauan reefs with contrasting bleaching, temperature and shading histories. Mol Ecol. 2004;**13**(8):2445-2458

[48] Mauricio R-L, Wendy P, Virginia W. Transcriptome analysis of a cnidarian-dinoflagellate mutualism reveals complex modulation of host gene expression. Bmc genomics. 2006;**7**(1):23

[49] Schwarz JA, Krupp DA, Weis VM. Late larval development and onset of symbiosis in the scleractinian coral *Fungia scutaria*. Biol Bull. 1999;**196**(1):70-79

[50] Weis VM, Reynolds WS, Krupp DA, others. Host-symbiont specificity during onset of symbiosis between the dinoflagellates *Symbiodinium spp*. and planula larvae of the scleractinian coral Fungia scutaria. Coral Reefs. 2001;**20**(3):301-308

[51] Logan DDK, LaFlamme AC, Weis VM, Davy SK. Flow-Cytometric Characterization of the Cell-Surface Glycans of Symbiotic Dinoflagellates (*Symbiodinium spp*.). J Phycol. 2010;**46**(3):525-33

[52] Weis VM. Cellular mechanisms of Cnidarian bleaching: stress causes the collapse of symbiosis. J Exp Biol. 2008;**211**(19):3059-3066

[53] Wood-Charlson EM, Hollingsworth LL, Krupp DA, Weis VM. Lectin/glycan interactions play a role in recognition in a coral/dinoflagellate symbiosis. Cell Microbiol. 2006;**8**(12):1985-93

[54] Freudenthal HD. *Symbiodinium* gen. nov. and *Symbiodinium microadriaticum* sp. nov., a Zooxanthella: Taxonomy, Life Cycle, and Morphology. J Protozool. 1962;**9**(1):45-52

[55] Jones AM, Berkelmans R, van Oppen MJ, Mieog JC, Sinclair W. A community change in the algal endosymbionts of a scleractinian coral following a natural bleaching event: field evidence of acclimatization. Proc R Soc B Biol Sci. 2008;**275**(1641):1359-1365

[56] Obura DO. Reef corals bleach to resist stress. Mar Pollut Bull. 2009;**58**(2):206-212

[57] Hoegh-Guldberg O. Climate change, coral bleaching and the future of the world's coral reefs. Mar Freshw Res. 1999;**50**(8):839-866

[58] Jokiel PL, Coles SL. Response of Hawaiian and other Indo-Pacific reef corals to elevated temperature. Coral Reefs. 1990;**8**(4):155-162

[59] McClanahan TR, Baird AH, Marshall PA, Toscano MA. Comparing bleaching and mortality responses of hard corals between southern Kenya and the Great Barrier Reef, Australia. Mar Pollut Bull. 2004;**48**(3):327-335

[60] Al-Naema N, Deb N, Saeed S, Dupont J, Ben-Hamadou R. Acute orthogonal stress driven by temperature, salinity and light intensity on Qatari Porites photosynthesis and growth. Qscience Proc. 2015;(5):17

[61] Gleason DF, Wellington GM. Ultraviolet radiation and coral bleaching. Nature. 1993;**365**(6449):836-838

[62] Hoegh-Guldberg O, Smith GJ. The effect of sudden changes in temperature, light and salinity on the population density and export of zooxanthellae from the reef corals *Stylophora pistillata* Esper and *Seriatopora hystrix* Dana. J Exp Mar Biol Ecol. 1989;**129**(3):279-303

[63] Lesser MP, Farrell JH. Exposure to solar radiation increases damage to both host tissues and algal symbionts of corals during thermal stress. Coral Reefs. 2004;**23**(3):367-377

[64] Yonge CM. The significance of the relationship between corals and zooxanthellae. Nature. 1931;**128**:309-311

[65] Glynn PW, D'croz L. Experimental evidence for high temperature stress as the cause of El Niño-coincident coral mortality. Coral Reefs. 1990;**8**(4):181-191

[66] Jokiel PL, Coles SL. Effects of temperature on the mortality and growth of Hawaiian reef corals. Mar Biol. 1977;**43**(3):201-208

[67] Steen RG, Muscatine L. Low temperature evokes rapid exocytosis of symbiotic algae by a sea anemone. Biol Bull. 1987;**172**(2):246-263

[68] Hoegh-Guldberg O, Jones RJ. Photoinhibition and photoprotection in symbiotic dinoflagellates from reef-building corals. Mar Ecol Prog Ser. 1999;**183**:73-86

[69] Lesser MP, Stochaj WR, Tapley DW, Shick JM. Bleaching in coral reef anthozoans: effects of irradiance, ultraviolet radiation, and temperature on the activities of protective enzymes against active oxygen. Coral Reefs. 1990;**8**(4):225-232

[70] Warner ME, Fitt WK, Schmidt GW. Damage to photosystem II in symbiotic dinoflagellates: a determinant of coral bleaching. Proc Natl Acad Sci. 1999;**96**(14):8007-8012

[71] Eakin CM, Morgan JA, Heron SF, Smith TB, Liu G, Alvarez-Filip L, et al. Caribbean corals in crisis: record thermal stress, bleaching, and mortality in 2005. PloS One. 2010;**5**(11):e13969

[72] Ban SS, Graham NAJ, Connolly SR. Evidence for multiple stressor interactions and effects on coral reefs. Glob Change Biol. 2014;**20**(3):681-97

[73] Fitt WK, Warner ME. Bleaching patterns of four species of Caribbean reef corals. Biol Bull. 1995;**189**(3):298-307

[74] Courtial L, Roberty S, Shick JM, Houlbrèque F, Ferrier-Pagès C. Interactive effects of ultraviolet radiation and thermal stress on two reef-building corals. Limnol Oceanogr. 2017;**62**(3):1000-13

[75] Drohan AF, Thoney DA, Baker AC. Synergistic effect of high temperature and ultraviolet-B radiation on the gorgonian *Eunicea tourneforti* (Octocorallia: Alcyonacea: Plexauridae). Bull Mar Sci. 2005;**77**(2):257-66

[76] Anthony KRN, Kline DI, Diaz-Pulido G, Dove S, Hoegh-Guldberg O. Ocean acidification causes bleaching and productivity loss in coral reef builders. Proc Natl Acad Sci. 2008;**105**(45):17442-6

[77] Hoegh-Guldberg O, Mumby PJ, Hooten AJ, Steneck RS, Greenfield P, Gomez E, et al. Coral reefs under rapid climate change and ocean acidification. Science. 2007; **318**(5857):1737-1742

[78] Pandolfi JM, Connolly SR, Marshall DJ, Cohen AL. Projecting coral reef futures under global warming and ocean acidification. Science. 2011;**333**(6041):418-22

[79] Brown BE. Coral bleaching: causes and consequences. Coral Reefs. 1997;**16**(1):S129-S138

[80] Glynn PW. Coral reef bleaching: ecological perspectives. Coral Reefs. 1993;**12**(1):1-17

[81] Goreau TJ, Hayes RL, McAlllister D. Regional patterns of sea surface temperature rise: implications for global ocean circulation change and the future of coral reefs and fisheries. World Resour Rev. 2005;**17**(3):350-370

[82] McWilliams JP, Côté IM, Gill JA, Sutherland WJ, Watkinson AR. Accelerating impacts of temperature-induced coral bleaching in the Caribbean. Ecology. 2005;**86**(8):2055-2060

[83] Hoegh-Guldberg O, Bruno JF. The impact of climate change on the world's marine ecosystems. Science. 2010;**328**(5985):1523-1528

[84] Alemu JB, Clement Y, others. Mass coral bleaching in 2010 in the southern Caribbean. PLoS One. 2014;**9**(1):e83829

[85] Mann ME, Rahmstorf S, Steinman BA, Tingley M, Miller SK. The likelihood of recent record warmth. Sci Rep. 2016;**6**:19831

[86] Heron SF, Maynard JA, van Hooidonk R, Eakin CM. Warming Trends and Bleaching Stress of the World's Coral Reefs 1985-2012. Sci Rep. 2016;**6**:38402

[87] Hill R, Brown CM, DeZeeuw K, Campbell DA, Ralph PJ. Increased rate of D1 repair in coral symbionts during bleaching is insufficient to counter accelerated photo-inactivation. Limnol Oceanogr. 2011;**56**(1):139-46

[88] Ross C, Ritson-Williams R, Olsen K, Paul VJ. Short-term and latent post-settlement effects associated with elevated temperature and oxidative stress on larvae from the coral *Porites astreoides*. Coral Reefs. 2013;**32**(1):71-79

[89] Yakovleva IM, Baird AH, Yamamoto HH, Bhagooli R, Nonaka M, Hidaka M. Algal symbionts increase oxidative damage and death in coral larvae at high temperatures. Mar Ecol Prog Ser. 2009;**378**:105-112

[90] Hume B, D'Angelo C, Burt J, Baker AC, Riegl B, Wiedenmann J. Corals from the Persian/Arabian Gulf as models for thermotolerant reef-builders: Prevalence of clade C3 *Symbiodinium*, host fluorescence and *ex situ* temperature tolerance. Mar Pollut Bull. 2013;**72**(2):313-322

[91] Takahashi S, Murata N. How do environmental stresses accelerate photoinhibition? Trends Plant Sci. 2008;**13**(4):178-182

[92] Hill R, Schreiber U, Gademann R, Larkum AW, Kühl M, Ralph PJ. Spatial heterogeneity of photosynthesis and the effect of temperature-induced bleaching conditions in three species of corals. Mar Biol. 2004;**144**(4):633-640

[93] Jones RJ, Hoegh-Guldberg O, Larkum AWD, Schreiber U. Temperature-induced bleaching of corals begins with impairment of the CO2 fixation mechanism in zooxanthellae. Plant Cell Environ. 1998;**21**(12):1219-30

[94] Lilley RM, Ralph PJ, Larkum AWD. The determination of activity of the enzyme Rubisco in cell extracts of the dinoflagellate alga *Symbiodinium* sp. by manganese chemiluminescence and its response to short-term thermal stress of the alga. Plant Cell Environ. 2010;**33**(6):995-1004

[95] Bhagooli R. Inhibition of Calvin-Benson cycle suppresses the repair of photosystem II in *Symbiodinium*: implications for coral bleaching. Hydrobiologia. 2013;**714**(1):183-90

[96] Perez S, Weis V. Nitric oxide and cnidarian bleaching: an eviction notice mediates breakdown of a symbiosis. J Exp Biol. 2006;**209**(14):2804-2810

[97] Hill R, Szabó M, ur Rehman A, Vass I, Ralph PJ, Larkum AWD. Inhibition of photosynthetic CO_2 fixation in the coral *Pocillopora damicornis* and its relationship to thermal bleaching. J Exp Biol. 2014;**217**(Pt 12):2150-62

[98] Wooldridge SA. Breakdown of the coral-algae symbiosis: towards formalising a linkage between warm-water bleaching thresholds and the growth rate of the intracellular zooxanthellae. Biogeosciences. 2013;**10**(3):1647-1658

[99] Tolleter D, Seneca FO, DeNofrio JC, Krediet CJ, Palumbi SR, Pringle JR, et al. Coral bleaching independent of photosynthetic activity. Curr Biol CB. 2013;**23**(18):1782-6

[100] Maor-Landaw K, Levy O. Survey of Cnidarian Gene Expression Profiles in Response to Environmental Stressors: Summarizing 20 Years of Research, What Are We Heading for? In: The Cnidaria, Past, Present and Future. Springer Netherlands. 2016:523-543

[101] Raina JB, Lutz A, Motti CA, Miller DJ, van Oppen MJH. Host Coenzyme Q Redox State Is an Early Biomarker of Thermal Stress in the Coral *Acropora millepora*. PLOS ONE. 2015;**10**(10):e0139290

[102] Buxton L, Takahashi S, Hill R, Ralph PJ. Variability in the primary site of photosynthetic damage in *Symbiodinium sp.*(dinophyceae) exposed to thermal stress1. J Phycol. 2012;**48**(1):117-126

[103] Bieri T, Onishi M, Xiang T, Grossman AR, Pringle JR. Relative Contributions of Various Cellular Mechanisms to Loss of Algae during Cnidarian Bleaching. PLOS ONE. 2016;**11**(4):e0152693

[104] Hoegh-Guldberg O, Poloczanska ES, Skirving W, Dove S. Coral Reef Ecosystems under Climate Change and Ocean Acidification. Front Mar Sci. 2017;**4**:158

[105] Logan CA, Dunne JP, Eakin CM, Donner SD. Incorporating adaptive responses into future projections of coral bleaching. Glob Change Biol. 2014;**20**(1):125-39

[106] Hughes TP, Baird AH, Bellwood DR, Card M, Connolly SR, Folke C, et al. Climate change, human impacts, and the resilience of coral reefs. science. 2003;**301**(5635):929-933

[107] Louis YD, Bhagooli R, Kenkel CD, Baker AC, Dyall SD. Gene expression biomarkers of heat stress in scleractinian corals: Promises and limitations. Comp Biochem Physiol Part C Toxicol Pharmacol. 2017;**191**:63-77

[108] Levas S, Grottoli AG, Schoepf V, Aschaffenburg M, Baumann J, Bauer JE, et al. Can heterotrophic uptake of dissolved organic carbon and zooplankton mitigate carbon budget deficits in annually bleached corals? Coral Reefs. 2016;**35**(2):495-506

[109] Tchernov D, Kvitt H, Haramaty L, Bibby TS, Gorbunov MY, Rosenfeld H, et al. Apoptosis and the selective survival of host animals following thermal bleaching in zooxanthellate corals. Proc Natl Acad Sci. 2011;**108**(24):9905-9909

[110] Aichelman HE, Townsend JE, Courtney TA, Baumann JH, Davies SW, Castillo KD. Heterotrophy mitigates the response of the temperate coral *Oculina arbuscula* to temperature stress. Ecol Evol. 2016;**6**(18):6758-69

[111] Tremblay P, Gori A, Maguer JF, Hoogenboom M, Ferrier-Pagès C. Heterotrophy promotes the re-establishment of photosynthate translocation in a symbiotic coral after heat stress. Sci Rep. 2016;**6**:srep38112

[112] Ainsworth TD, Heron SF, Ortiz JC, Mumby PJ, Grech A, Ogawa D, et al. Climate change disables coral bleaching protection on the Great Barrier Reef. Science. 2016;**352**(6283):338-42

[113] Salih A, Cox G, Szymczak R, Coles SL, Baird AH, Dunstan A, et al. The role of host-based color and fluorescent pigments in photoprotection and in reducing bleaching stress in corals. In: Proc 10th Int Coral Reef Symp. 2006:746-756

[114] Coles SL, Brown BE. Coral bleaching-capacity for acclimatization and adaptation. Adv Mar Biol. 2003;**46**:183-223

[115] Palumbi SR, Barshis DJ, Traylor-Knowles N, Bay RA. Mechanisms of reef coral resistance to future climate change. Science. 2014;**344**(6186):895-898

[116] LaJeunesse TC, Smith RT, Finney J, Oxenford H. Outbreak and persistence of opportunistic symbiotic dinoflagellates during the 2005 Caribbean mass coral 'bleaching' event. Proc R Soc Lond B Biol Sci. 2009;**276**(1676):4139-48

[117] Kemp DW, Hernandez-Pech X, Iglesias-Prieto R, Fitt WK, Schmidt GW. Community dynamics and physiology of *Symbiodinium spp*. before, during, and after a coral bleaching event. Limnol Oceanogr. 2014;**59**(3):788-797

[118] Dove SG, Lovell C, Fine M, Deckenback J, Hoegh-Guldberg O, Iglesias-Prieto R, et al. Host pigments: potential facilitators of photosynthesis in coral symbioses. Plant Cell Environ. 2008;**31**(11):1523-1533

[119] Shick JM, Dunlap WC. Mycosporine-like amino acids and related gadusols: biosynthesis, accumulation, and UV-protective functions in aquatic organisms. Annu Rev Physiol. 2002;**64**(1):223-262

[120] Hofmann GE, Todgham AE. Living in the now: physiological mechanisms to tolerate a rapidly changing environment. Annu Rev Physiol. 2010;**72**:127-145

[121] Császár NBM, Seneca FO, Van Oppen MJH. Variation in antioxidant gene expression in the scleractinian coral *Acropora millepora* under laboratory thermal stress. Mar Ecol Prog Ser. 2009;**392**:93-102

[122] Sampayo EM, Ridgway T, Bongaerts P, Hoegh-Guldberg O. Bleaching susceptibility and mortality of corals are determined by fine-scale differences in symbiont type. Proc Natl Acad Sci. 2008;**105**(30):10444-10449

[123] Banaszak AT. Photoprotective physiological and biochemical responses of aquatic organisms. UV Eff Aquat Org Ecosyst. 2003;329-356

[124] Flower J, Ortiz JC, Chollett I, Abdullah S, Castro-Sanguino C, Hock K, et al. Interpreting coral reef monitoring data: A guide for improved management decisions. Ecol Indic. 2017;**72**:848-69

[125] Hill R, Ralph P. Post-bleaching viability of expelled zooxanthellae from the scleractinian coral *Pocillopora damicornis*. ýMar. Ecol. Prog. Ser. 2007;**352**:137-144

[126] Foster T, Short JA, Falter JL, Ross C, McCulloch MT. Reduced calcification in Western Australian corals during anomalously high summer water temperatures. J Exp Mar Biol Ecol. 2014;**461**:133-43

[127] Maynard J, van Hooidonk R, Eakin CM, Puotinen M, Garren M, Williams G, et al. Projections of climate conditions that increase coral disease susceptibility and pathogen abundance and virulence. Nat Clim Change. 2015;**5**(7):688-94

[128] Rosenberg E, Koren O, Reshef L, Efrony R, Zilber-Rosenberg I. The role of microorganisms in coral health, disease and evolution. Nat Rev Microbiol. 2007;**5**(5):355-362

[129] Pandolfi JM. Incorporating Uncertainty in Predicting the Future Response of Coral Reefs to Climate Change. Annu Rev Ecol Evol Syst. 2015;**46**(1):281-303

[130] Wild C H-GO, Loya Y, Bythell J, Fitt W, Gates R, Iglesias-Prieto R, et al. Bleaching and Relate Ecological Factors. Melbourne, Australia. Ed. Currie Communications; 2009

[131] Cheal AJ, MacNeil MA, Cripps E, Emslie MJ, Jonker M, Schaffelke B, et al. Coral-macroalgal phase shifts or reef resilience: links with diversity and functional roles of herbivorous fishes on the Great Barrier Reef. Coral Reefs. 2010;**29**(4):1005-1015

[132] Hughes TP, Graham NA, Jackson JB, Mumby PJ, Steneck RS. Rising to the challenge of sustaining coral reef resilience. Trends Ecol Evol. 2010;**25**(11):633-642

[133] Huang D. Threatened reef corals of the world. PLoS One. 2012;**7**(3):e34459

Chapter 6

Dynamics of Stony Coral and Octocoral Juvenile Assemblages Following Disturbance on Patch Reefs of the Florida Reef Tract

Lucy A. Bartlett, Vanessa I.P. Brinkhuis,
Rob R. Ruzicka, Michael A. Colella,
Kathleen Semon Lunz, Erin H. Leone and
Pamela Hallock

Additional information is available at the end of the chapter

http://dx.doi.org/10.5772/intechopen.71606

Abstract

In January 2010, reefs in the Upper and Middle Florida Keys experienced prolonged exposure to extremely cold water temperatures, below lethal thresholds for many reef organisms including corals. We examined post-disturbance juvenile assemblages of stony corals and octocorals on eight patch reefs, four of which were categorized as high impact and four as low impact, based on declines in stony-coral cover following disturbance. We established permanent quadrats to conduct field surveys in spring and fall of 2012 and 2013. Overall, juvenile abundances of both stony corals and octocorals were greater on low-impact sites, suggesting that those sites had higher recruitment and juvenile survival than high-impact sites. Juvenile assemblages also showed a regional pattern, with more stony corals on Middle Keys sites and more octocorals on Upper Keys sites. The stony-coral juvenile assemblage was dominated by *Siderastrea siderea* (46%) and *Porites astreoides* (19%), whereas previously abundant species such as *Orbicella annularis* were nearly absent (<3%). Octocoral juveniles were dominated by *Antillogorgia* spp. (25%), *Gorgonia* spp. (21%), *Eunicea* spp. (19%) and *Erythropodium caribaeorum* (14%). Overall, post-disturbance juvenile assemblages displayed a wide range of octocoral genera, but only a few select stony-coral species, which exhibited either opportunistic or hardy life-history characteristics.

Keywords: recruitment, *Siderastrea siderea*, *Porites astreoides*, *Antillogorgia*, *Gorgonia*, *Eunicea*

1. Introduction

Coral reefs around the world continue to decline in response to compounding stressors, including those caused by humans and those occurring naturally. Often, natural disturbances such as hurricanes and thermal anomalies (especially those associated with El Niño Southern Oscillations (ENSOs) or cold-water events) can cause mass mortality in coral communities [1, 2]. For the past several decades, chronic anthropogenic disturbances, including reduced water quality, pollution, terrestrial runoff, and coastal development, have been exacerbating the impacts of natural disturbances by contributing to stressful environmental conditions that inhibit reef-community recovery [3–5]. Corals are thereby threatened by local anthropogenic stressors coupled with the challenges of warming seas and ocean acidification [6]. The additive and synergistic effects of these stressors can prevent reef-building corals from recovering following natural disturbances [1, 4, 5, 7]. This complex disturbance regime is readily evident in the history of coral communities of the Florida Keys [8–13].

The Florida reef tract was historically dominated by stony-coral species such as the branching *Acropora* spp. and the massive *Orbicella* spp., which have undergone dramatic declines in the last several decades [10, 14, 15]. In many documented cases around the world [16–19], macroalgae have taken over reefs following decline in stony-coral cover. In the Florida reef tract, octocorals and sponges, rather than macroalgae, have assumed the role of spatial opportunists and are replacing stony corals on many reefs [13, 20]. In other areas of the Caribbean where stony corals have declined, octocorals have maintained similar densities [21].

Octocorals, like macroalgae, are opportunistic, with higher recruitment and growth rates than stony corals [22–25]. Although octocorals do not confer the same ecosystem services as stony corals, they do offer more structural attributes than macroalgae. Octocorals found in the Florida Keys span a range of functional groups. Some grow as fans and plumes, creating canopies that provide habitat and protection for reef organisms [26], while encrusting species protect reef framework from bioeroding grazers such as parrot fish [27].

The Florida reef tract, off southeast Florida and the Florida Keys, is near the northern limit of coral-reef development and occasionally experiences environmental extremes, like the cold-water event that occurred in January 2010 [28–30]. In addition, nearshore patch reefs of the Florida reef tract are exposed to fluctuations in temperature and salinity and to greater variability in turbidity and light attenuation [31, 32]. Their proximity to land also potentially exposes these reefs to more direct anthropogenic effects such as terrestrial runoff, pollution [33–35], and fishing, boating and diving [31]. Despite this, prior to the 2010 cold-water event, patch reefs demonstrated the highest coral cover, highest growth rates and lowest incidences of partial mortality on the reef tract and hence had been identified as resilient to a variety of acute and chronic stressors [32].

In January 2010, patch reef communities were profoundly impacted when a prolonged influx of Arctic air caused sea-surface temperature to drop below the lethal threshold for many benthic organisms. Water temperature fell well below the 16°C lethal threshold for stony corals and remained low for several days [28, 36–38]. The event caused extreme, yet spatially inconsistent mortality among stony corals of the nearshore patch reefs in the Upper and Middle Florida Keys [28].

In the present study, we quantified juvenile abundance to determine if coral-recruitment success and post-settlement survivorship would provide early indications of recovery on these patch reefs. We chose four patch reefs that were severely impacted by the cold-event (high-impact sites) and four patch reefs that were essentially unchanged in community composition (low-impact sites) to create a balanced study design. The study compared juvenile communities of stony corals (comprising the orders Scleractinia and Anthomedusae, specifically family Milleporidae) and octocorals on patch reefs that endured significant losses of coral cover (high impact) with those that were less impacted (low impact). We included octocoral juveniles in the study to assess for susceptibility of patch reefs to documented increases in octocoral cover concurrent with declines in stony-coral cover occurring throughout the Florida reef tract. In addition, we used Coral Reef Evaluation and Monitoring Project (CREMP) data to compare macroalgal cover before and after the event to investigate the prospective role of macroalgae in stony-coral decline and recovery in this setting. We had four overarching questions: (1) Does coral recruitment differ between high-impact and low-impact sites? (2) Does coral recruitment differ between stony corals and octocorals? (3) Is there a difference in recruitment between the Upper and Middle Florida Keys? and (4) Did macroalgal cover increase following the cold-water event and if so, did it influence coral recruitment?

2. Methods

2.1. Site selection and survey protocol

Eight patch reef sites were selected for this study: four in the Upper Keys region (Admiral, Porter, Burr and Turtle) and four in the Middle Keys region (Dustan Rocks, West Turtle, Rawa and Thor) (**Figure 1**). All sites were selected from annual monitoring locations in the Coral Reef Evaluation and Monitoring Project (CREMP), for which historical demographic and benthic cover data were available for adult stony corals and macroalgae [13, 15, 39]. In each region, two reefs were categorized as high impact and two reefs as low impact (reference sites) based on the difference in percent stony-coral cover between 2009 and 2010 (**Figure 2**). A site was considered to be high impact if there was a 6% decrease in absolute stony-coral cover from summer 2009 to summer 2010, whereas low-impact sites decreased in cover by less than 1% or had slight gains during this time period [37].

Although all eight study sites were categorized as patch reef habitat, they differed in certain characteristics, including structural complexity, topography, depth and species composition. For example, Admiral reef was at least 3 m shallower than any other sites and was unique in that nearly all coral cover was previously comprised of massive *Orbicella annularis* colonies, which were decimated by the cold-water event [28]. Porter patch reef was structurally less complex than most of the other sites, with little rugosity and higher octocoral cover. The remaining sites, Dustan Rocks, Rawa, West Turtle, Thor, Burr and Turtle reefs, had similar depths and structural complexity and were representative of typical mid-channel patch reefs located in the Florida Keys.

Thirty-two randomly placed, permanent 0.25 m^2 quadrats were established at each patch reef. Sixteen quadrats were placed along each 22-m-long by 2-m-wide CREMP monitoring

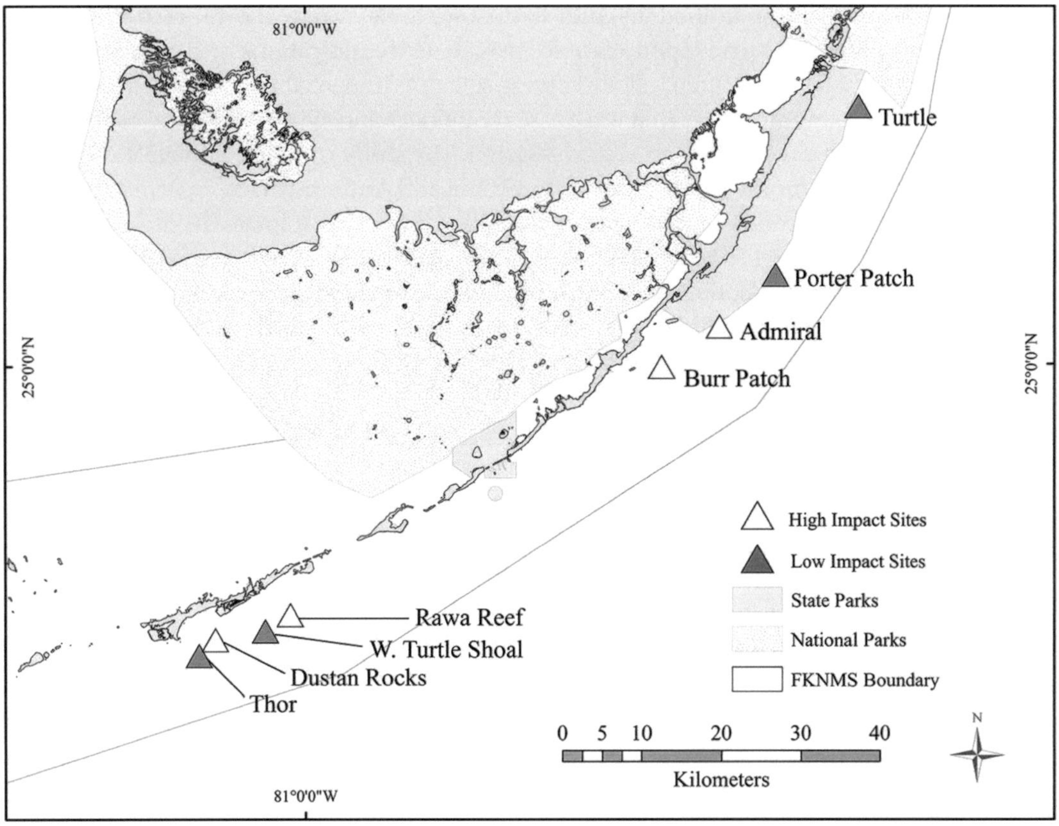

Figure 1. Map of the Middle and Upper Florida Keys with locations of recruitment study sites. White triangles represent high-impact sites, and gray triangles represent low-impact sites.

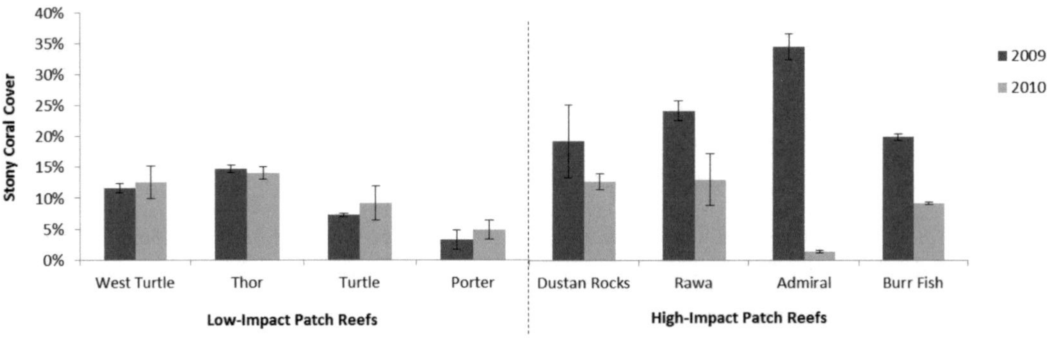

Figure 2. Percentage of stony coral cover in 2009 and 2010 at eight patch reef sites (error bars are standard error of the means). Extreme loss (≥6%) of stony coral cover was recorded on four patch reefs (high-impact sites). Cover values from CREMP data.

transect [40], and two transects were used per site. Quadrat locations within transects were randomly determined by subdividing the belt transect into 0.5 m by 0.5 m squares, numbering them (1–176), and using a random number generator to choose quadrat location. To avoid placing quadrats in locations where the substrate was unsuitable for coral settlement, more than 50% of the quadrat could not be comprised of living coral or sand. When a quadrat location did not meet these criteria, the next random number was selected. Three small (~7.6 cm long) masonry nails were inserted into the substrate and marked with flagging tape to delineate three corners of each quadrat to aid in relocation.

Photographs were taken of each quadrat and used as a visual aid for relocation. For the survey, all stony corals having a maximum diameter ≤4 cm and all octocoral juveniles with a maximum height ≤4 cm were identified, measured, mapped and photographed in March/April and September/October in 2012 and 2013. The 4-cm dimension for scleractinians was selected based on studies that found 4 cm to be an appropriate size cutoff for juveniles [41–43]. For octocorals, a 4-cm-height cutoff was applied because several gorgonian species can exceed 4 cm in height in a single year [24, 25]. This conservative threshold was applied so that octocorals under 4 cm in height were considered less than a year old to aid in distinguishing between year classes (e.g., juveniles identified in the first year of surveys would presumably be larger than 4 cm in the second year). Stony corals were identified to species, and octocorals were identified to genus. When identification was not possible, the colony was designated as either unknown stony coral (UNKS) or unknown octocoral (UNKO). The unknown categories were used by observers when a juvenile was too small to identify confidently, was obscured by another benthic organism, or was unidentifiable for some other reason.

2.2. High-impact vs. low-impact reefs

To compare the juvenile abundance of stony corals and octocorals at high-impact and low-impact sites, we built a generalized linear mixed model assuming a negative binomial distribution in SAS v9.3 (SAS Inc., Cary, NC). We included coral type (stony and octocoral) and site category (high-impact vs. low-impact) as well as their interaction as fixed factors. The negative binomial distribution was used in place of Poisson to account for over-dispersion (i.e., variance being unequal to mean) [44]. To account for the subsampling nature of data collection, we included random effects for multiple transects per site, as well as multiple quadrats per transect. As sites were repeatedly sampled, we incorporated a repeated-measures variance structure into the statistical model to account for potential autocorrelation, but removed it when the effect was estimated to be zero. We used a means comparison to determine which relationships were significant. Comparisons resulting in a p value <0.05 were considered statistically significant. Sites (n = 8) and survey periods (n = 4) were pooled, and relative frequencies of stony-coral and octocoral juvenile colonies were calculated to illustrate differences in high-impact versus low-impact site-frequency distributions as well as to highlight the high frequency of quadrats with very few or zero recruits found within them.

We employed a Before-After-Control-Impact analysis framework [45–47] to assess the impact of the 2010 cold-water event on macroalgae (including cyanobacteria) using the CREMP long-term monitoring database from 2007 to 2012. The 2007–2009 samples represented the "before" dates, while 2010–2012 represented "after." We built a linear mixed model to incorporate period (before vs. after), impact class (high impact vs. low impact), and their interaction, as well as random terms for year, site, and the year-by-site interaction. Model residuals were examined to ensure they met the assumptions of normality and homogeneity of variance. All analyses were performed in SAS v9.4.

To assess the impact of percent cover of macroalgae on juvenile abundance, we used the CREMP long-term monitoring data for each of the sites, and the spring juvenile abundance counts. Spring counts were used because they were done at about the same time of year as the CREMP surveys. We summed juvenile abundance across quadrats to enable us to compare to percent cover data at the transect level. We built generalized linear mixed models assuming a negative binomial distribution, separately for octocoral and stony-coral juvenile abundance. We included site category (high or low impact) as a categorical fixed factor and macroalgal percent cover as a continuous fixed factor.

2.3. Spatial patterns

To test for relationships among sites and to elucidate any regional differences in community assemblages, we used PRIMER v6 (PRIMER-E Ltd., Plymouth, UK) for the following multivariate analyses. We performed a CLUSTER analysis on abundances of stony-coral and octocoral juveniles using a Bray-Curtis resemblance matrix on untransformed data. We applied dispersion weighting to account for differences in variance structure of species counts due to spatial clumping [48]. We conducted a SIMPROF analysis to check for structure in the data and identify where clusters were significant. We then created a nonmetric multidimensional scaling (MDS) plot to visually display the relationships among sites. Nonmetric MDS plots map the sites in two-dimensional space, with proximity between sites representing similarity and a stress level of <0.2 indicating a useful representation of the relationship [49]. We used an analysis of similarity (ANOSIM) test to demonstrate where the dissimilarities were strongest among sites.

To assess relationships between adult and juvenile stony-coral assemblages, we used similar procedures to those outlined above for juvenile abundance. We acquired adult stony-coral densities (colonies/m^2) for each of the eight patch reef sites from 2011 and 2012 CREMP demographic survey data. We omitted octocoral data from this analysis since adult octocoral data were not available for all sites. We calculated stony-coral juvenile densities by dividing the total abundance by the study area (4 m^2/station). We then performed an MDS analysis on Bray-Curtis similarities of square-root-transformed, dispersion-weighted and averaged data to create a two-dimensional spatial visUalization of relationships among adult and juvenile assemblage densities of stony corals. We used the PRIMER v6 RELATE test to compare the adult and juvenile Bray-Curtis similarity matrices. The RELATE test uses the Spearman Rank Correlation Coefficient (Rho), with a Rho of +1 or −1 representing the highest possible correlation and with a p value of <0.05 representing a significant result. SIMPER analysis showed which sites had the greatest overall similarity between adult and juvenile assemblages.

2.4. Species compositions

To determine which species of stony corals and genera of octocorals were most abundant for all quadrats, data were pooled for each site and survey period to calculate overall taxonomic composition (**Table 1**).

http://dx.doi.org/10.5772/intechopen.71606

Stony coral species	% of total	Octocoral species	% of total
Siderastrea siderea	46.2	*Antillogorgia* spp.	24.5
Porites astreoides	19.1	*Gorgonia ventalina*	20.7
Siderastrea radians	6.3	*Eunicea* spp.	18.9
Agaricia agaricites	4.7	*Erythropodium caribaeorum*	14.3
Stephanocoenia intercepta	4.6	*Muriceopsis* spp.	4
Porites porites	4.4	*Pseudoplexaura* spp.	3
Montastraea cavernosa	3.7	*Briareum asbestinum*	3
Millepora alcicornis	2.7	*Muricea* spp.	1.8
Orbicella spp.	2.2	*Plexaura* spp.	1.5
Dichocoenia stokesii	0.94	*Plexaurella* spp.	1.3
Scolymia cubensis	0.85	*Pterogorgia* spp.	0.14
Colpophyllia natans	0.83	Unknown octocoral	6.8
Eusmilia fastigiata	0.74		
Pseudodiploria strigosa	0.41		
Mycetophyllia lamarckiana	0.37		
Oculina diffusa	0.27		
Diploria labyrinthiformis	0.24		
Agaricia fragilis	0.07		
Favia fragum	0.07		
Solenastrea bournoni	0.04		
Unknown stony coral	1.3		

Table 1. Juvenile stony coral and octocoral total percent species composition.

3. Results

3.1. High-impact vs. low-impact reefs

We found a significant interaction between coral type (stony corals vs. octocorals) and impact class (high impact vs. low impact) for juvenile abundance ($F_{1.1982}$ = 27.1, $P<0.001$). Both octocorals and stony corals were present in greater abundance at the low-impact sites (**Figure 3**). The difference in juvenile abundance between high- and low-impact sites was more pronounced for octocorals than that for stony corals. For both stony corals and octocorals, median frequencies for low-impact sites were higher than those for high-impact sites as well (**Figure 4**). The most frequent abundance value recorded within a single quadrat was zero. There were three to four times more quadrats (approximately 1 m^2) where

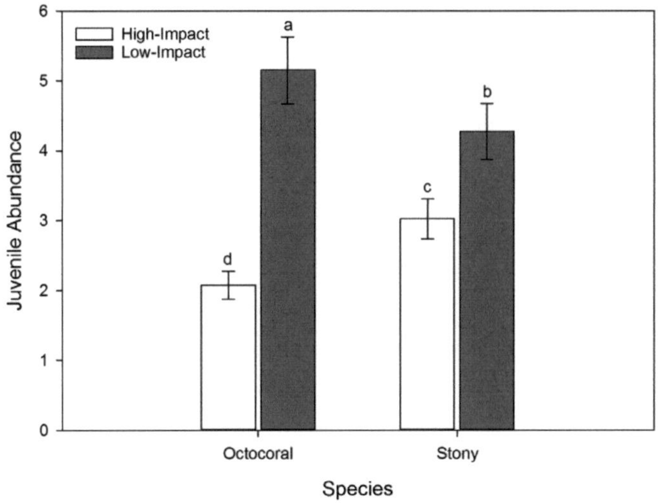

Figure 3. Juvenile coral abundance (least-squares mean ± SEM) per quadrat (0.25 m^2) by coral type (stony coral vs. octocoral). White bars are high-impact sites, and gray bars are low-impact sites. Means sharing any common letters are not significantly different (P<0.001): octocoral and stony coral abundance was significantly lower on high-impact sites.

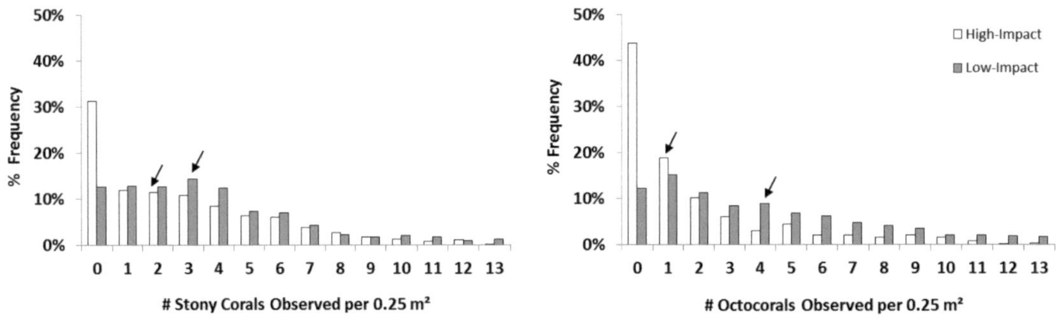

Figure 4. Relative-frequency distributions for stony coral (left) and octocoral (right) abundance by 0.25 m^2 quadrat (all sites and survey periods pooled). Both distributions are skewed to the right, with highest frequencies occurring at lower abundance values. Abundance data are not normally distributed (Shapiro-Wilk, p<0.001). White bars are high-impact sites, and gray bars are low-impact sites. Arrows indicate median values.

no juveniles were recorded on high-impact sites than on low-impact sites. Otherwise, frequency distributions of stony corals and octocorals were similar for high-impact and low-impact sites.

The BACI analysis revealed macroalgal cover increased following the 2010 cold-water event. There was a clear influence of time period ($F_{1,86}$ = 15.6, P = 0.0002), indicating that macroalgal cover was more abundant post-disturbance. However, there was no difference between impact classes ($F_{1.86}$ = 1.87, P = 0.175), suggesting that this increase was not influenced by a level of coral mortality. The lack of a significant interaction term also indicates that changes in macroalgal cover were influenced more by reef-tract wide processes and high annual variability, which overshadowed any differences that would have been associated with post-event mortality ($F_{1.86}$ = 1.10, P = 0.298, **Figure 5**). Additionally, we found no effect of macroalgal cover on juvenile coral abundance for either stony corals (**Table 2**) or octocorals (**Table 3**).

3.2. Spatial patterns

The three northernmost sites (Turtle, Porter and Admiral reefs) had different juvenile assemblages than the remaining sites (**Figure 6**, SIMPROF $P<0.05$). On these three sites, juvenile assemblages were dominated by octocorals, whereas assemblages on the remaining sites were dominated by stony corals. The stony coral *S. siderea* was the most abundant juvenile on all four Middle Keys sites as well as the southernmost Upper Keys site, Burr Patch. Juvenile assemblages on the three northernmost sites (Turtle, Porter and Admiral) were dominated by three different octocoral genera *Antillogorgia, Eunicea* and *Gorgonia,* respectively (**Table 4**). Cluster analysis demonstrated that the spatial pattern was based on regional differences rather than site-impact category.

Similarities among post-disturbance adult and juvenile scleractinian assemblages by site were evaluated using CREMP demographic data for 2011 and 2012. Site clustering was very similar to the juvenile MDS plot (RELATE Spearman Rank Correlation Rho = 0.7, $p<0.001$), with all Middle Keys sites and Burr displaying similar adult and juvenile scleractinian assemblages,

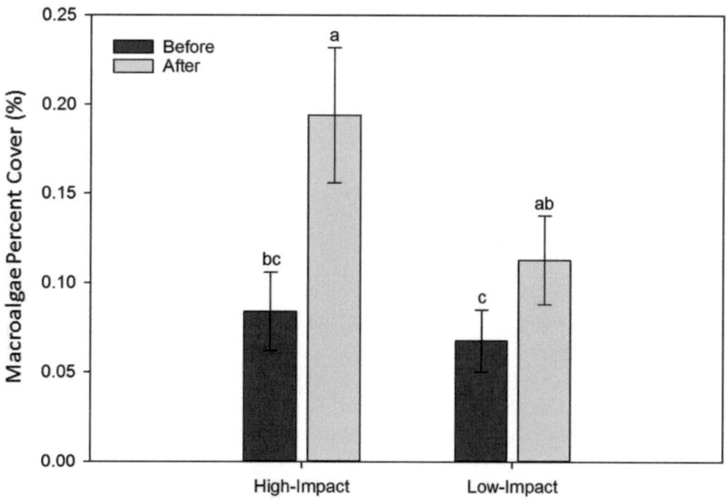

Figure 5. Macroalgae percent cover (least-squares mean ± 95% CI) by impact class (high vs. low impact) and time period (before and after 2010 bleaching event). Means sharing any common letters are not significantly different ($P > 0.05$).

Effect	Num DF	Den DF	Estimate	Standard error	F value	Pr > F
Site category	1	15	–	–	0.21	0.655
Macroalgae	1	15	−0.880	0.895	97	0.341

Site category (high impact vs. low impact) included as a categorical fixed factor and macroalgal cover as a continuous fixed factor.

Table 2. Model results showing impact of macroalgal cover on stony coral juvenile abundance.

Effect	Num DF	Den DF	Estimate	Standard error	F value	Pr > F
Site category	1	15	–	–	3.29	0.0896
Macroalgae	1	15	0.264	0.996	0.07	0.794

Site category (high impact vs. low impact) included as a categorical fixed factor and macroalgal cover as a continuous fixed factor.

Table 3. Model results showing impact of macroalgal cover on octocoral juvenile abundance.

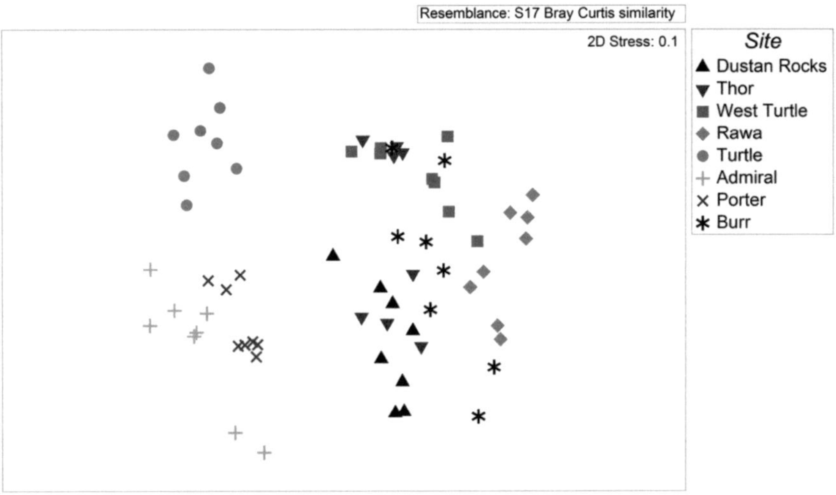

Figure 6. MDS plot of Bray-Curtis similarities of juvenile stony coral and octocoral abundances among sites, survey periods pooled (eight samples per site).

while Porter and Admiral displayed different assemblages (**Figure 7**). A main difference that resulted from removing octocorals from the analysis was that Turtle grouped with the larger cluster, instead of remaining distinct. This was true for both juveniles and adults, suggesting that the stony-coral assemblage on Turtle reef was actually quite similar to that at the Middle Keys sites. Juvenile and adult assemblages from the same sites tended to be similar. SIMPER analysis showed that within-site similarity between adults and juveniles ranged from 30 to

http://dx.doi.org/10.5772/intechopen.71606

Site	SSID	PAST	SINT	PPOR	MCAV	AAGA	SRAD	ANTI	GORG	EUNI	ERYT	PSEU	BRIA	MURO
Turtle	6	5	0	1	0	1	0	**23**	5	5	1	5	2	2
Porter	1	1	0	0	0	1	1	2	2	**6**	2	1	0	0
Admiral	1	1	0	0	0	0	0	3	**12**	1	0	0	0	0
Burr	**9**	3	0	0	0	1	1	1	1	1	1	0	0	2
Rawa	**8**	2	1	1	2	0	0	0	0	0	0	0	0	0
W. Turtle	**11**	6	2	2	2	2	1	2	1	2	1	0	0	0
Dustan	**6**	1	1	0	0	0	1	3	1	2	0	0	0	0
Thor	**12**	4	3	1	0	0	0	2	1	2	3	0	1	0

Bolded values are the dominant species for that site. Sites are listed from northernmost to southernmost with high-impact sites in italics. Species and genera coding is the first letter of the genus and the first three letters of the species for stony corals (SSID-*Siderastrea siderea*, PAST-*Porites astreoides*, SINT-*Stephanocoenia intersepta*, PPOR-*Porites porites*, MCAV-*Montastraea cavernosa*, AAGA-*Agaricia agaricites*, SRAD-*Siderastrea radians*) and the first four letters of the genus for octocorals (ANTI-*Antillogorgia*, GORG-*Gorgonia*, EUNI-*Eunicea*, ERYT-*Erythropodium*, PSEU-*Pseudoplexaura*, BRIA-*Briareum*, MURO-*Muriceopsis*).

Table 4. Median densities (colonies per m^2, per site) for top seven juvenile stony coral species (left side) and top seven juvenile octocoral genera (right side).

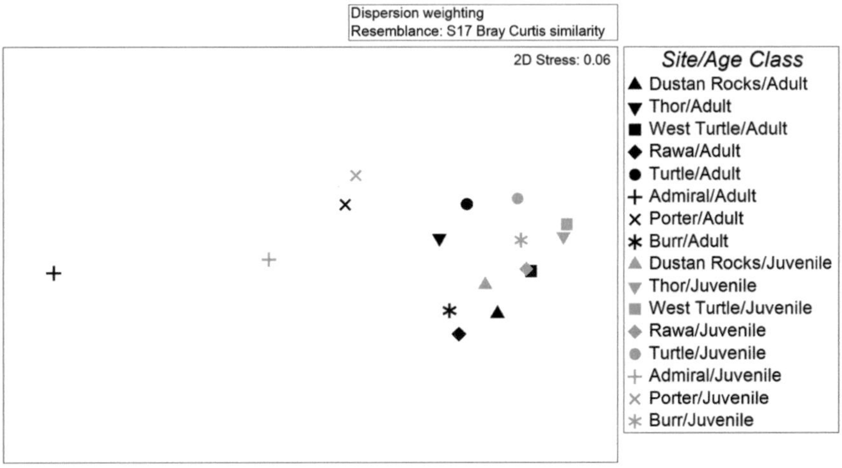

Figure 7. MDS plot of stony-coral adult and juvenile species densities among sites with survey periods pooled (juveniles—eight samples per site, adults—four samples per site). Data were dispersion-weighted, square-root-transformed and averaged.

72%. West Turtle and Porter reefs showed the greatest similarity between adult and juvenile assemblages, while Admiral showed the least similarity (**Table 5**).

3.3. Species compositions

The most abundant stony-coral species were *Siderastrea siderea* (46.2%), *Porites astreoides* (19.1%), *Siderastrea radians* (6.3%), *Agaricia agaricites* (4.7%), *Stephanocoenia intersepta* (4.6%), *Porites porites* (4.4%) and *Montastraea cavernosa* (3.7%) (**Table 1**). *Siderastrea siderea* made up nearly half of all stony corals observed, and *P. astreoides* made up almost one-fifth. Combined, these two species comprised over 65% of the stony-coral juvenile assemblage, in the study. Ten percent of the stony-coral assemblage was made up of other species, each accounting for

Site	%Similarity
W. Turtle	72
Porter	67
Turtle	67
Dustan	66
Burr	61
Rawa	60
Thor	55
Admiral	30

Table 5. Percent similarity between adult and juvenile stony-coral densities on each site (SIMPER test in PRIMER v6).

<1% of the total, with the exception of fire coral, *Millepora alcicornis*, which made up 2.7%, and *Orbicella* spp., which made up 2.1% (**Table 1**).

The most abundant octocoral genera were *Antillogorgia* (formerly *Pseudopterogorgia*, 24.5%), *Gorgonia* (20.7%), *Eunicea* (19.0%), *Erythropodium* (14.3%), *Muriceopsis* (4.0%), *Pseudoplexaura* (3.0%) and *Briareum* (3.0%) (**Table 1**). The four top octocoral genera (*Antillogorgia, Gorgonia, Eunicea* and *Erythropodium*) made up almost 80% of all octocorals seen, but unlike the stony corals, percent composition was more evenly distributed among these four genera. The unknown octocoral category included 6.8% of all juvenile octocoral observations and was included in analyses comparing total stony-coral abundance to total octocoral abundance, but was not included in species and genera analyses.

4. Discussion

With increasing frequency and intensity of perturbations impacting coral reefs around the world, an understanding of recovery processes is crucial. While coral-reef response to disturbances involving warm-water bleaching events, hurricanes and nutrient stress is well documented [4, 50, 51], comparatively less is known of recovery from cold-stress events [29, 30]. This study examined juvenile assemblages following the cold-water mass-mortality event in 2010, which resulted in one of the greatest losses of reef-building corals in the Florida Keys in modern time [28, 36, 38]. We found fewer stony coral and octocoral juveniles on the high-impact sites (**Figures 3** and **4**), suggesting that recruitment to severely degraded reefs was impaired. Unlike in many documented cases of abated coral recovery, here, macroalgae was not implicated for impairing coral recovery as we found no significant difference in macroalgal cover between high- and low-impact sites following the event (**Figure 5**). In addition, we observed a regional trend in which juvenile stony corals were more abundant on the Middle Keys reefs, while juvenile octocorals were more abundant on the northernmost Upper Keys reefs (**Figure 6**, **Table 4**). Moulding [43] found similar results in 2004, with higher stony-coral recruit density in the Middle Keys than in the Upper Keys. During our study, only *Siderastrea siderea* and *Porites astreoides* accounted individually for >10% of the juveniles recorded and together totaled 65%. Of the juvenile octocorals identified, four genera accounted for >10% individually and together totaled 78% of those recorded (**Table 1**). Results suggest that recovery will be slower in areas where disturbance impacts were greatest and that post-mortality species composition will be influenced by regional differences in settlement and survival, as well as by differences in species life-history characteristics.

4.1. High-impact vs. low-impact reefs

The level of mortality on a reef negatively correlated with recruitment. Partial and complete colony mortality on high-impact sites [28] appeared to have dampened recruitment and/or juvenile survival. Higher abundance and cover of living coral are important drivers of coral recruitment [52–54]. However, percent cover on high-impact sites was comparable to cover on low-impact sites following the cold-water event. This suggests that the magnitude of mortality and post-event conditions on a reef had more influence on recruitment than total percent cover. Thus, regardless of the similarities in coral cover following the disturbance, severe mortality on high-impact sites placed these reefs at a clear reproductive disadvantage for recruitment of both octocorals and stony corals.

Massive scleractinian species often found on reefs of the Florida reef tract, such as *Orbicella* (formerly *Montastraea*) spp., *Diploria* spp., *Colpophyllia natans* and *Siderastrea siderea,* reproduce via broadcast spawning, releasing gametes only once a year in synchrony with the late summer lunar cycle [55]. The success of sexual reproduction via spawning is a function of reef connectivity and is highly dependent on physical characteristics of the water column, including tides and currents [56]. Nonmassive coral species, such as *Porites* spp., *Agaricia* spp. and *Siderastrea radians,* are internal brooders, reproducing many times a year [43, 55]. Brooded planulae are larger and more developed upon release into the water column, settle sooner and presumably, closer to the parent colonies than larvae produced by broadcast spawning corals, which require the fertilization of gametes in the water column after release [57]. The reduced size of the overall coral population in addition to the suboptimal condition of surviving adult corals at the high-impact reefs would have likely lowered successful reproduction, while low-impact sites may have benefited from the local supply of brooded planulae from the relatively healthy colonies still present on those reefs. In addition, coral larvae may actively select suitable substratum for settlement via chemical signals produced by adult colonies and associated flora such as crustose coralline algae (CCA) [58–60], and this process may have been disturbed on high-impact sites. It is unclear which stage of the reproduction cycle was influenced most by the mortality event; however, our results show that stony-coral percent cover alone did not explain the observed differences in juvenile abundances.

While macroalgal cover increased on high-impact sites in the years following the cold-water anomaly, there was no difference in macroalgal cover between high- and low-impact sites, suggesting that this increase reflected the natural variability in macroalgal cover along the Florida reef tract [13] and that the production of macroalgae is independent of coral-mortality event-driven processes [61, 62]. Additionally, we found no effect of macroalgal cover on juvenile coral abundance for either stony corals or octocorals. Our ability to discern differences in macroalgal cover between high- and low-impacts sites was likely limited by low replication (e.g., only two CREMP transects per site were used to match pooled quadrat data) and high annual variability in cover (e.g., high inter-site variability). Macroalgae can interfere with coral recruitment through either competition for space or more direct methods involving allelopathy [63, 64]. Macroalgae colonize a reef quickly following disturbance, taking advantage of the vacant space [17–19]. However, here the increase in macroalgal cover on high-impact sites did not differ significantly from cover on low-impact sites and did not influence juvenile assemblages; thus, it seems factors other than macroalgal or stony-coral cover drove recruitment in this study. Potential factors include the magnitude of the loss of live coral cover, the post-event condition of the surviving reef community, and the regional location of a particular reef, all of which could influence larval supply, settlement rates and post-settlement survival.

4.2. Spatial trends

A strong regional trend was observed, in which juvenile stony corals dominated the Middle Keys sites and Burr Patch (the southernmost site in the Upper Keys), whereas juvenile octocorals dominated the three northernmost sites, Turtle, Admiral and Porter. While this trend was unanticipated, there are certain environmental factors that differ between the Middle and

Upper Keys regions that could have influenced this pattern. Middle Keys patch reefs lie close to tidal connections with Florida Bay and so experience greater turbidity, greater fluctuations in temperature and salinity, and higher inputs of terrigenous materials [65–67]. Connections to Florida Bay become less frequent toward the Upper Keys. Burr Patch is the southwestern-most Upper Keys site in this study and is the closest of all Upper Keys sites to tidal connections. Its location and proximity to the northernmost connection to Florida Bay may explain its similarity in juvenile assemblage to the Middle Keys sites. It is still unclear, however, how such a hydrodynamic difference between regions would facilitate such a pattern. We were unable to compare adult to juvenile octocoral assemblages, due to lack to adult demographic data, but such a comparison may help to explain the observed spatial patterns in the data.

The regional pattern we observed in the site clustering is a strong indication that post-disturbance juvenile abundance was influenced by regional differences in hydrodynamics. Site-specific differences in structural complexity, topography and reef structure also may have influenced these results. While each of the eight study sites are designated as patch reefs, such reefs are inherently diverse, with different depths, structural complexities and species compositions such that site-level differences undoubtedly influenced results. In fact, species compositions of adult assemblages did prove similar to juvenile species compositions at certain sites (**Figure 7**, **Table 4**), including West Turtle, Porter and Turtle (72, 67 and 67% similarity among adult and juvenile assemblages, respectively). In 2010, Kuffner et al. [68] found a strong influence of "reef effect" on variance of the biological community structure in Biscayne National Park and suggested prior residents on a reef, or "priority" effects were important in determining future benthic-community structure. It seems likely similar effects may be at play here.

Additionally, the location of the Florida reef tract near the northern latitudinal limit for coral-reef accretion may help explain the reduction in stony-coral juveniles and the proliferation of octocoral juveniles on the northernmost study sites. Results corroborate regional trends observed on the Florida reef tract [43, 69], where the density of juvenile stony corals increased in a southwesterly direction down the reef tract. This is an important finding, and further investigation is needed to elucidate the causes behind this pattern.

4.3. Species composition

Species-abundance data suggest that specific life-history traits play a major role in which species recruit to and survive on patch reefs. Known ecological traits of *Siderastrea siderea* and *P. astreoides* likely influenced successful recruitment and survival. *Siderastrea siderea* can tolerate a wide range of temperature and environmental fluctuations [28, 36, 37] but it is slow-growing, can reach massive sizes and reproduces typically once a year via broadcast spawning [55, 70]. In this study, the high tolerance of environmental fluctuations clearly outweighs the slow growth and reproductive mode of *S. siderea,* making it the most abundant stony-coral juvenile observed on study patch reefs. *Porites astreoides* is a smaller, weedy, brooding species that can reproduce prolifically, multiple times a year, and reaches maturity sooner but that is less tolerant of temperature fluctuations [41, 71–73]. Even though *P. astreoides* was nearly extirpated at some of the high-impact sites [28] during the cold-water event, the species can quickly recover after disturbances [74]. Our results are consistent with those of van Woesik et al. [75] which indicated that *S. siderea* and *P. astreoides* have also had high recruitment rates on outer bank reefs in the Florida Keys. In this study, each species demonstrated a different, yet successful, method of recruitment to patch reefs.

The overall juvenile assemblage was essentially devoid of *Orbicella* spp. (<3% of the juvenile stony-coral species composition). *Orbicella* spp. are very sensitive to temperature fluctuations [70, 76, 77]. The few small *Orbicella* colonies found were most likely surviving tissue remnants of older, larger colonies that had undergone near-complete mortality. In recent years, recruitment by *Orbicella* spp. on the Florida reef tract has been poor [43, 69, 75]. *Orbicella* spp. reproduce only once per year in synchrony with the late summer lunar cycle via broadcast spawning [55]. Their breeding method and life history characteristics do not support high recruitment rates, but rather high survival rates of settled juveniles [43, 55]. This reproduction method is apparently not the best for present conditions along Florida reef tract.

Octocorals were an important part of the coral juvenile assemblage on all of our patch reef sites. Relative to stony corals, less is known about reproduction in octocorals. Many gorgonians are thought to be gonochoric brooders [78]. Brooding can occur internally, as occurs with stony corals, or externally, on the surface of the female colony as occurs with the species *Briareum asbestinum* [79]. Others are broadcast spawners, releasing both sperm and egg into the water column for fertilization as occurs with *Gorgonia ventalina* [80, 81]. As with stony corals, octocorals spawn in synchrony with the lunar cycle, usually following a full moon, though not necessarily synchronous across species [80].

Relative to stony corals, octocorals are weedy, fast-growing, opportunistic organisms that can quickly colonize newly available substrate, outcompeting stony corals for suitable settlement space [24, 80]. The flexible-branching morphology of octocorals facilitates asexual reproduction through a number of methods, giving octocorals a unique advantage following disturbance events. The ability for vegetative growth allows octocorals to aggregate in high densities and to colonize space quickly [82, 83]. This can provide an advantage following a disturbance event, allowing them to outcompete more slowly reproducing organisms, such as spawning stony corals, for the newly available space. Octocoral extension rates are also orders of magnitude higher than those of stony corals [22–25].

The most abundant octocoral genus that we recorded was *Antillogorgia* (25%), which includes several species, many of which have fast growth rates [23] and often colonize quickly following disturbances. Octocorals were most abundant on Admiral, Porter and Turtle, the northernmost sites in the study. Overall, octocorals exhibited a wider range of successfully recruiting genera than did stony corals, which may promote higher species diversity in future assemblages. The successful recruitment of octocorals may further promote the ongoing shift in community structure to an octocoral-dominated state, which has been observed in adult assemblages throughout the Florida Keys [13]. More research is needed on the changing role of octocorals in reef ecosystems of the Florida reef tract.

5. Conclusions

Our results support previous findings that Florida stony-coral populations struggle to recover after major perturbations. Patch reefs were considered one of the last refuges for corals on the Florida reef tract that could potentially serve as a larval source for repopulating the highly depauperate outer bank reefs [32]. The 2010 cold-water event decimated many major framework-building corals that had previously shown resistance to a variety of acute and chronic stressors impacting Florida Keys reefs. Short-term recovery thus far has been defined by recruitment and survival of opportunistic and hardy stony-coral species and octocorals. If early results persist, composition of the scleractinian assemblages on these patch reefs will be homogenized by the few coral species that are successfully recruiting in the Florida Keys, or, as documented elsewhere in Florida and the Caribbean, modern reefs will be dominated by associated coral-reef fauna (e.g., octocorals) interspersed with the few corals that are resilient to disturbances. A better understanding of local hydrodynamics, reproductive timing and dispersal mechanisms of both octocorals and stony corals is needed to provide insight into potential drivers of the spatial patterns observed in this study as well as the changing ecological role of octocorals on the Florida reef tract.

Acknowledgements

We thank everyone who contributed to the many grueling hours of diving with our faces pressed close to the substratum, searching endlessly for tiny juvenile corals. Funding to support this work came from the US EPA Water Quality Protection Program (X7-97468002). We thank Dr. Ilsa B. Kuffner for constructive advice in improving the manuscript. The statements, conclusions and recommendations expressed herein are those of the authors and do not necessarily reflect the views of the State of Florida, the US Environmental Protection Agency or the University of South Florida.

Author details

Lucy A. Bartlett[1,2,3], Vanessa I.P. Brinkhuis[1], Rob R. Ruzicka[1], Michael A. Colella[1], Kathleen Semon Lunz[1], Erin H. Leone[1] and Pamela Hallock[2]*

*Address all correspondence to: pmuller@usf.edu

1 Florida Fish and Wildlife Conservation Commission Wildlife Research Institute, Saint Petersburg, FL, USA

2 University of South Florida, Saint Petersburg, FL, USA

3 USGS St. Petersburg Coastal and Marine Science Center, Saint Petersburg, FL, USA

References

[1] Baker AC, Glynn PW, Riegl B. Climate change and coral reef bleaching: An ecological assessment of long-term impacts, recovery trends and future outlook. Estuarine, Coastal and Shelf Science. 2008;**80**:435-471

[2] Edmunds PJ, Witman JD. Effect of Hurricane Hugo on the primary framework of a reef along the south shore of St. John, US Virgin Islands. Marine Ecology Progress Series. 1991;**78**:201-204

[3] Gardner TA, Cote IM, Gill JA, Grant A, Watkinson AR. Long-tern region-wide declines in Caribbean corals. Science. 2003;**301**:958-960

[4] Graham NAJ, Nash KL, Kool JT. Coral reef recovery dynamics in a changing world. Coral Reefs. 2011;**30**:283-294

[5] Hughes TP, Connell JH. Multiple stressors on coral reefs: A long-term perspective. Limnology and Oceanography. 1999;**44**:932-940

[6] Bruno JF, Valdivia A. Coral reef degradation is not correlated with local human population density. Scientific Reports. 2016;**6**:29778

[7] Wakeford M, Done TJ, Johnson CR. Decadal trends in a coral community and evidence of changed disturbance regime. Coral Reefs. 2008;**27**:1-13

[8] Downs CA, Fauth JE, Robinson CE, Curry R, et al. Cellular diagnostics and coral health: Declining coral health in the Florida Keys. Marine Pollution Bulletin. 2005;**51**:558-569

[9] Miller M, Bourque A, Bohnsack J. An analysis of the loss of acroporid corals at Looe Key, Florida, USA: 1983-2000. Coral Reefs. 2002;**21**:179-182

[10] Palandro DA, Andréfouët S, Hu C, Hallock P, et al. Quantification of two decades of shallow-water coral reef habitat decline in the Florida Keys National Marine Sanctuary using Landsat data (1984-2002). Remote Sensing of Environment. 2008;**112**:3388-3399

[11] Porter J, Dustan P, Jaap W, Patterson K, et al. Patterns of spread of coral disease in the Florida Keys. In: Porter J, editor. The Ecology and Etiology of Newly Emerging Marine Diseases. Vol. 159. Netherlands: Springer; 2001. pp. 1-24

[12] Porter JW, Lewis SK, Porter KG. The effect of multiple stressors on the Florida Keys coral reef ecosystem: A landscape hypothesis and a physiological test. Limnology and Oceanography. 1999;**44**:941-949

[13] Ruzicka R, Colella M, Porter J, Morrison J, et al. Temporal changes in benthic assemblages on Florida Keys reefs 11 years after the 1997/1998 El Niño. Marine Ecology Progress Series. 2013;**489**:125-141

[14] Burman SG, Aronson RB, van Woesik R. Biotic homogenization of coral assemblages along the Florida reef tract. Marine Ecology Progress Series. 2012;**467**:89

[15] Somerfield PJ, Jaap WC, Clarke KR, Callahan M, et al. Changes in coral reef communities among the Florida Keys, 1996-2003. Coral Reefs. 2008;**27**:951-965

[16] Hoegh-Guldberg O. Climate change, coral bleaching and the future of the world's coral reefs. Marine and Freshwater Research. 1999;**50**:839-866

[17] Hughes TP. Catastrophes, phase shifts, and large-scale degradation of a Caribbean coral reef. Science. 1994;**256**:1547-1551

[18] Hughes TP, Rodrigues MJ, Bellwood DR, Ceccarelli D, et al. Phase shifts, herbivory, and the resilience of coral reefs to climate change. Current Biology. 2007;**17**:360-365

[19] McCook L. Macroalgae, nutrients and phase shifts on coral reefs: Scientific issues and management consequences for the Great Barrier Reef. Coral Reefs. 1999;**18**:357-367

[20] McMurray SE, Henkel TP, Pawlik JR. Demographics of increasing populations of the giant barrel sponge Xestospongia muta in the Florida Keys. Ecology. 2010;**91**:560-570

[21] Lenz EA, Bramanti L, Lasker HR, Edmunds PJ. Long-term variation of octocoral populations in St. John, US Virgin Islands. Coral Reefs. 2015;**34**:1099-1109

[22] Cary LR. Observations upon the growth-rate and ecology of gorgonians. Carnegie Institute Washington Publisher. 1914;**182**:79-90

[23] Yoshioka BB. Aspects of the ecology of *Pseudopterogorgia americana* and *Pseudopterogorgia acerosa* [MS Thesis]. University of Puerto Rico; 1979

[24] Yoshioka PM, Yoshioka BB. A comparison of the survivorship and growth of shallow-water gorgonian species of Puerto Rico. Marine Ecology Progress Series. 1991;**69**:253-260

[25] Brazeau DA, Lasker HR. Growth rates and growth strategy in a clonal marine invertebrate, the Caribbean octocoral Briareum asbestinum. The Biological Bulletin. 1992;**183**: 269-277

[26] Kumagai NH. Role of food source and predator avoidance in habitat specialization by an octocoral-associated amphipod. Oecologia. 2008;**155**:739-749

[27] Kuffner IB, Toth LT. A geological perspective on the degradation and conservation of western Atlantic coral reefs. Conservation Biology. 2016;**30**:706-715

[28] Colella M, Ruzicka R, Kidney J, Morrison J, Brinkhuis V. Cold-water event of January 2010 results in catastrophic benthic mortality on patch reefs in the Florida Keys. Coral Reefs. 2012;**31**:621-632

[29] Hudson JH, Shinn EA, Halley RB, Lidz B. Sclerochronology: A tool for interpreting past environments. Geology. 1976;**4**:361-364

[30] Porter JW, Battey JF, Smith GJ. Perturbation and change in coral reef communities. Proceedings of the National Academy of Sciences of the United States of America. 1982;**79**:1678-1681

[31] Ginsburg RN, Gischler E, Kiene WE. Partial mortality of massive reef-building corals: An index of patch reef condition, Florida reef tract. Bulletin of Marine Science. 2001;**69**:1149-1173

[32] Lirman D, Fong P. Is proximity to land-based sources of coral stressors an appropriate measure of risk to coral reefs? An example from the Florida Reef Tract. Marine Pollution Bulletin. 2007;**54**:779-791

[33] Chiappone M, Sullivan K. Distribution, abundance and species composition of juvenile scleractinian corals in the Florida reef tract. Bulletin of Marine Science. 1996;**58**:555-569

[34] Fabricius KE. Effects of terrestrial runoff on the ecology of corals and coral reefs: Review and synthesis. Marine Pollution Bulletin. 2005;**50**:125-146

[35] Lapointe BE, O'Connell JD, Garrett GS. Nutrient couplings between on-site sewage disposal systems, groundwaters, and nearshore surface waters of the Florida Keys. Biogeochemistry. 1990;**10**:289-307

[36] Kemp DW, Oakley CA, Thornhill DJ, Newcomb LA, Schmidt GW, Fitt WK. Catastrophic mortality on inshore coral reefs of the Florida Keys due to severe low-temperature stress. Global Change Biology. 2011;**17**:3468-3477

[37] Kemp DW, Colella MA, Bartlett LA, Ruzicka RR, Porter JW, Fitt WK. Life after cold death: Reef coral and coral reef responses to the 2010 cold water anomaly in the Florida Keys. Ecosphere. 2016;**7**(6)

[38] Lirman D, Schopmeyer S, Manzello D, Gramer LJ, et al. Severe 2010 cold-water event caused unprecedented mortality to corals of the Florida Reef Tract and reversed previous survivorship patterns. PLoS One. 2011;**e23047**:6

[39] Porter JW, Kosmynin V, Patterson KL, Porter KG, et al. Detection of coral reef change by the Florida Keys coral reef monitoring project. In: The Everglades, Florida Bay, and Coral Reefs of the Florida Keys: An Ecosystem Sourcebook. Boca Raton: CRC Press. 2001. pp. 749-769

[40] Ruzicka R, Colella M, Semon K, Brinkhuis V, et al. CREMP 2009 Final Report. St. Petersburg, FL: Fish & Wildlife Research Institute/Florida Fish & Wildlife Conservation Commission; 2010

[41] Bak RPM, Engel MS. Distribution, abundance and survival of juvenile hermatypic corals (Scleractinia) and the importance of life history strategies in the parent coral community. Marine Biology. 1979;**54**:341-352

[42] Loya Y. Recolonization of Red Sea corals affected by natural catastrophes and man-made perturbations. Ecology. 1976;**57**:278-289

[43] Moulding AL. Coral recruitment patterns in the Florida Keys. Revista de Biología Tropical. 2005;**53**:75-82

[44] Bliss CI, Fisher RA. Fitting the negative binomial distribution to biological data. Biometrics. 1953;**9**:176-200

[45] Stewart-Oaten A, Murdoch WW, Parker KR. Environmental Impact Assessment: "Pseudoreplication" in Time? Ecology. 1986;**67**:929-940

[46] Stewart-Oaten A, Bence JR, Osenberg CW. Assessing effects of unreplicated perturbations: No simple solutions. Ecology. 1992;**73**:1396-1404

[47] Stewart-Oaten A, Bence JR. Temporal and spatial variation in environmental impact assessment. Ecological Monographs. 2001;**71**:305-339

[48] Clarke KR, Chapman MG, Somerfield PJ, Needham HR. Dispersion-based weighting of species counts in assemblage analyses. Marine Ecology Progress Series. 2006;**320**:11-27

[49] Clarke K, Warwick R. Change in Marine Communities: An Approach to Statistical Analysis and Interpretation. Plymouth, UK: PRIMER-E; 2001

[50] Connell JH, Hughes TP, Wallace CC. A 30-year study of coral abundance, recruitment, and disturbance at several scales in space and time. Ecological Monographs. 1997;**67**:461-488

[51] Vega Thurber RL, Burkepile DE, Fuchs C, Shantz AA, McMinds R, Zaneveld JR. Chronic nutrient enrichment increases prevalence and severity of coral disease and bleaching. Global Change Biology. 2014;**20**:544-554

[52] Gibbs DA, Hay ME. Spatial patterns of coral survivorship: Impacts of adult proximity versus other drivers of localized mortality. PeerJ. 2015;**3**:e1440

[53] Lasker HR. Recruitment and Resilience of a Harvested Caribbean Octocoral. PLoS One. 2013;**8**:e74587

[54] Vermeij M. Substrate composition and adult distribution determine recruitment patterns in a Caribbean brooding coral. Marine Ecology Progress Series. 2005;**295**:123-133

[55] Szmant AM. Reproductive ecology of Caribbean reef corals. Coral Reefs. 1986;**5**:43-53

[56] Richmond RH, Hunter CL. Reproduction and recruitment of corals: comparisons among the Caribbean, the tropical Pacific, and the Red Sea. Marine Ecology Progress Series. 1990;**60**:185-203

[57] Richmond RH. Reproduction and recruitment in corals: Critical links in the persistence of reefs. In: Life and Death of Coral Reefs. New York: Chapman & Hall; 1997. pp. 175-197

[58] Fadlallah YH. Sexual reproduction, development and larval biology in scleractinian corals. Coral Reefs. 1983;**2**:129-150

[59] Morse AN, Morse DE. Flypapers for coral and other planktonic larvae. Bioscience. 1996;**46**:254-262

[60] Ritson-Williams R, Arnold SN, Fogarty ND, Steneck RS, Vermeij MJ, Paul VJ. New perspectives on ecological mechanisms affecting coral recruitment on reefs. Smithsonian Contributions to the Marine Sciences. 2009;(38):437-457

[61] Suchley A, McField MD, Alvarez-Filip L. Rapidly increasing macroalgal cover not related to herbivorous fishes on Mesoamerican reefs. PeerJ. 2016;**4**:e2084

[62] Toth LT, van Woesik R, Murdoch TJT, Smith SR, Ogden JC, Precht WF, Aronson RB. Do no-take reserves benefit Florida's corals? 14 years of change and stasis in the Florida Keys National Marine Sanctuary. Coral Reefs. 2014;**33**:565-577

[63] Kuffner IB, Walters LJ, Becerro MA, Paul VJ, Ritson-Williams R, Beach KS. Inhibition of coral recruitment by macroalgae and cyanobacteria. Marine Ecology Progress Series. 2006;**323**:107-117

[64] Tanner JE. Competition between scleractinian corals and macroalgae: An experimental investigation of coral growth, survival and reproduction. Journal of Experimental Marine Biology and Ecology. 1995;**190**:151-168

[65] Ginsburg RN, Shinn EA. Distribution of the reef-building community in Florida and the Bahamas. American Association of Petroleum Geologists Bulletin. 1964;**48**:527

[66] Montague CL, Ley JA. A possible effect of salinity fluctuation on abundance of benthic vegetation and associated fauna in northeastern Florida Bay. Estuaries. 1993;**16**:703-717

[67] Vega-Rodriguez M, Müller-Karger F, Hallock P, Quiles-Perez G, et al. Influence of water-temperature variability on stony coral diversity in Florida Keys patch reefs. Marine Ecology Progress Series. 2015;**528**:173-186

[68] Kuffner IB, Grober-Dunsmore R, Brock JC, Hickey TD. Biological community structure on patch reefs in Biscayne National Park, FL, USA. Environmental Monitoring and Assessment. 2010;**164**:513-531

[69] Miller MW, Weil E, Szmant AM. Coral recruitment and juvenile mortality as structuring factors for reef benthic communities in Biscayne National Park, USA. Coral Reefs. 2000;**19**:5-123

[70] Mayor AG. The Effects of Temperature Upon Tropical Marine Animals. Washington: Carnegie Institution of Washington Publishing Department, Marine Biology Papers of the Tortugas Marine Lab. 1914;**6**(183):1-24

[71] Chornesky EA, Peters EC. Sexual reproduction and colony growth in the scleractinian coral *Porites astreoides*. The Biological Bulletin. 1987;**172**:161-177

[72] Edmunds PJ. Population biology of Porites astreoides and *Diploria strigosa* on a shallow Caribbean reef. Marine Ecology Progress Series. 2010;**418**:87-104

[73] Soong K. Colony size as a species character in massive reef corals. Coral Reefs. 1993;**12**: 77-83

[74] Green DH, Edmunds PJ, Carpenter RC. Increasing relative abundance of *Porites astreoides* on Caribbean reefs mediated by an overall decline in coral cover. Marine Ecology Progress Series. 2008;**359**:1-10

[75] van Woesik R, Scott WJ, Aronson RB. Lost opportunities: Coral recruitment does not translate to reef recovery in the Florida Keys. Marine Pollution Bulletin. 2014;**88**:110-117

[76] Roberts H, Rouse Jr LJ, Walker ND, Hudson J. Cold-water stress in Florida Bay and northern Bahamas: A product of winter cold-air outbreaks. Journal of Sedimentary Research. 1982;**52**:145-155

[77] Voss GL. Sickness and death in Floridas coral reefs. Natural History. 1973;**82**:41-47

[78] Brazeau DA, Lasker HR. The reproductive cycle and spawning in a Caribbean gorgonian. The Biological Bulletin. 1989;**176**:1-7

[79] Brazeau DA, Lasker HR. Sexual reproduction and external brooding by the Caribbean gorgonian *Briareum asbestinum*. Marine Biology. 1990;**104**:465-474

[80] Fitzsimmons-Sosa K, Hallock P, Wheaton J, Hackett KE, Callahan MK. Annual cycles of gonadal development of six common gorgonians from Biscayne National Park, Florida, USA. Caribbean Journal of Science. 2004;**40**:144-150

[81] Kahng SE, Benayahu Y, Lasker HR. Sexual reproduction in octocorals. Marine Ecology Progress Series. 2011;**443**:265-283

[82] Benayahu Y, Loya Y. Long-term recruitment of soft corals (Octocorallia: Alcyonacea) on artificial substrata at Eilat (Red Sea). Marine Ecology Progress Series. 1987;**38**:161-167

[83] Lasker HR. Vegetative reproduction in the octocoral *Briareum asbestinum* (Pallas). Journal of Experimental Marine Biology and Ecology. 1983;**72**:157-169

Chapter 7

The Fate of Corals: Will They Overcome Competition with Algae and Cyanobacteria in a Changing Environment?

Monica Puyana

Additional information is available at the end of the chapter

http://dx.doi.org/10.5772/intechopen.71568

Abstract

Coral reefs provide substantial ecological and economic services to coastal communities in the tropics. Hence, there is a great concern about the permanence of these ecosystems and the consequent loss of the economic, ecological and social services coral reefs provide due to their susceptibility to natural and anthropogenic threats. Large-scale processes such as strong El Niño Southern Oscillation events, global warming and ocean acidification represent significant challenges for coral reefs. Benthic cyanobacteria and seaweed have substantially increased in reef areas, facilitated by excessive nutrient input, reduction in herbivore populations and global warming. This review briefly describes the currently known aspects of coral, algae and cyanobacterial interactions, as well as the local and global environmental and ecological aspects that have caused the increase of algae and cyanobacteria in detriment to reef corals. Reef communities will keep changing in the light of large-scale events and anthropogenic influences. As short-term measures, ambitious programs for grazer reintroduction could help curb population growth of algae and cyanobacteria. Medium- to long-term measures should be oriented at limiting nutrient input to water bodies.

Keywords: cyanobacteria, algae, coral reefs, competition, phase changes

1. Introduction

Coral reefs constitute one of the most important, diverse and productive ecosystems in the planet. These ecosystems provide a great number of goods as well as economic and ecological services. Coral reefs protect coastal areas such as seagrasses and mangroves from erosion [1–3].

Coral reefs supply different food sources to coastal communities and also provide habitat for fish, molluscs and crustaceans of commercial interest [4–6]. Coral reefs also harbor a large diversity of microorganisms, algae and invertebrates that have provided researchers, the pharmaceutical, nutraceutical and cosmeceutical industries with many interesting bioactive molecules [7, 8].

Currently, the oceans are suffering remarkable changes evidenced by the increase and intensity of diseases, mass mortalities and blooms of several photosynthetic organisms. All these events have occurred in both disturbed and undisturbed areas around the globe (reviewed in [9]). Among some of the causes that have been linked to these events are global warming, increased UV radiation, overfishing, pollution and coastal eutrophication and oxygen depletion. However, some of all these factors may act in a synergistic manner rather than in isolation, considering that there are large spatial and temporal scales involved in all those processes [9–11].

Degradation of coral reefs has increased considerably world-wide, at least for which there are records of the phenomenon, since there is no consensus of what a pristine reef is [12]; therefore, it is very hard to assess the magnitude of changes and impacts compared to "base line" conditions [13]. Reef degradation has been linked to human activities and impacts; hence, there is a great concern about the permanence of these ecosystems and the consequent loss of the economic, ecological and social services coral reefs provide [14–16]. Moreover, large-scale processes such as the incidence of strong El Niño Southern Oscillation [ENSO] events, global warming and ocean acidification represent an enormous challenge for coral reefs to survive and remain as we know them today [17].

2. Cyanobacteria in reefs

Cyanobacteria are eubacteria with photosynthetic capabilities considered as "primitive microalgae" for a long time. These are ancient organisms dating back to 3.5 billion years as evidenced by their fossil record, particularly in stromatolites [18]. Recent molecular data have revealed that cyanobacteria are a polyphyletic group, and taxa, traditionally grouped by having similar morphologies, have different phylogenetic affiliations [19–22]. Cyanobacteria have a wide distribution in terrestrial and aquatic environments. Some taxa are involved in close symbiotic relationships with diatoms, sponges, corals, lichens and plants such as water ferns and cycads [23–26]. Many cyanobacteria are able to fix atmospheric nitrogen, making a significant contribution of this element into environments where it is usually limiting [27–30].

Marine benthic cyanobacteria thrive in a wide variety of habitats including rocky coastlines, sandy beaches, mangroves, marshes and swamps. The distribution of cyanobacterial mats is influenced by sediment type, tidal exposure and wave energy [31]. In coral reefs, given their abundance at certain times and locations, cyanobacteria may play a similar role to algae in terms of primary production and interactions with herbivores [32].

Cyanobacteria have long been regarded as regular elements of tropical marine communities and for that reason included in floristic lists from a number of tropical regions worldwide [33–35]. However, blooms of marine benthic cyanobacteria have become frequent in coastal areas receiving significant nutrient input from runoff and groundwater, as well as a result of large-scale climatic and hydrological changes [10, 11, 26, 36, 37]. Cyanobacterial blooms, toxin production and derived health risks are currently considered a public health hazard [38–43].

Blooms of marine benthic cyanobacteria grow on the substrate forming dark flimsy mats, usually red, purple, brown or black. There are several reports of marine benthic cyanobacteria blooms in several sites in the Pacific, in the island of Guam [44–46], in some Micronesian atolls and also in the Hawaiian Islands [50]. In the Atlantic, cyanobacterial blooms are a recurrent event strongly subject to seasonality in the state of Florida [51, 52]. In the Caribbean, there are reports from Puerto Rico [53], the Bahamas, Belize and the Virgin Islands [52], and the Colombian Caribbean [35, 54, 55]. Cyanobacterial blooms not only occur in coral reefs but also in seagrass meadows [56–58], mangroves [56, 57], estuaries and coastal lagoons [59].

Blooms in Guam and Australia, due to their extension, persistence and toxicity, have caused mass mortalities of fishes, as well as significant economic losses and health problems in residents or visitors to coastal areas. Marine turtles and manatees are also impacted by these blooms due to toxin exposure [60, 61]. Additionally, benthic cyanobacteria are often the dominant organisms that colonize coral skeletons after bleaching events [62]. Although several species of cyanobacteria grow in microbial mats, these tend to be dominated by a single species, although the reasons by which a species dominates over the other(s) are not yet entirely clear [63]. Additionally, cyanobacteria and turf algae may prevent or inhibit the settlement of coral larvae [50, 64–67]. The increasing occurrence, frequency and intensity of benthic cyanobacterial blooms, is now considered a great threat to coral reefs around the world [51]. Benthic cyanobacteria most commonly linked to bloom formation are included in the orders Chroococcales and Oscillatoriales, with the most common genera being *Schizothrix*, *Hormothamnion*, *Oscillatoria* and *Lyngbya* [68]. *Lyngbya*, in particular, is probably the most studied genus for which there is a great number of bloom reports and whose chemistry and chemical ecology are better documented [44, 45, 69–75]. However, recent studies have reassessed the phylogenetic affiliation of this genus finding that it is a rather diverse complex of species now reassigned into the genera *Moorea*, *Okeania* or *Lyngbya*, which may explain the vast chemical complexity of what used to be considered a single genus [19–22]. Although the taxonomy of the group has changed substantially, for the purpose of this review, I will refer to the names that were originally assigned in the papers where they were described.

Blooms of benthic cyanobacteria develop fast, covering vast areas in a period of few weeks. In those events, cyanobacteria grow on the substrate forming mats that can smother benthic organisms. As the bloom progresses, it can turn adjacent sediment and waters anoxic [49]. Cyanobacterial detachment from the sediment, either by flotation or wave action, may promote their dispersion [49, 68].

3. Algae in reefs

Algae are very important members of reef communities where they provide food and refuge to many organisms. They are also the basis of the reef food web [76]. Due to their different phylogenetic affiliations, varying forms and functions, algae are classified into three large functional groups based on growth form and size, ecological characteristics, toughness, susceptibility to herbivores and photosynthetic abilities. Functional groups are used to characterize algal communities in coral reefs and to understand their distribution and responses to environmental factors [76].

Filamentous and small algae with fast growth rates are included in the turf algae category. Multispecies assemblages of turf algae can be very abundant in reefs despite their low biomass. They are an attractive food source for several herbivores, preventing their overgrowth. Cyanobacteria are also included in this category [76]. The next category, fleshy macroalgae, is commonly known as seaweeds and includes the most well-known forms. Seaweed can be very abundant in reefs and have several adaptations to avoid herbivory [77–79]. A third category, coralline or crustose algae, secretes calcium carbonate and is important in cementing the reef framework but also in attracting coral recruits [80, 81].

4. Factors favoring the growth of algae and cyanobacteria in reefs

Competition for light and space between benthic algae and corals affects coral resilience and is essential when assessing the degree of reef health [65, 76]. In healthy coral reefs, reduced nutrient availability and high grazing pressure are the most important factors controlling the growth of algal turf and fleshy algae [82]. Phase shifts or phase changes are evident by a decrease in the cover and recruitment of corals compared to the colonization and substrate cover by turf algae, seaweed and cyanobacteria [14, 65, 83]. Phase shifts are common in many degraded reefs due to disparity in coral-algal interactions [15, 76, 86]. Under stressful conditions, favored by bleaching events, partial mortality or reduced grazing pressure, algae and cyanobacteria become competitively superior to corals, eventually overgrowing and killing them [15, 87].

Alleged causes linked to cyanobacterial blooms in fresh and brackish waters have been thoroughly studied (reviewed in [88]). However, blooms of marine benthic cyanobacteria are not understood that well. Normally, they occur in shallow waters, and environmental factors such as high temperatures, reduced wave action and the availability of phosphorus, nitrogen and iron have been linked to their formation [48, 49, 66, 89, 90]. Individual strains of cyanobacteria vary greatly in their bloom dynamics, which in some cases may be controlled by physical disturbances rather that nutrient availability or competition with macroalgae [91]. Several studies, however, have established correlations between bloom formation and mild wave action and increases in water temperature, phosphate levels and/or iron bioavailability. Bloom persistence, however, has been correlated to the low palatability of cyanobacteria to most generalist herbivores [45, 50, 70, 88, 92, 93].

Algal and cyanobacterial blooms may be difficult to explain because as multispecies consortia (at least in the case of cyanobacterial blooms) the dynamics in bloom formation may be a result of species-specific adaptations or preferences (reviewed in [94]). Also, the resulting increased growth may be a result of nutrient loading and/or reduction or absence of grazing [94]. Hence, food web dynamics may be important to understand this phenomenon. However, it is a no-brainer that the removal of large and small herbivores [12] has certainly favored the growth of cyanobacteria and algae, which have taken advantage of the increased nutrient availability and decreased grazing pressure.

5. Water temperature and cyanobacterial blooms

Sustained water temperatures over 28°C, alterations of flow and time of residence, decrease in water viscosity, large-scale mixing processes and small-scale turbulence are physical factors that correlate with cyanobacterial bloom formation [10, 88, 89, 95].

Thorough monitoring of *Lyngbya majuscula* blooms in the coasts of Puerto Rico from September 1998 until February 2000 showed that during September 1998 and September 2000 to December 2000, *Lyngbya* achieved the greatest cover coinciding with the highest temperatures recorded. While mean cover values ranged between 7 and 82% year-round, *Lyngbya* cover reached up a 100% during the warmest periods [53]. In Rosario Islands, Colombia, a similar pattern was observed in September 2010 and 2011, where the cover of cyanobacterial mats reached a 100%, concurring with sustained water temperatures above 30°C [93].

In Australia, benthic cyanobacterial blooms are recognized as an environmental health hazard [47, 49, 96]. Monitoring of *L. majuscula* mats in eastern Australia during 2005 showed that during the cool winter months blooms did not develop, but between the months of October and November, when water temperature increased, there was a fast development of *Lyngbya* mats. Blooms began gradually at the end of October, when total bottom cover of *L. majuscula* did not surpass 10% in an area of 49 hectares. During November, the total bottom cover of *L. majuscula* reached values close to 40% in an approximate area of 329 hectares. At the same time, water temperature increased from 23.5°C in October to 30.8°C in December. By the end of December, the bloom covered almost a 100% of the total available substrate in an area of 529 hectares [89]. I calculated an approximate wet biomass of 5000 Tons when the bloom reached its peak. As temperature decreased, so did *Lyngbya* cover. Similar patterns were registered in the Colombian Caribbean. Mats of *L. majuscula* and *L. sordida* covered extensions close to a 100% of the available reef substrate in Rosario Islands in September 2010 when the water temperature surpassed 30°C [93]. Also, mats of *Symploca hydnoides* and *Phormidium submembranaceum* were very conspicuous in Old Providence Island during October 2009 and July 2010. Again, those blooms took place during the warmest months recorded in the San Andres Archipelago [above 28°C] [55, 93].

6. Nutrient input favors the growth of algae and cyanobacteria

Water quality degradation as a result of increased nutrient input promotes the development and persistence of algal blooms and is one of the reasons that explains their expansion worldwide [94]. Although assessing the influence of nutrients is far more complicated than correlating these with increased water temperatures, experts agree that nutrients play a significant role in favoring blooms. Regular or pulsated nutrient delivery promotes bloom development; the overall composition—not just concentration—of nutrient input impacts blooms and high-biomass blooms may be partly sustained by exogenous nutrient sources [94].

Several studies have shown that cyanobacterial blooms are stimulated by the increase in nutrient concentration in water bodies. Nutrient increase may be a result of sewage disposal, animal husbandry or agricultural activities [26, 41, 49, 56, 57, 88, 89, 95, 97]. Rains also stimulate bloom formation due to increased runoff and river flow providing extra amounts of nutrients from terrestrial environments to coastal areas [49, 55, 98]. Groundwater and atmospheric deposition may also influence bloom formation and dynamics [94]. Nutrients that do have a stimulating growth effect on cyanobacteria are iron, phosphorus and nitrogen [26, 36, 49, 57, 89, 90, 99]. In the Colombian Caribbean, I have detected an apparent synergistic effect of high temperatures, the onset of rainy seasons and nutrient increase with incidence of marine cyanobacteria blooms at various sites [55, 93].

Anthropogenic activities such as the use of fertilizers and the dumping of urban and industrial waste waters provide significant amounts of nutrients into water bodies [26]. The input of key nutrients favor the growth of phototrophic organisms such as plants, algae and cyanobacteria, largely controlling aquatic primary production in marine environments where nitrogen is usually limiting. These blooms, with oxygen consumption during the hours of darkness, generate anoxia and consequent mortality of fish and other aquatic organisms [56].

7. The effect of cyanobacteria and algae on corals

There is significant evidence that reef degradation has increased on a global scale. Just in the Caribbean region, at least 20% of live coral coverage has been lost per decade [6]. This loss has been attributed to the increase in human population, dumping of waste water, soil erosion and subsequent sediment input by rivers, eutrophication, imbalance of food chains, proliferation of macroalgae [65, 100, 101], diseases and climate change [101, 102], as well as the removal or decimation of top predators and large herbivores [12].

Competition is a process that determines the structure, composition and diversity of benthic communities in coral reefs [5, 103–107]. In coral reefs, competition between sessile organisms such as benthic algae and corals has become very significant for the resilience of corals due to the dominance and vast cover of algae and cyanobacteria in affected coral reefs [65]. Reef deterioration has caused a significant replacement of live coral cover by cyanobacteria and macroalgae [59, 83, 108].

The strong competition for space between coral, algae and marine cyanobacteria can determine the structure, composition and abundance of these three groups in coral reefs [59, 109]. Overgrowth of algae or cyanobacteria on corals can cause deleterious effects on their health [50, 54, 55, 66, 110]. The vast occurrences of benthic algae and cyanobacteria in coral reefs are certainly an indication of the prevalent ecological conditions and may serve as indicators of coral reef health and local ecological imbalances [51, 111].

Coral reefs have a great diversity of herbivores that can exert a strong pressure on communities of primary producers such as macroalgae and cyanobacteria [2, 15]. In tropical areas, the most important herbivores in terms of consumption and impact on macroalgae are fish and sea urchins [70, 72, 77]. These herbivores can consume between 60 and 100% of the algal biomass daily [112–114]. Herbivory and nutrient input are two more determinant factors that define the relationship between algae and corals in reefs [85, 115] and the success of algae in other coastal ecosystems [116].

However, the generalized decrease in the populations of reef herbivores is an indirect cause of the increase in algal coverage and consequent phase shifts [2, 15, 50, 83]. Overfishing in reef areas has decimated the populations of herbivorous fish causing imbalances in the populations of macroalgae and their consequent overgrowth on corals and other substrates [2, 83, 84]. Occasionally, other herbivores such as sea urchins can sometimes increase in response to the decrease in herbivorous fishes and control algal populations. In the Caribbean basin, however, mass mortalities of sea urchins, particularly *Diadema antillarum*, triggered an excessive increase of algae in Jamaica and other locations [83].

Benthic cyanobacteria are efficient colonizers in a wide array of substrates such as coral skeletons, live coral, sand and even macroalgae. Cyanobacteria increase the magnitude of the phase shifts [51]. Benthic cyanobacteria, as well as macroalgae, are favored by bottom-up (increase in nutrients in the water bodies) and top-down effects (such as the decline of herbivores in reefs) [85]. Various studies have shown that the most important factor in controlling algae populations in coral reef areas is herbivory and the lack of it increases the consequences of phase changes more so than eutrophication, especially in the Caribbean [117].

Seaweed or macroalgae, compared to terrestrial plants, are mostly foliage, lacking dense structural material such as lignin, cellulose and hemicellulose, being more susceptible to herbivore consumption [118]. Benthic cyanobacterial mats strongly resemble macroalgae because they may exhibit large biomass and could become a potential food source for reef herbivores [44]. During blooms, cyanobacteria can surpass macroalgae in terms of available biomass [44, 72, 93].

Herbivory in reef areas generates a strong pressure on macroalgae and cyanobacteria. In "healthy" reefs, grazing rates may reach 100% of the produced biomass [77, 85, 112, 119]. Macroalgae and cyanobacteria, however, minimize grazing by means of spatial or temporal escapes, tolerating herbivory by compensating tissue loss with fast growth rates or investing in structural or chemical defenses [77–79, 113]. The production of feeding deterrent compounds, in particular, is well documented in these groups and plays an important role against grazing, enabling the growth and persistence of algae and cyanobacteria in coral reef areas [78, 119]. Besides deterring herbivores, chemically defended cyanobacteria could be favored by selective fish and urching grazing over palatable macroalgae, removing potential competitors and favoring their growth and expansion in reef areas [26].

Massive colonization of hard substrates by algae and benthic cyanobacteria can determine competitive networks, structure, composition and abundance of these three groups locally [51, 109]. In events where the abundance of cyanobacteria or macroalgae greatly increases, there is also an increase in overgrowth interactions with corals. The overgrowth of cyanobacteria and algae over corals may cause deleterious effects on their health [50, 66, 110]. In the Colombian Caribbean, the growth of benthic cyanobacterial mats over different species of scleractinian corals and soft corals eventually results in the death of the coral tissue that underlies those mats [54, 93] (**Figure 1**). In the Florida Keys, [50] determined that the brown seaweeds *Dictyota* and *Lobophora*, as well as several species of *Lyngbya*, not only inhibited the larval settlement of the hard coral *Porites astreoides* and the soft coral *Briareum asbestinum*, but also showed increased recruit mortality when the larvae came into direct contact with these organisms. Hence, seaweed and cyanobacteria may be considered competitively superior to corals not just by taking up space but also by inhibiting their recruitment.

Competition through allelopathic mechanisms by benthic marine cyanobacteria causes harmful effects on corals and soft corals. This has been observed in the field and tested in laboratory experiments in various sites such as the Bahamas, Belize, Florida and Panama [52], Florida [120], Hawaii [121] and the Colombian Caribbean [54, 93]. Experiments have been performed over coral embryos and larvae [50, 66] or adult corals [110].

[66] compared the recruitment and survival of embryos of the hard corals *Pocillopora damicornis* and *Acropora surculosa*, in the presence of the cyanobacterium *L. majuscula*. Coral embryos got tangled in the dense Lyngbya filaments and died. Cyanobacterial mats may trap coral embryos but also promote sediment deposition and accumulation surrounding benthic microbial mats [122]. Under these mats, anoxic conditions may develop, which may favor nitrogen fixation but are potentially deadly to the very sensitive coral embryos [66]. Since recruitment is a key process in the maintenance and recovery of coral reef ecosystems [50, 123], any alteration in this process has negative implications and will affect the persistence and resilience of corals.

The fact that allelopathy is so hard to prove experimentally does not mean that it does not occur in reef environments. Cyanobacteria, in particular, while being in direct contact with corals, could release allelopathic compounds as a result of abrasion by water motion with consequent cell rupture. Many studies have shown that toxin release from cyanobacteria is a result of cell lysis due to abrasion, stress and cell death [124–126]. Additionally, environmental factors such as temperature may elicit active toxin liberation in these microorganisms [124, 126].

Competitive interactions between the hard coral *Porites lutea*, a brown alga *Dictyota dichotoma* and the cyanobacterium *Lyngbya bouillonii* were evaluated in Sesoko Island, Japan. While coral growth is compromised by direct contact and abrasion by the alga, mats of the cyanobacterium *L. bouillonii* are able to kill live coral tissue upon direct contact [110].

Interactions between cyanobacteria, a hard coral and a soft coral in Rosario Islands, Colombian Caribbean, were evaluated *in situ* in order to identify deleterious effects potentially related to allelopathic mechanisms. Cyanobacterial extracts were incorporated into Phytagel™ gels and these were placed in direct contact with the hard coral *Porites porites* and the soft coral *B. asbestinum* [127]. HPLC chromatographic profiles of zooxanthellae in coral tissues were evaluated after 24, 48 and 72 hours of exposure. Extracts from *Lyngbya* spp. showed a clear effect on the zooxanthellae chromatographic profiles evidenced by an increase in pheophytin, a degradation product from chlorophyll. The effect was greater with longest exposure time. These results suggest that cyanobacteria may compete against corals due to their fast growth rates, defenses against herbivory and allelopathic mechanisms. Further evidence was

http://dx.doi.org/10.5772/intechopen.71568

Figure 1. Benthic cyanobacteria from Colombian Caribbean reefs. (A) Mats of *Moorea producens* growing at the base of a *Eunicea* soft coral. The soft coral retracts its polyps due to abrasion and possibly to avoid contact with the mat. Rosario Islands, 2016. (B) Filamentous mats of *Caldora* sp. This cyanobacterium grows profusely over various kinds of seaweeds. Old Providence Island, 2008. (C) Unidentified red filamentous mats overgrowing hard corals and green algae. Old Providence Island, 2008. (D and E) Multispecies cyanobacterial mats overgrowing soft corals, causing polyp necrosis and tissue death. The remaining gorgonin skeletons continue to be colonized by other cyanobacteria and filamentous algae. Old Providence Island, 2016. (F) Puffs of filamentous cyanobacteria may overgrow live corals causing bleaching and/or necrosis. In this case, they are growing over gorgonin skeletons. Old Providence Island, 2016. Photos A, D–F: Monica Puyana; B–C: Julian Prato.

obtained while testing the potential allelopathic effects of two organic extracts from marine cyanobacteria over live corals. Organic extracts from two different microbial consortia, a consortium of *S. hydnoides* and *P. submembranaceum* and a consortium of *Lyngbya cf. semiplena, L. majuscula, Oscillatoria nigroviridis* and *O. margaritifera,* were tested at 0.002 ppm and 0.011 ppm, both below their natural concentration. In order to do so, fragments of the hard coral *Madracis mirabilis* were placed in individual containers, and once acclimated, with their polyps fully expanded, cyanobacterial extracts were resuspended in ethanol and seawater and coral behavior was registered. Both extracts generated the retraction of coral polyps with variable speed and intensity. Ethanol controls, on the other hand, presented a rapid recovery and a minimum shrinkage of polyps. Fragments exposed to cyanobacterial extracts showed 80 to 99% of polyp retraction, whereas solvent controls did not exceed 25% of polyp retraction. No extract proved to be lethal, and after 20 hours, corals extended their polyps displaying full recovery [93].

In another assay, the toxicity of four cyanobacterial extracts to embryos of *Montastrea annularis,* obtained during the mass spawning event in Rosario Islands in September, 2011, was evaluated. In this assay, we assessed embryo mortality 6, 18 and 24 hours after acute exposure to cyanobacterial extracts in concentrations of 1000, 500, 100 and 10 ppm. All tested extracts were toxic at concentrations of 500 and 1000 ppm, causing 100% embryo mortality within 6 hours. Controls with seawater and ethanol did not affect coral embryos. The extract that showed greater toxicity was obtained from a mixed *Lyngbya* assemblage, causing a 90% embryo mortality at the lowest concentration tested (10 ppm) after only 6 hours of exposure.

In summary, competition between corals, cyanobacteria and/or algae may take place either against coral embryos or larvae or during their adult stages. Some of the mechanisms include allelopathic inhibition on adult corals and embryos, negative effects of abrasion by direct physical contact, drastic decrease in oxygen levels near and under cyanobacterial mats, reduction of available space in reef substrates affecting larval recruitment and mortality of embryos entangled in cyanobacterial mats. All these effects strongly suggest that blooms of algae and benthic cyanobacteria pose a risk for the recruitment and development of reef builders and other reef organisms such as soft corals [50, 51, 66, 93, 110, 120, 121].

8. Climate change, ocean acidification and future of reefs

Resilience or the ability to withstand and recover from the negative phenomena affecting coral reefs is an essential role for their persistence. Recovery after each disturbance involves the re-establishment of coral coverage, the growth of surviving coral fragments, reproductive success and the subsequent settlement and survival of coral embryos and larvae [16, 123]. However, the recovery process in coral reefs is nowadays more difficult because the substrate in these ecosystems is frequently dominated by algae and cyanobacteria. These photosynthetic organisms efficiently colonize available substrates following disturbances and become dominant in degraded reefs [15, 65, 83, 123], reducing the resilience of coral reefs overall

[2, 15]. Algal dominance can become a serious bottleneck for the re-establishment of corals, depending on the characteristics of the dominant algal assemblages in each place [65, 123].

Large-scale hydrological disturbances and environmental changes due to global warming and ocean acidification are recognized as growing threats to coral reefs worldwide [17, 123, 128–130]. Global warming is also believed to facilitate the development, frequency and expansion of cyanobacterial blooms [10, 11, 26, 37]. These large-scale disturbances may have a more direct effect on corals, either by reducing growth rates and calcification regimes under acidic seawater conditions or by the onset of bleaching events under warmer conditions [17, 123]. Unless there is a significant reduction in CO_2 atmospheric levels, reefs with reduced populations of grazers and increased nutrient input are predicted from shift from predominantly coral-dominated to predominantly algal-dominated states [17, 130]. This pattern has been recognized for some time now, particularly in the Caribbean and the Eastern Pacific [131–135]. Management options such as restoring herbivore populations and limiting nutrient input to reef areas may only be effective under controlled CO_2 input to seawater [17, 123, 130]. Sea urchins are important grazers of noxious algae [136]; therefore, restoration of grazers such as urchins is believed to alleviate the problem of excessive algal biomass in coral reefs. In Hawaii, juveniles of the local urchin *Tripneustes gratilla* are raised in hatcheries and reintroduced in reef areas in order to remove excessive biomass of invasive algal species and restore the natural ecosystem function [137]. However, control of cyanobacterial populations by grazers seems very unlikely [26].

Approximately two-thirds of coral reefs in the Caribbean are threatened by human activities such as coastal development, wastewater and sediment input, pollution and overfishing [138]. The economic impact of coral bleaching, coral diseases and cyanobacterial blooms has not been quantified, but it is clear that these have caused significant changes in Caribbean reef communities [138]. Therefore, it is important to assess the impact of cyanobacterial blooms at local scales in order to understand their causes and consequences in order to address significant monitoring and management measures. The relationships between nutrient input and bloom formation are obscured by changes in food webs, habitat alterations and climate change. Nutrient enrichment has several effects in food webs, predator-prey interactions and overall nutrient dynamics. However, experts agree that management of nutrient inputs to watershed may be the most important measure to prevent excessive growth of algae and cyanobacteria [26, 94]. The Australian government, for instance, has invested significant funds to reduce nitrogen inputs into Moreton Bay to prevent blooms of *L. majuscula* [90].

9. Conclusions

In order to answer the question whether corals will overcome competition with algae and cyanobacteria in environments, the evidence shows that reef communities have changed over time and will keep changing in the light of large-scale events and anthropogenic influences. Algae and cyanobacteria are thriving as the total bottom cover of reef corals is reduced. So even if corals survive, their communities will not be the same. We do need to gain a better understanding of causes, at local and regional scales, that support bloom formation. Also, it is crucial to convince managers that only long-term studies and periodic surveys will help to understand and manage algae and cyanobacteria in reefs and other marine ecosystems. As short-term measures, ambitious programs for grazer reintroduction could help curb population growth of algae and cyanobacteria. Medium- to long-term measures should be oriented at limiting nutrient input to water bodies.

Acknowledgements

I want to acknowledge COLCIENCIAS, Universidad Jorge Tadeo Lozano, Universidad Nacional de Colombia sede Bogotá (DIB and Facultad de Ciencias), Fundación para la Promoción de la Investigación y la Tecnología del Banco de la República and Fundación Mariano Ospina Pérez-ICETEX for providing financial and/or administrative support for my past and ongoing research on marine cyanobacteria and also for supporting some undergraduate and graduate students over the years. Students are fundamental to our field and laboratory work, so I want to thank Julian Prato and Felipe Nieto who carried out several of the experiments presented in this chapter. Valuable discussions and exchange with my colleagues Alberto Acosta, Universidad Javeriana; and Freddy Ramos, Leonardo Castellanos, Ernesto Mancera and Brigitte Gavio, Universidad Nacional de Colombia, are also acknowledged. Many people have collaborated with me over the years: Uriah Steele, David Ryan Steele, Jerónimo Vásquez at Old Providence Island; Enrique Pomare in San Andrés Island; Elvira Alvarado, Valeria Pizarro and Diving Planet in Cartagena. Also, undergraduate and graduate students Nicolás Restrepo, Isabel Hernández, Laura Rodríguez, Juliana Vanegas, Laura Becerra, Lina Bayona, Jairo Quintana, Farja Ayala and Fredy Duque have helped in many aspects over the years. Professor Edisson Tello from Universidad de la Sabana and Professor Rafael Barragán from Universidad Santo Tomás also helped during field work. I would like to thank Rafael Vieira, Jaime Rojas, Paola Pinzón and all the personnel at Oceanario de las Islas del Rosario for lodging us and supporting our field work in Rosario Islands. Professor Valerie Paul from Smithsonian Marine Station at Fort Pierce has guided us through the murky waters of the taxonomy, chemistry and molecular biology of marine cyanobacteria. Last but not least, I would like to thank Comunidad de Orika, Isla Grande, Islas del Rosario for allowing us to perform research in their territory and also to the Raizal community of Old Providence Island, particularly Miss Francia D'Armas and Arelis Howard and families who always helped us and made us feel at home.

Author details

Monica Puyana

Address all correspondence to: monica.puyana@utadeo.edu.co

Department of Biological and Environmental Sciences, Jorge Tadeo Lozano University, Bogotá, Colombia

References

[1] Moberg F, Folke C. Analysis: Ecological goods and services of coral reef ecosystems. Ecological Economics. 1999;**29**:215-233

[2] Hughes TP, Baird AH, Bellwood DR, Card M, Connolly SR, Folke C, Grosberg R, Hoegh-Guldberg O, Jackson JBC, Kleypas J, Lough JM, Marshall P, Nyström M, Palumbi SR, Pandolfi JM, Rosen B, Roughgarden J. Climate change, human impacts, and the resilience of coral reefs. Science. 2003;**301**:929-933

[3] Ritson-Williams R, Arnold SN, Fogarty ND, Steneck RS, Vermeij MJA, Paul VJ. New perspectives on ecological mechanisms affecting coral recruitment on reefs. In: Lang MA, MacIntyre IG, Rützler K, editors. Proceedings of the Smithsonian Marine Science Symposium. Vol. 2009. Washington, DC: Smithsonian Institution Scholarly Press; 2009. p. 437-457

[4] Jennings S, Polunin NVC. Impacts of fishing on tropical reef ecosystems. Ambio. 1996; **25**:44-49

[5] Birkeland C. Life and Death of Coral Reefs. New York: Chapman and Hall; 1997. 536 p

[6] Mumby PJ, Steneck R. Coral reef management and conservation in the light of rapidly evolving ecological paradigms. Trends in Ecology and Evolution. 2008;**23**:555-563

[7] Molinski TF, Dalisay DS, Lievens SL, Saludes JP. Drug development from marine natural products. Nature Reviews-Drug Discovery. 2009;**8**:69-85

[8] Leal MC, Puga J, Serôdio J, Gomes NCM, Calado R. Trends in the discovery of new marine natural products from invertebrates over the last two decades-where and what are we bioprospecting? PLoS One. 2012;**7**:e30580. DOI: 10.1371/journal.pone.0030580

[9] Hayes ML, Bonaventura J, Mitchell TP, Prospero JM, Shinn EA, Van Dolah F, Barber RT. How are climate and marine biological outbreaks functionally linked? Hydrobiologia. 2001;**460**:213-220. DOI: 10.1023/A:1013121503937

[10] Paerl HW, Huisman J. Blooms like it hot. Science. 2008;**320**:57-58

[11] Paerl HW, Huisman J. Climate change: A catalyst for global expansion of harmful algal blooms. Environmental Microbiology Reports. 2009;**1**:27-37

[12] Jackson JBC. What was natural in the coastal oceans? Proceedings of the National Academy of Sciences. 2001;**98**:5411-5418

[13] Bruno JF, Precht WF, Vroom PS, Aronson RB. Coral reef baselines: How much macroalgae is natural? Marine Pollution Bulletin. 2014;**80**:24-29. DOI: 10.1016/j.marpolbul.2014.01.010

[14] Pandolfi JM, Bradbury RH, Sala E, Hughes TP, Bjorndal KA, Cooke RG, McArdle D, McClenachan L, Newman MJH, Paredes G, Warner RR, Jackson JBC. Global trajectories of the long-term decline of coral reef ecosystems. Science. 2003;**301**:955-958

[15] Hughes TP, Rodrigues MJ, Bellwood DR, Ceccarelli D, Hoegh-Guldberg O, McCook L, Moltschaniwskyj N, Pratchett MS, Steneck RS, Willis B. Phase shifts, herbivory, and the resilience of coral reefs to climate change. Current Biology. 2007;**17**:360-365

[16] Birrell CL, McCook LJ, Willis BL, Diaz-Pulido GA. Effects of Benthic algae on the replenishment of corals and the implications for the resilience of coral reefs. Oceanography and Marine Biology: An Annual Review. 2008;**46**:25-63

[17] Anthony KRN, Maynard JA, Diaz-Pulido G, Mumby PJ, Marshall PA, Cao L, Hoegh-Guldberg O. Ocean acidification and warming will lower coral reef resilience. Global Change Biology. 2011;**17**:1798-1808. DOI: 10.1111/j.1365-2486.2010.02364.x

[18] Whitton BA, Potts M. Introduction to the Cyanobacteria. In: Whitton BA, Potts M, editors. The Ecology of Cyanobacteria. Dordrecht: Kluwer Academic Publishers; 2000. p. 1-11

[19] Engene N, Coates RC, Gerwick WH. 16S rRNA gene heterogeneity in the filamentous marine cyanobacterial genus *Lyngbya*. Journal of Phycology. 2010;**46**:591-601

[20] Engene N, Choi H, Esquenazi E, Rottacker E, Ellisman MH, Dorrestein PC, Gerwick WH. Underestimated biodiversity as a major explanation for the perceived rich secondary metabolite capacity of the cyanobacterial genus *Lyngbya*. Environmental Microbiology. 2011;**13**:1601-1610

[21] Engene N, Rottacker EC, Kasvtovsky J, Byrum T, Choi H, Ellisman MH, Komárek J, Gerwick WH. *Moorea producens* gen. nov., sp. nov. and *Moorea bouillonii* comb. nov., tropical marine cyanobacteria rich in bioactive secondary metabolites. International Journal of Systematic and Evolutionary Microbiology. 2012;**62**:1171-1178

[22] Engene N, Paul V, Byrum T, Gerwick W, Thor A, Ellisman M. Five chemically rich species of tropical marine cyanobacteria of the genus *Okeania* gen. nov. [Oscillatoriales, Cyanoprokaryota]. Journal of Phycology. 2013;**69**:1095-1106

[23] Adams DG. Cyanobacteria in symbiotic interactions. In: Whitton BA, Potts M, editors. The Ecology of Cyanobacteria. Vol. 2000. Dordrecht: Kluwer Academic Publishers; 2000. p. 523-561

[24] Cox PA, Banack SA, Murch SJ. Biomagnification of cyanobacterial neurotoxins and neurodegenerative disease among the Chamorro people of Guam. Proceedings of the Natural Academy of Sciences. 2003;**100**:13380-13383

[25] Usher KM. The ecology and phylogeny of cyanobacterial symbionts in sponges. Marine Ecology. 2008;**29**:178-192

[26] Paerl HW, Paul VJ. Climate change: Links to global expansion of harmful cyanobacteria. Water Research. 2012;**49**:1349-1363

[27] Corredor JE, Wilkinson CR, Vicente VP, Morell JM, Otero E. Nitrate release by Caribbean reef sponges. Limnology and Oceanography. 1988;**33**:114-120

[28] Capone DG, Zehr JP, Paerl HW, Bergman B, Carpenter EJ. *Trichodesmium*, a globally significant marine cyanobacterium. Science. 1997;**276**:1221-1229

[29] Diaz MC, Ward BB. Sponge mediated nitrification in tropical benthic communities. Marine Ecology Progress Series. 1997;**156**:97-107

[30] Ferris MJ, Palenik B. Niche adaptation in ocean cyanobacteria. Nature. 1998;**396**:226-228

[31] Hoffman L. Marine cyanobacteria in tropical regions: Diversity and ecology. European Journal of Phycology. 1999;**34**:371-379

[32] Paul VJ, Cruz-Rivera E, Thacker R. Chemical mediation of macroalgal-herbivore interactions. In: McClintock J, Baker B, editors. Marine Chemical Ecology. Boca Raton: CRC Press; 2001. p. 227-265

[33] Littler DS, Littler MM. 2000. Caribbean Reef Plants. An Identification Guide to the Reef Plants of the Caribbean, Bahamas, Florida and Gulf of Mexico. Offshore Graphics: Washington; 2000. 542 p

[34] Diaz-Pulido G, Diaz-Ruiz M. Diversity of benthic marine algae of the Colombian Atlantic. Biota Colombiana. 2003;**4**:203-246

[35] Reyes V, Gavio B, Velasquez H. Notes on the marine algae of the International Biosphere Reserve *Seaflower*, III. New records of Cyanophyta for the Caribbean coast of Colombia. Nova Hedwigia. 2013;**97**:349-360

[36] Ahern KS, Ahern CR, Udy JW. In situ field experiment shows *Lyngbya majuscula* [cyanobacterium] growth stimulated by added iron, phosphorus and nitrogen. Harmful Algae. 2008;**7**:389-404

[37] Paerl HW, Fulton RS. Ecology of harmful cyanobacteria. In: Graneli E, Turner J, editors. Ecology of Harmful Marine Algae. Berlin: Springer-Verlag; 2006. p. 95-107

[38] Paerl HW. A comparison of cyanobacterial bloom dynamics in freshwater, estuarine and marine environments. Phycologia. 1996;**35**:25-35

[39] Sivonen K. Cyanobacterial toxins and toxin production. Phycologia. 1996;**35**(6 suppl): 12-24

[40] Sellner KG. Physiology, ecology, and toxic properties of marine cyanobacterial blooms. Limnology and Oceanography. 1997;**42**(5, part 2):1089-1104

[41] Metcalf JS, Codd GA. Cyanobacterial Toxins in the Water Environment: A Review of Current Knowledge. UK: Foundation for Water Research. 2004. 36 p

[42] Miller MA, Kudela RM, Mekebri A, Crane D, Oates SC, Tinker MT, Staedler M, Miller WA, Toy-Choutka S, Dominik C, Hardin D, Langlois G, Murray M, Ward K, Jessup DA. Evidence for a novel marine harmful algal bloom: Cyanotoxin (microcystin) transfer from land to sea otters. PLoS One. 2010;**5**:e12576. DOI: 10.1371/journal.pone.0012576

[43] Leao PN, Engene N, Antunes A, Gerwick WH, Vasconcelos V. The chemical ecology of cyanobacteria. Natural Products Reports. 2012;**29**:372-391

[44] Pennings S, Pablo SR, Paul VJ. Chemical defenses of the tropical, benthic marine cyanobacterium *Hormothamnion enteromorphoides*: Diverse consumers and synergisms. Limnology and Oceanography. 1997;**42**:911-917

[45] Nagle D, Paul VJ. Chemical defense of a marine cyanobacterial bloom. Journal of Experimental Marine Biology and Ecology. 1998;**225**:29-38

[46] Thacker RW, Ginsburg DW, Paul VJ. Effects of herbivore exclusion and nutrient enrichment on coral reef macroalgae and cyanobacteria. Coral Reefs. 2001;**19**:318-329

[47] Dennison WC, O'Neil JM, Duffy EJ, Oliver PE, Shaw GR. Blooms of the cyanobacterium *Lyngbya majuscula* in coastal waters of Queensland, Australia. Bulletin de l'Institut océanographique [Monaco]. 1999;**19**:501-506

[48] O'Neil JM, Shaw GR, Dennison WC. Blooms of the toxic cyanobacteria *Lyngbya majuscula* in coastal Queensland waters. In: Abstracts 9th International Conference on Algal Blooms. Tasmania, 2000

[49] Albert S, JM O'N, Udy JW, Ahern KS, CM O'S, Dennison WC. Blooms of the cyanobacterium *Lyngbya majuscula* in coastal Queensland, Australia: Disparate sites, common factors. Marine Pollution Bulletin. 2005;**51**:428-437

[50] Kuffner IB, Walters LJ, Becerro MA, Paul VJ, Ritson-Williams R, Beach KS. Inhibition of coral recruitment by macroalgae and cyanobacteria. Marine Ecology Progress Series. 2006;**323**:107-117

[51] Paul VJ, Thacker R, Banks K, Golubic S. Benthic cyanobacterial blooms impact the reef of south Florida [Broward County, USA]. Coral Reefs. 2005;**24**:693-697

[52] Ritson-Williams R, Paul VJ, Bonito V. Marine benthic cyanobacteria overgrow coral reef organisms. Coral Reefs. 2005;**24**:629

[53] Stielow S, Ballantine D. Benthic cyanobacterial, *Microcoleus lyngbyaceus*, blooms in shallow, inshore Puerto Rican seagrass habitats, Caribbean Sea. Harmful Algae. 2003;**2**:127-133. DOI: 10.1016/S1568-9883(03)00007-6

[54] Puyana M, Prato J. Overgrowth of reef organisms by benthic cyanobacteria in the Colombian Caribbean. Mutisia. 2013;**3**:58-60

[55] Puyana M, Acosta A, Bernal-Sotelo K, Velásquez-Rodríguez T, Ramos F. Spatial scale of cyanobacterial blooms in Old Providence Island, Colombian Caribbean. Universitas Scientiarum. 2015;**20**:83-105. DOI: 10.11144/Javeriana.SC20-1.sscb

[56] Paerl HW. Controlling eutrophication along the freshwater–marine continuum: Dual nutrient (N and P) reductions are essential. Estuaries and Coasts. 2009;**32**:593-601

[57] Pittman SJ, Pittman KM. Short-term consequences of a benthic cyanobacterial bloom (*Lyngbya majuscula* Gomont) for fish and penaeid prawns in Moreton Bay [Queensland, Australia]. Estuarine, Coastal and Shelf Science. 2005;**63**:619-632

[58] Conley D, Paerl H, Howarth R, Boesch D, Seitzinger S, Havens K, Lancelot C, Likens G. Controlling eutrophication: Nitrogen and phosphorus. Science. 2009;**323**:1014-1015. DOI: 10.1126/science.1167755

[59] Charpy L, Casareto B, Langlade M-J, Suzuki Y. Cyanobacteria in coral reef ecosystems: A review. Journal of Marine Biology. 2012;**2012**:1-9. DOI: 10.1155/2012/259571

[60] Arthur KE, Limpus CJ, Roelfsema CM, Udy JW, Shaw GR. A bloom of *Lyngbya majuscula* in Shoalwater Bay, Queensland, Australia: An important feeding ground for the green turtle (*Chelonia mydas*). Harmful Algae. 2006;**5**:251-265

[61] Harr KE, Szabo NJ, Cichra M, Phlips EJ. Debromoaplysiatoxin in *Lyngbya*-dominated mats on manatees (*Trichechus manatus latirostris*) in the Florida King's Bay ecosystem. Toxicon. 2008;**52**:385-388

[62] Diaz-Pulido G, McCook LJ. The fate of bleached corals: Patterns and dynamics of algal recruitment. Marine Ecology Progress Series. 2002;**232**:115-128

[63] Stal LJ. Physiological ecology of cyanobacteria in microbial mats and other communities. The New Phytologist. 1995;**131**:1-32

[64] Edmunds PJ, Carpenter RC. Recovery of *Diadema antillarum* reduces macroalgal cover and increases abundance of juvenile corals on a Caribbean reef. Proceedings of the National Academy of Sciences of the United States of America. 2001;**98**:5067-5071. DOI: 10.1073/pnas.071524598

[65] McCook LJ, Jompa J, Diaz-Pulido G. Competition between corals and algae on coral reefs: A review of evidence and mechanisms. Coral Reefs. 2001;**19**:400-417

[66] Kuffner IB, Paul V. Effects of the benthic cyanobacterium *Lyngbya majuscula* on larval recruitment of the reef corals *Acropora surculosa* and *Pocillopora damicornis*. Coral Reefs. 2004;**23**:455-458

[67] Vermeij M, van Moorselaar I, Engelhard S, Hörnlein C, Vonk S, Visser P. The effects of nutrient enrichment and herbivore abundance on the ability of turf algae to overgrow coral in the Caribbean. PLoS One. 2010;**5**(12):e14312. DOI: 10.1371/journal.pone.0014312

[68] Tan LT, Goh BPL. Chemical ecology of marine cyanobacterial secondary metabolites: A mini review. Journal of Coastal Development. 2009;**13**:1-9

[69] Pennings SC, Weiss AM, Paul VJ. Secondary metabolites of the cyanobacterium *Microcoleus lyngbyaceus* and the sea hare *Stylocheilus longicauda*: Palatability and toxicity. Marine Biology. 1996;**126**:735-743

[70] Thacker R, Nagle D, Paul VJ. Effects of repeated exposures to marine cyanobacterial secondary metabolites on feeding by juvenile rabbitfish and parrotfish. Marine Ecology Progress Series. 1997;**147**:21-29

[71] Thacker RW, Becerro MA, Lumbang WA, Paul VJ. Allelopathic interactions between sponges on a tropical reef. Ecology. 1998;**79**:1740-1750

[72] Nagle D, Paul VJ. Production of secondary metabolites by filamentous tropical marine cyanobacteria: Ecological functions of the compounds. Journal of Phycology. 1999;**35**:1412-1421

[73] Nagle DG, Paul VJ, Roberts AM. Ypaoamide, a new broadly acting feeding deterrent from the marine cyanobacterium *Lyngbya majuscula*. Tetrahedron Letters. 1996;**37**:6263-6266

[74] Nagle DG, Camacho FT, Paul VJ. Dietary preferences of the opisthobranch mollusc *Stylocheilus longicauda* for secondary metabolites produced by the tropical cyanobacterium *Lyngbya majuscula*. Marine Biology. 1998;**132**:267-273

[75] Luesch H, Harrigan GG, Goetz G, Horgen FD. The cyanobacterial origin of potent anticancer agents originally isolated from sea hares. Current Medicinal Chemistry. 2002; **9**:1791-1806

[76] Díaz-Pulido G, McCook LJ. Macroalgae [Seaweeds]. In: Chin A, editor. The State of the Great Barrier Reef On-line. Townsville: Great Barrier Reef Marine Park Authority; 2008 http://www.gbrmpa.gov.au/corp_site/info_services/publications/sotr/downloads/SORR_Macroalgae.pdf Accessed: 2017-03-30

[77] Duffy JE, Hay ME. Seaweed adaptations to herbivory. Bioscience. 1990;**40**:368-376

[78] Hay ME. Marine-terrestrial contrasts in the ecology of plant chemical defenses against herbivores. Trends in Ecology and Evolution. 1991;**6**:362-365

[79] Hay ME, Kappel QE, Fenical W. Synergisms in plant defenses against herbivores: Interactions of chemistry, calcification, and plant quality. Ecology. 1994;**75**:1714-1726

[80] Littler MM, Littler DS. Models of tropical reef biogenesis: The contribution of algae. Progress in Phycological Research. 1984;**3**:323-364

[81] Heyward AJ, Negri AP. Natural inducers of coral larval metamorphosis. Coral Reefs. 1999;**18**:273-279

[82] Littler MM, Littler DS, Brooks BL. Harmful algae on tropical coral reefs: Bottom-up eutrophication and top-down herbivory. Harmful Algae. 2006;**5**:565-585

[83] Hughes TP. Catastrophes, phase shifts, and large-scale degradation of a Caribbean coral reef. Science. 1994;**265**:1547-1551

[84] Mumby PJ, Steneck RS, Edwards AJ, Ferrari R, Coleman R, Harborne AR, Gibson PJ. Fishing down a Caribbean food web relaxes trophic cascades. Marine Ecology Progress Series. 2012;**445**:13-24

[85] Littler MM, Littler DS, Brooks BL. Herbivory, nutrients, stochastic events, and relative dominances of benthic indicator groups on coral reefs: A review and recommendations. Smithsonian Contributions to the Marine Sciences. 2009;(38):401-414

[86] McCook LJ. Macroalgae, nutrients and phase shifts on coral reefs: Scientific issues and management consequences for the Great Barrier Reef. Coral Reefs. 1999;**18**:357-367

[87] Jompa J, McCook LJ. The effects of nutrients and herbivory on competition between a hard coral (*Porites cylindrica*) and a brown alga (*Lobophora variegata*). Limnology and Oceanography. 2002;**47**:527-534

[88] Paerl H, Hall N, Calandrino E. Controlling harmful cyanobacterial blooms in a world experiencing anthropogenic and climatic-induced change. Science of the Total Environment. 2011;**409**:1739-1745. DOI: 10.1016/j.scitotenv.2011.02.001

[89] Ahern KS, Ahern CR, Savige GM, Udy JW. Mapping the distribution, biomass and tissue nutrient levels of a marine benthic cyanobacteria bloom (*Lyngbya majuscule*). Marine and Fresh Water Research. 2007;**58**:883-904

[90] Ahern KS, Ahern CR, Udy JW. Nutrient additions generate prolific growth of *Lyngbya majuscula* (cyanobacteria) in field and bioassay experiments. Harmful Algae. 2007; **6**:134-151

[91] Thacker R, Paul VJ. Are benthic cyanobacteria indicators of nutrient enrichment? Relationships between cyanobacterial abundance and environmental factors on the reef flats of Guam. Bulletin of Marine Science. 2001;**69**:497-508

[92] Thacker R, Paul VJ. Morphological, chemical, and genetic diversity of tropical marine cyanobacteria *Lyngbya* spp. and *Symploca* spp. (Oscillatoriales). Applied and Environmental Microbiology. 2004;**70**:3305-3312

[93] Prato J. Afloramientos de cianobacterias marinas bentónicas en San Andrés, Providencia y las Islas del Rosario (Caribe colombiano): Caracterización y evaluación de su posible papel ecológico. [Master's Thesis]. Bogota: Universidad Nacional de Colombia; 2013

[94] Heisler J, Glibert PM, Burkholder JM, Anderson DM, Cochlan W, Dennison WC, Dortch Q, Gobler CJ, Heil CA, Humphries E, Lewitus A, Magnien R, Marshall HG, Sellner K, Stockwell DA, Stoecker DK, Suddleson M. Eutrophication and harmful algal blooms: A scientific consensus. Harmful Algae. 2008;**8**:3-13

[95] Watkinson AJ, JM O'N, Dennison WC. Ecophysiology of the marine cyanobacterium, *Lyngbya majuscula* (Oscillatoriaceae) in Moreton Bay, Australia. Harmful Algae. 2005;**4**:697-715

[96] Osborne NJT, Webb PM, Shaw GR. The toxins of *Lyngbya majuscula* and their human and ecological health effects. Environment International. 2001;**27**:381-392

[97] Paerl HW, Piehler MF. Nitrogen and marine eutrophication. In: Capone DG, Bronk DA, Mulholland MR, Carpenter EJ, editors. Nitrogen in the Marine Environment. 2nd ed. Oxford: Elsevier; 2008. p. 529-567. DOI: 10.1016/B978-0-12-372522-6.00011-6

[98] Vivas-Aguas L, Tosic M, Sánchez J, Narváez S, Cadavid B, Bautista P, Betancourt J, Parra J, Echeverri L, Espinosa L. Diagnóstico y evaluación de la calidad ambiental marina en el Caribe y Pacífico colombiano. Red de vigilancia para la conservación y protección de las aguas marinas y costeras de Colombia-REDCAM. Informe técnico 2010. INVEMAR: Santa Marta; 2010. 208 p

[99] O'Neil JM, Davis TW, Burford MA, Gobler CJ. The rise of harmful cyanobacteria blooms: The potential roles of eutrophication and climate change. Harmful Algae. 2012;**14**:313-334. DOI: 10.1016 j.hal.2011.10.027

[100] Garzón-Ferreira J, Díaz JM. The Caribbean coral reefs of Colombia. In: Cortés J, editor. Latin American Coral Reefs. Amsterdam: Elsevier Science; 2003. p. 275-300

[101] Aronson R, Precht W. Conservation, precaution, and Caribbean reefs. Coral Reefs. 2006;**25**:441-450. DOI: 10.1007/S00338-006-0122-9

[102] Mora C. A clear human footprint in the coral reefs of the Caribbean. Proceedings of the Royal Society of London B. 2008;**275**:767-773. DOI: 10.1098/rspb.2007.1472

[103] Logan A. Interespecific aggression in hermatypic corals from Bermuda. Coral Reefs. 1984;**3**:131-138

[104] Lang JC, Chornesky EA. Competition between scleractinian reef corals: A review of mechanisms and effects. In: Dubinsky Z, editor. Coral Reefs. Amsterdam: Elsevier; 1990. p. 209-252

[105] López-Victoria M, Zea S, Weil E. Competition for space between encrusting excavating Caribbean sponges and other coral reef organisms. Marine Ecology Progress Series. 2006;**312**:113-121

[106] Chaves-Fonnegra A, Zea S. Coral colonization by the encrusting excavating Caribbean sponge *Cliona delitrix*. Marine Ecology. 2010;**32**:1-12. DOI: 10.1111/j.1439-0485.2010.00416.xs

[107] Chadwick NE, Morrow KM. Competition among sessile organisms on coral reefs. In: Dubinsky Z, Stambler N, editors. Coral Reefs: An Ecosystem in Transition. Heidelberg: Springer; 2011. p. 347-371

[108] INVEMAR. 2005. Informe del los ambientes marinos y costeros en Colombia: Año 2005. Serie de publicaciones periódicas No. 8. INVEMAR: Santa Marta; 2005 360 p

[109] Díaz-Pulido G, McCook LJ. The fate of bleached corals: Patterns and dynamics of algal recruitment. Marine Ecology Progress Series. 2002;**232**:115-128

[110] Titlyanov E, Yakovleva I, Titlyanova T. Interaction between benthic algae (*Lyngbya bouillonii, Dictyota dichotoma*) and scleractinian coral *Porites lutea* in direct contact. Journal of Experimental Marine Biology and Ecology. 2007;**342**:282-291

[111] Golubic S, Abed RMM, Palinska K, Pauillac S, Chinain M, Laurent D. Marine toxic cyanobacteria: Diversity, environmental responses and hazards. Toxicon. 2010;**56**:836-841

[112] Carpenter RC. Partitioning herbivory and its effects on coral reef algal comunities. Ecological Monographs. 1986;**56**:345-363

[113] Hay ME, Fenical W. Marine plant-herbivore interactions: The ecology of chemical defense. Annual Review of Ecology and Systematics. 1988;**19**:111-145

[114] Valentine JF, Heck KL, Kirsch KD, Webb D. Role of sea urchin *Lytechinus variegatus* grazing in regulating subtropical turtlegrass *Thalassia testudinum* meadows in the Florida Keys [USA]. Marine Ecology Progress Series. 2000;**200**:213-222

[115] Hay ME. The role of seaweed chemical defenses in the evolution of feeding specialization and in the mediation of complex interactions. In: Paul VJ, editor. Ecological Roles for Marine Natural Products. Ithaca: Comstock Press; 1992. p. 93-118

[116] Vaz Pinto F, Olabarria C, Arenas F. Role of top-down and bottom-up forces on the invasibility of intertidal macroalgal assemblages. Journal of Sea Research. 2013;**76**:178-186

[117] Burkepile DE, Hay ME. 2006. Herbivore versus nutrient control of marine primary producers: Context-dependent effects. Ecology. 2006;**87**:3128-3139

[118] Hay ME, Steinberg PD. The chemical ecology of plant-herbivore interactions in marine versus terrestrial communities. In: Rosenthal J, Berenbaum M, editors. Herbivores: Their Interaction with Secondary Metabolites, Evolutionary and Ecological Processes. San Diego: Academic Press; 1992. p. 371-413

[119] Paul VJ, Hay ME. Seaweed susceptibility to herbivory: Chemical and morphological correlates. Marine Ecology Progress Series. 1986;**33**:255-264

[120] Kiryu Y, Landsberg JH, Peters EC, Tichenor E, Burleson C, Perry N. Pathological effects of cyanobacteria on sea fans in southeast Florida. Journal of Invertebrate Pathology. 2015;**129**:13-27. DOI: 10.1016/j.jip.2015.04.007

[121] Smith JE, Kuwabara J, Flanagan K, duPlessis S, Coney J, Beets J, Takabayashi M, Barnes S, Turner J, Brown D, Griesemer BK, Stanton F. An unusual cyanobacterial bloom in Hawai'i. Coral Reefs. 2008;**27**:851

[122] Birrell CL, McCook LJ, Willis BL. Effects of algal turfs and sediment on coral settlement. Marine Pollution Bulletin. 2005;**51**:408-414

[123] Diaz-Pulido G, Harii S, McCook LJ, Hoegh-Guldberg O. The impact of benthic algae on the settlement of a reef-building coral. Coral Reefs. 2010;**29**:203-208

[124] Belov AP. A model of phycotoxin release by cyanobacterial cells. Ecological Modelling. 1998;**110**:105-117

[125] Ross C, Santiago-Vazquez L, Paul V. Toxin release in response to oxidative stress and programmed cell death in the cyanobacterium *Microcystis aeruginosa*. Aquatic Toxicology. 2006;**78**:66-73

[126] Preubel K, Wessel G, Fastner J, Chorus I. Response of cylindrospermopsin production and release in *Aphanizomenon flos-aquae* (Cyanobacteria) to varying light and temperature conditions. Harmful Algae. 2009;**8**:645-650

[127] Nieto CF. Cianobacterias bentónicas arrecifales: Evaluación de sus interacciones alelopáticas mediante ensayos in situ en comunidades coralinas Islas del Rosario, Caribe Colombiano [Undergraduate thesis]. Bogota: Universidad Jorge Tadeo Lozano; 2012

[128] Paerl HW, Gardner WS, Havens KE, Joyner AR, McCarthy MJ, Newell SE, Qin B, Scott JT. Mitigating cyanobacterial harmful algal blooms in aquatic ecosystems impacted by climate change and anthropogenic nutrients. Harmful Algae. 2016;**54**:213-222. DOI: /10.1016/j.hal.2015.09.009

[129] Ainsworth CH, Mumby PJ. Coral–algal phase shifts alter fish communities and reduce fisheries production. Global Change Biology. 2015;**21**:165-172. DOI: 10.1111/gcb.12667

[130] Diaz-Pulido G, Gouezo M, Tilbrook B, Dove SG, Anthony KRN. High CO_2 enhances the competitive strength of seaweeds over corals. Ecology Letters. 2010;**14**:156-162. DOI: 10.1111/j.1461-0248.2010.01565.x

[131] Gardner TA, Cote IM, Gill JA, Grant A, Watkinson AR. Long-term region-wide declines in Caribbean corals. Science. 2003;**301**:958-960

[132] Martinez S, Acosta A. Cambio temporal en la estructura de la comunidad coralina del área de Santa Marta-Parque Nacional Natural Tayrona [Caribe Colombiano]. Boletín de Investigaciones Marinas Y Costeras. 2005;**34**:161-191

[133] Rodríguez-Ramírez A, Reyes-Nivia MC, Zea S, Navas-Camacho R, Garzón-Ferreira J, Bejarano S, Orozco C. Recent dynamics and condition of coral reefs in the Colombian Caribbean. Revista de Biología Tropical. 2010;**58**(Suppl 1):107-131

[134] Schutte VGW, Selig ER, Bruno JF. Regional spatio-temporal trends in Caribbean coral reef benthic communities. Marine Ecology Progress Series. 2010;**402**:115-122

[135] Zapata FA, Rodríguez-Ramírez A, Caro-Zambrano C, Garzón-Ferreira J. Mid-term coral-algal dynamics and conservation status of a Gorgona Island (Tropical Eastern Pacific) coral reef. Revista de Biología Tropical. 2010;**58**(Suppl 1):81-94

[136] Lodeiros C, García N. The use of sea urchins to control fouling during suspended culture of bivalves. Aquaculture. 2004;**231**:293-298

[137] Westbrook CE, Ringang RR, Cantero SMA, HDAR & TNC Urchin Team, Toonen RJ. Survivorship and feeding preferences among size classes of outplanted sea urchins, *Tripneustes gratilla*, and possible use as biocontrol for invasive alien algae. PeerJ. 2015; **3**:e1235. DOI: 10.7717/peerj.1235

[138] Burke L, Maidens J. Reefs at Risk in the Caribbean. Executive Summary. World Resources Institute: Washington; 2004. 16 p

Chapter 8

Coral Microcosms: Challenges and Opportunities for Global Change Biology

Patrick Schubert and Thomas Wilke

Additional information is available at the end of the chapter

http://dx.doi.org/10.5772/intechopen.68770

Abstract

Well-maintained coral-microcosm systems provide a good opportunity for performing global-change simulations under controlled conditions and allow long-term experiments while avoiding problems with natural fluctuations. However, despite rapid technical progress over the last few years in maintaining corals, microcosm experiments remain demanding and challenging. Therefore, this paper focuses on problems and opportunities associated with maintaining corals for global-change experiments, and the pitfalls associated with simulating natural and anthropogenic disturbances. We start in Section 1 with a brief assessment of the global situation of coral reefs and discuss problems and challenges associated with microcosm experiments. Section 2 covers the technical setup of coral-aquarium systems in respect to the necessary hardware and safety precautions. Section 3 provides information on coral-species selection, coral-propagation techniques, and the choice of associated fauna and flora. Problems with maintaining controlled conditions are deliberated in Section 4, including water chemistry as well as pest and disease control. The paper closes with conclusions for global-change studies in coral-microcosm systems (Section 5). As this review provides important insights into the rapidly developing field of coral-microcosm experiments, it might be of particular interest for graduate and post-graduate students in marine sciences, for global-change researchers, as well as for administrators and technicians interested in maintaining corals under fully-controlled conditions.

Keywords: aquaculture, coral propagation, environmental change, experimental design, reef aquarium, Scleractinia, simulation studies

1. Introduction

Coral reefs belong to the most diverse and valuable ecosystems on Earth. They offer habitats for ca. 100,000 to >500,000 described species [1, 2] and the actual number might be higher by one magnitude [3]. Coral reefs also provide crucial ecosystem services as an important source of food for humans, as natural coastal defence, and as a recreational resource. Moreover, the biodiversity in coral reefs is seen as an important source for drug discovery [4].

Reef-building corals cover a total surface area of 260,000–600,000 km^2 [2]. They are typically restricted to latitudes between 25°N and 25°S. The optimal water temperature for most species is 23–29°C, and the optimal salinity is 32–42 parts per thousand (ppt). In addition, zooxanthellate corals require abundant light, restricting them to the euphotic zone of the oceans [5].

Reef-building corals are very sensitive to environmental change, both natural and anthropogenic, and it is estimated that around 50% of the world's reefs are threatened by human activities and that about 20% of the reefs have been destroyed already [6]. Major threats include destructive fishing practices and overfishing, habitat destruction, pollution, eutrophication, changes in food webs, unsustainable tourism, sedimentation, global warming and ocean acidification (e.g. [7, 8]).

There is ample empirical evidence for the adverse effects of these stressors on particular coral species or even entire reefs, and some of the causal links between environmental disturbances and biological responses of corals, such as increasing water temperature and coral bleaching or decreasing ocean-water pH and reduced calcification rates [9–11], are well understood.

However, as most adverse effects are multifactorial, a precise assessment of the individual contribution of stressors in natural systems, particularly those related to global change, remains challenging [12, 13]. Besides additive effects, multiple stressors could also act synergistically or antagonistically [14]. In fact, for developing effective management strategies, the individual contribution of stressors is of the utmost concern, enabling stakeholders to identify the most important parameters in a particular system. Therefore, scientists are interested in quantifying both the individual and combined effects of stressors acting on reef-building corals.

Although numerous field observations are being carried out to address these problems, studies in natural systems are typically confounded by the presence of variables other than those of interest. The relationship can thus be characterized, at best, as correlative, and a direct inference of cause and effect remains difficult [15]. In particular, environmental problems at the global scale can typically not be addressed using traditional scientific experiments [16]. The latter authors also argued that microcosms experiments (i.e. "experimental ecological systems at a small spatial scale") using model organisms could be a suitable methodology for addressing global problems, such as ecosystem responses to climate change.

Microcosms enable the manipulation of a single or few variables, and to compare the effects on organisms over time against control conditions. However, unlike natural systems, microcosm experiments are an abstraction from reality, and no single setup might explain the complex impact of global change on populations, species, and communities. Instead, each setup may help answer a specific question [15]. Besides generating such specific knowledge, microcosm studies can also help develop theories and meaningful policy implications [16].

Coral-microcosm experiments are a relatively new approach. Only some 30 years ago, several technical breakthroughs were achieved, enabling researchers to keep corals healthy in closed tanks [17]. However, controlled laboratory experiments add a level of complexity to keeping corals in aquarium systems, particularly in long-term microcosm studies. To ensure a stable growth of corals and to avoid a potential bias introduced by unintended variations of system parameters, a broad spectrum of environmental factors has to be regulated [18, 19]. Given the complex chemical nature of seawater and the dynamics of biological consumption processes, this remains a challenging task. Growth rates of corals, for example, are largely controlled by the Ca^{2+} content and alkalinity of the seawater. As growth processes deplete the water of Ca^{2+}, differential growth rates also have a differential feedback effect on the Ca^{2+} level and alkalinity of the water. These problems are of particular concern for global-change studies involving manipulations of CO_2 and pH levels [15, 18]. Similar problems are of concern for the choice of associated animal (e.g. herbivorous fish) and plant species (e.g. coralline algae) to be maintained in the microcosm system for enabling a healthy growth of corals. Further challenges lie in the selection of the general experimental setup (e.g. size of tanks, natural vs. synthetic seawater, single vs. multiple water-circulation systems) and in the choice of the technical equipment (e.g. type of lighting, circulation and control systems).

All these considerations may have a profound impact on the quality of the data generated, on associated costs, on the maximum possible duration of the experiment, and on its susceptibility to failure. Moreover, microcosm experiments are increasingly being designed for long-term durations to enable an assessment of evolutionary adaptations of corals. Finally, a wealth of technical novelties has been introduced to the market in recent years. Therefore, complex decisions have to be made by the experimenter prior to the setup of coral-microcosm experiments.

For these reasons, this article aims at discussing the challenges and opportunities of utilizing coral microcosms for global-change studies. Based on literature reviews, expert interviews, and our own >15 years of experiences with maintaining stony corals, we inform about the technical setup of coral microcosms in Section 2, provide information on the study organisms in Section 3, discuss problems of maintaining controlled conditions in Section 4, and finish with conclusions concerning setup and operation of coral microcosms for global-change studies (Section 5).

The insights provided might be of particular interests for graduate and postgraduate students in marine sciences, for global-change researchers, for technicians and animal keepers, as well as for decision makers responsible for the administrative planning of coral-microcosm facilities.

2. Setup of coral-microcosm systems

2.1. Experimental design

Maintaining stony corals in tanks is a challenging task. Conducting controlled (long-term) microcosm experiments adds another level of complexity. Besides comprehensive technical, biological, and chemical knowledge, extensive experiences with experimental design and the manipulation of environmental variables are required in order to perform these experiments in a way that compelling conclusions can be drawn from the data generated [17, 18, 20, 21].

Many technical (e.g. size and number of tanks, choice of technical equipment) and biological decisions (e.g. study species, associated fauna and flora) have to be made prior to the setup of microcosm experiments. However, the first and most critical step is the selection of the general experimental design based on the study question, the study species chosen, and the intended duration of the experiment. Several key decisions have to be made. They include the choice of (i) natural versus synthetic seawater, (ii) open versus semi-closed versus closed systems and iii) number of water-circulation systems to be implemented.

Natural and synthetic seawater differ in various characteristics of significance for coral-microcosm experiments (see **Table 1**). Of particular interest are availability, overall quality, consistency, and toxicity. Particularly in coastal areas, natural seawater is readily available. The chemical composition of off-shore seawater is usually highly consistent. However, the overall quality strongly varies with source, mean of transportation (e.g. hygiene of ballast tanks, containers and delivery pumps), and subsequent treatment. A principal problem of natural seawater is chemical and biological contamination. In particular, the high abundance and diversity of bacteria, viruses, archaea, algae and fungi are of concern. Whereas for standard marine aquarium purposes, natural microbial communities could jumpstart the biological cycle in the system and can be an important source of food for vertebrates and invertebrates, the adverse impact on coral global-change experiments in microcosm setups could be considerable. Corals are holobionts that can adjust the composition of their microbial endosymbionts depending on environmental conditions. Therefore, natural seawater makes it more

	Synthetic seawater	Natural seawater
Availability	Always available	Depends on location and infrastructure available
Quality	Usually high, variation in consistency possible, low toxicity	Depends on the source of water and the transport process, consistency typically high, water often contaminated
Treatment	Only basic treatment required (dissolution in deionized or reverse-osmosis purified water, control/adjustment of pH, temperature and salinity)	Often complex treatment necessary (e.g. ultra-filtration, dark-treatment, ultraviolet sterilization, chlorination)
Costs	Medium	Low to high, depends on location, source and treatment
Pros for microcosm experiments	High quality, no contamination, readily available	Natural and consistent chemical composition, enables studies with natural microbial communities
Cons for microcosm experiments	Often variable chemical composition, deionized or reverse-osmosis water required for preparation	Biological and chemical contamination possible, often requires filtration and decontamination, cannot be stored over long periods of time

Table 1. Properties of synthetic and natural seawater.

difficult to maintain controlled conditions throughout the experiment. It requires ultrafiltration as well as extensive decontamination prior to use (e.g. chemical decontamination, dark or ultraviolet treatment). This, however, creates a new set of problems (e.g. the need for dechlorination after chlorine treatment). Moreover, due to its higher toxicity, natural seawater may perform inferior as a culture medium for juvenile invertebrates, compared to synthetic seawater [22, 23]. Besides a low toxicity, another advantage of synthetic seawater is its ready availability and overall high quality. However, chemical consistency may vary among brands, and sometimes even within lots and individual packages. Moreover, high-quality deionized or reverse-osmosis water is required for preparing synthetic seawater.

The second major decision to be made with regard to the planning of coral-microcosm setups is the question of closed versus semi-closed versus open systems (reviewed in [24]). According to the latter author, closed microcosm systems are barred from exchange of food energy, seawater, as well as associated fauna and flora. They only allow gas exchange, freshwater refill to replace evaporation and exchange of light and heat energy. By contrast, semi-closed systems also allow for partial seawater exchange to maintain water quality by replacing inorganic nutrients and trace elements. Finally, open systems permit gas and seawater exchange, but also moderate inputs of supplemental food for associated faunas as well as the replacement of dead plants and animals [17, 25, 26]. The major goal of coral-microcosm experiments is not self-sufficiency of the system but the manipulation of a single or few variables by maintaining all other variables as constant as possible. Therefore, closed systems are, in many cases, impractical as a constant water quality for the demanding stony corals cannot be assured (e.g. metabolized trace elements have to be replaced). Possible exceptions are short-term experiments. Whether semi-closed or open microcosm systems are more appropriate largely depends on the duration of the experiment, the study species, as well as the associated fauna and flora. Maintaining near-natural and relatively constant conditions in coral microcosms over a long time often requires the addition of fishes and other animals, which typically depend on external food supplies. Moreover, deceased animals and plants have to be replaced. Thus, long-term coral-microcosm experiments are mostly designed as open systems (*sensu* [24], see also **Figure 1**).

The third principal decision concerns the number of water-circulation systems to be implemented. This, in turn, depends on the statistical design of the experiment and the study question. In most cases, a design with a single water-circulation system is preferred in order to keep all but one or few target variables constant. This enables, for example, the exchange of planktonic organisms (including pathogens and parasites) throughout the system. Moreover, the associated larger water volume makes the system less susceptible to unintended fluctuations of water parameters. However, particularly for experiments studying the effects of biotic factors (e.g. the composition of microbial communities in seawater or the impact of pathogens) or chemical parameters (e.g. toxins), a single water-circulation system may be impractical. Another possibility for stabilizing water circulation is to integrate a larger 'buffer' tank containing fish for nutrient intake and a deep-sand bed for biological filtration (**Figure 1**). Moreover, an algae filter with an inverse lighting regime might compensate for diurnal fluctuations of pH values [27, 28].

Recommendations: The experimental setup is largely determined by the scientific question of interest, study species, associated fauna and flora (if applicable), and the intended duration of

Figure 1. 'Ocean 2100' global-change simulation experiment at Justus Liebig University Giessen. Nine experimental coral tanks (three hidden) are connected via a technical tank (not shown) to a 'buffer' reef tank with live rocks and a deep-sand bed (right). This long-term setup (intended duration 10 years) has been designed as an open microcosm system with synthetic seawater and a single water-circulation system. The replicate system (not shown here) has its own water-circulation system and 'buffer' tank.

the experiment. In fact, the longer the duration of the experiment, the more detailed planning is needed. Unless natural seawater of high quality is readily available or the composition of its natural microbial community is of interest, synthetic seawater is favoured in microcosm experiments. In that case, high-quality products should be preferred and whole packages must always be used for preparing the water [17, 25]. Moreover, important parameters (e.g. pH, alkalinity, salinity) have to be checked prior to use. Closed coral-microcosm systems are typically only applicable for short-term experiments (over few weeks) without the need for associated faunas (such as herbivorous fish). For medium-term experiments (several months) without associated faunas, semi-closed setups are preferable. Long-term experiments (several years) or setups that require supplementary food supply are typically designed as open microcosm systems (**Figure 1**). For semi-closed and open setups, a seawater exchange of at least 20% per month is recommended [28]. As for the water-circulation systems in coral microcosms, the least number with the largest effective volume should be chosen for each experimental system (for a review of the statistical needs in global-change experiments, see [18]).

2.2. Lighting

Light is fundamental to all photosynthetic processes and thus crucial for zooxanthellate corals [5]. Defined and controlled light conditions are also important for assuring reproducible results obtained from coral-microcosm experiments. Among others, light (1) affects density and photosynthetic activity of hermatypic corals [29], (2) increases calcification rates in hermatypic corals [30], (3) influences the activity of associated faunas such as diurnal fishes [31, 32], (4) affects the metabolic efficiency of corals and thus survival [33, 34] and (5) influences the phenotype of scleractinian corals [35].

Although most stony corals require abundant light with a broad spectrum, conditions are often species-specific [36]. Under controlled microcosm conditions, it is therefore important to meet the requirements of the study species for achieving near-natural growth rates and physiological responses [37]. Too little light may, for example, decrease the metabolic efficiency and growth rates in stony corals, or may cause shifts in phenotype morphology [35]. Too much light could burn the zooxanthellae or cause coral bleaching [38].

For most coral-microcosm setups, no sufficient natural light is available. Therefore, artificial lighting has to be used with the appropriate intensity and colour spectrum. This is a challenging task as these characteristics, for example, change with water depth. Today, four popular artificial lighting systems are available, which differ in some of the main characteristics of relevance for global-change experiments in stony corals (see **Table 2** and **Figure 2**): T5 fluorescent lamps, metal-halide lamps, LED lamps and light-emitting plasma lamps.

The widespread introduction of metal-halide lamps into reef aquariums some 30 years ago made it possible to maintain stony corals with comparatively low effort, and for many years they have been the standard lighting equipment. They are well suited for high water columns, can be fitted to suit a wide range of tank sizes, are available with different colour temperatures and have, in general, a well-balanced spectrum. Disadvantages are their relatively low energy efficiency, a short lifetime and a high heat production.

The decline of metal-halide lamps over the past 10 years is mostly due to improvements in T5 fluorescent lighting, making the latter very popular for reef aquariums [39]. It is more energy efficient than metal-halide lamps, comes in a wide range of colour temperatures and its spectral characteristic is relatively good. Moreover, the spread of light is comparatively even, enabling relatively constant conditions across experimental tanks. However, similar to metal-halide lamps, T5 tubes have a comparably short lifetime. Moreover, light intensity and spectrum change over time, and spectral characteristics are affected by ambient temperature. T5 lighting is only suitable for shallow tanks.

Very recently, LED lighting has advanced to the point where it can be used to maintain stony corals, as long as the quality of light meets the requirement of the study organism [40]. High-quality LED lighting combines excellent energy efficiency with long-term stability of spectrum and intensity. The spread of light can be controlled by lenses for individual LEDs. In sophisticated systems, intensity and colour temperature can be adjusted electronically, though achieving natural spectra remains a problem. As the respective colour spectrum is produced by an array of individual LEDs, partial failure of LEDs, which is often difficult to detect, changes the spectrum. Moreover, some LED lighting systems require active cooling, making them vulnerable to humidity and salt deposits. Finally, some coral species appear to be sensitive to LED light (Schubert, unpublished data).

	T5 fluorescent lamp	Metal-halide lamp	LED lamp	Plasma lamp
Technology	Gas-discharge lamp that uses internal electrodes	Metal-halide lamp that produces light by an electric arc	Light-emitting diode lamp	Gas-discharge lamp that uses an electric or magnetic field
Acquisition cost	Low	Medium	Medium to high	High
Maintenance requirements	Medium	Medium	Medium	Low
Energy efficiency	Medium	Low	High	High
Reliability	Medium	Medium	Low to medium	High
Lifetime	Ca. 10 months	Ca. 10 months	>48 months*	>48 months*
Spectral coverage	Good	Good	Good	Good
Spectral and brightness stability over lifetime	Change over time	Change over time	Relatively constant over time*	Relatively constant over time
Effective water column height	Up to ca. 60 cm	Up to several meters	Up to several meters	Up to several meters
Spread of light	Very even	Even	Depends on the lenses used, lamps may flicker	Even
Pros for microcosm experiments	Various types with different characteristics available	Common and well established	Some LED lamps allow an individual adjustment of spectral composition, little waste heat released to the water	Various types with different characteristics available, little waste heat released to the water, IP68 housings available
Cons for microcosm experiments	Spectrum affected by ambient temperature and age of bulb, air-cooled lamps often affected by humidity and salt	Much waste heat released to the water, decreasing number of manufacturers	High ambient temperature, humidity and mineral deposits decrease reliability and lifetime of the lamp, failure of individual LEDs often hard to detect	So far, no long-term experiences available

*High-quality lamps

Table 2. Properties of popular artificial lighting systems used for coral-microcosm experiments.

The very latest editions to reef-aquarium lighting systems are modern plasma lamps. They are highly energy efficient, show a long-term stability of spectrum and intensity, and spectrum and colour-temperature can be custom-tailored by the manufacturer. Moreover, the spread of light is very even and typically no active cooling is necessary, allowing for the construction of housings according to the IP66 or IP68 standards. Though the equipment is still very expensive, energy and maintenance costs are very low. Thus, lifetime costs might be the lowest of

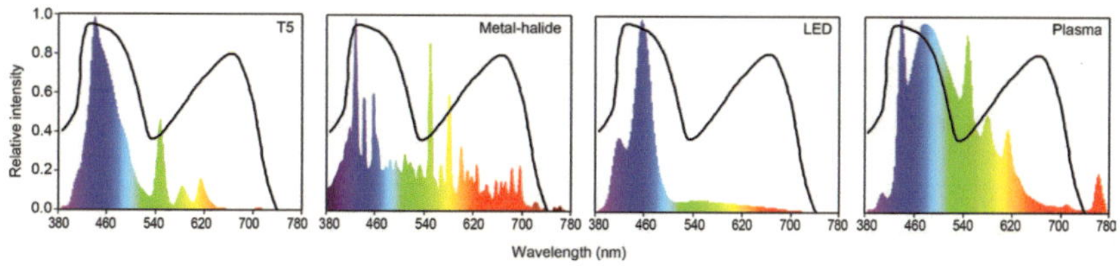

Figure 2. Irradiance spectra of common light sources. T5: ATI AquaBlueSpezial and BluePlus (ATI, Hameln, Germany); Metal-halide: BLV 12,000 K, 250 W (BLV, Steinhoring, Germany); LED: Orphek Atlantik V4 (Orphek, Sao Paulo, Brazil); Plasma: TGS029, 150 W (Aqua ECO Store, Kaltenkirchen, Germany). Spectra were generated using a Lighting Passport Spectrometer (Asensetek, New Taipei City, Taiwan). Black curves indicate PAR reference.

all lighting systems discussed. However, due to their recent introduction, so far no long-term experiences exist for the application of plasma lamps to coral-microcosm systems [36, 40].

Recommendations: The choice of lighting system for coral-microcosm experiments largely depends on the specific parameters investigated, the study species and the intended duration of the experiment. In general, high-quality T5, LED or plasma lamps should be considered. Some lighting systems are optimized to enhance coral growth and to 'improve' the colour intensity of the corals maintained. They are thus not suitable for most global-change experiments. All lamps/tubes used in a system should be at the same stage of lifetime. Open lighting systems have to be protected against heat, mineral deposits and water. All systems should be maintained regularly, which may also include the control of light intensity and spectrum. This is particularly important for LED lamps and respective hand-held LED testers and light metres are available on the market.

2.3. Water movement

Water movement in aquarium systems is crucial to the vitality of stony corals [41]. A controlled movement is also critical for obtaining reproducible results in global-change studies using microcosm setups. Among others, water movement increases the exchange rate of gases and thus photosynthetic efficiency [42], increases mass-transfer of materials across the tissue-water interface [43, 44], increases food capture and thus energy supply to the coral [43, 45], facilitates cleaning of corals and prevents build-up of detritus [46], and influences the phenotype of scleractinian corals [35].

Most stony corals are adapted to strong water movement and/or wave action [47, 48]. Insufficient water movement may, for example, enhance detritus and sediment build-ups, and could thus cause unintended and unpredictable local processes in nutrient balance (nitrification and denitrification). It may also foster the emergence of anaerobic zones in tanks, affects the biological filtration rate of the system and thus facilitates uneven growth rates of corals across experimental tanks. Excessive and/or strongly concentrated water movement, on the contrary, may increase the stress level of some corals, damage sensitive species and cause atypical growth forms.

Three popular systems for generating water movement in coral microcosms are available: (1) water-flow systems where pumps create a laminar or a turbulent water movement,

(2) water-oscillation systems (e.g. Wavebox®), which set the entire water body in motion and (3) water-spill systems where a water bucket equipped with a tilting mechanism creates a regular wave motion (**Table 3**).

Of these systems, the water-flow system is most commonly used. One or more pumps either create a laminar (i.e. streamlined) or turbulent (i.e. irregular or mixed) flow. Turbulent flows are typically found in oceans in water depths of less than 12–15 m, and laminar flows in depths more than that [49]. For generating the water flow, radial-flow and axial-flow velocity pumps are typically used [50]. The latter are preferred because the water flow is more uniform. Water-flow systems are relatively cost-efficient and can be installed in most tank systems. Disadvantages are that the direction and intensity of water movement vary across the tank. Moreover, flow velocity will be higher at the periphery of a bushy coral compared to its centre.

In recent years, another water-movement technology, the water-oscillation system (such as the Wavebox; Tunze, Penzberg, Germany), has made its way into coral-microcosm systems. A Wavebox consists of one or more axial-flow-pulsing pumps and a controller. Determined by tank resonance, the intermittent operation of the system sets the entire water body in motion, assuring water movement in all parts of the tank [51]. Maximal displacement at either end of the tank is several centimetres. Another advantage of the oscillating nature of water movement is the uniform growth morphometry of the corals seen in such systems. Disadvantages are the robustness of the construction required due to the resonance generated and the possible interferences with other tanks, the need to place the water overflow in the central part of the tank, and the need for additional pumps in larger systems to create a linear flow.

The third approach, water-spill systems, is less common and typically used for specific purposes [50]. It is a wave machine that usually consists of a bucket equipped with a tilting mechanism. The bucket is filled with water and once the water level reaches a certain level, it tips over and releases the water to create a spill. Water-spill systems are ideally suited to simulate

	Water-flow system	**Water-oscillation system**	**Water-spill system**
Technology	Radial- or axial-flow pumps	Pulsing pump with controller	Tilting water bucket
Acquisition cost	Low to medium	High	Low to high
Maintenance requirements	Low to medium	Low to medium	Low
Efficiency	Medium to high	High	High
Reliability	Medium to high	Medium to high	High
Pros for microcosm experiments	Applicable to tanks of various shapes and sizes of 20 to >20,000 L	High efficiency, optimal water movement, near-natural growth morphology of corals	Adaptable to all tank sizes
Cons for microcosm experiments	Low energy efficiency, high amount of waste heat	Restricted to rectangular, medium-sized tanks (0.6–3.5 m length)	Mostly individually manufactured

Table 3. Properties of popular water-movement systems for coral-microcosm experiments.

wave actions in littoral zones. They can be adapted to all tank sizes. As only few commercial products are available, custom-made solutions are typically required.

Recommendations: The choice of water-movement systems for coral-microcosm experiments depends on the specific parameters investigated, the study species, and the size and shape of the tanks used. The water-flow system, though widely distributed, might not be suitable for most questions related to global change due to difficulties in ensuring relatively constant conditions throughout the tank. A possible exception is the study of corals that live in water depths characterized by laminar water movements. The application of water-spill systems is also restricted to specific research questions. They are mainly suitable for studying processes in coral species that live in the littoral zone. By contrast, water-oscillation systems are applicable to a wide range of questions and species. They produce a relatively homogeneous water movement, thus preventing a potential bias of the study results. Though there are some size restrictions on the tanks (see **Table 3**), most coral-microcosm setups might fall within the suitable range. For larger systems, several Waveboxes can be combined and/or complemented with axial-flow pumps to increase water flow.

2.4. Miscellaneous supporting hardware

Coral reefs are characterized by relatively stable water conditions, and most inhabitants react very sensitively to sudden changes in environmental parameters [52]. Moreover, fluctuations in water chemistry may be caused by the reef organisms themselves due to metabolic processes. In natural systems, the physiological impact of organisms on water parameters is limited due to the comparatively low ratio of biomass and water volume [26]. However, compensating for these problems in microcosm systems remains challenging (see also Section 2.1.). This concerns the water chemistry but also other 'tank effects' such as temperature fluctuations, microbial contamination and the accumulation of pollutants.

With the growing popularity of seawater aquariums in general and reef aquariums in particular, the selection of devices and methods for controlling and maintaining healthy conditions for stony corals has significantly increased. However, some of the available commercial solutions and products are not suitable for a precise control of parameters; others may cause harmful side effects in coral-microcosm setups, and yet others have efficiency and reliability issues. Therefore, the appropriate equipment/method typically has to be chosen based on the design and specific goal of the microcosm experiment.

One of the most significant challenges in coral-microcosm experiments is to assure near-natural calcification conditions. In particular, Ca^{2+} and Mg^{2+} levels have to be maintained, and alkalinity has to be stabilized [27, 53, 54]. Three approaches are commonly used: the calcium reactor, the Balling method and the Kalkwasser stirrer (**Table 4**). A calcium reactor is filled with $CaCO_3$ material (such as coral rubble), which slowly dissolves when the pH value is lowered through the addition of CO_2 [28, 55]. The efficiency of a calcium reactor largely depends on the type and size of the reactor, flow-through rate, type and grain size of the substrate used, as well as the pH value set. An alternative approach is the 'Balling method', that is, the individual addition of pre-mixed solutions of $CaCl_2$, $MgCl_2$ and $NaHCO_3$ [17, 56]. The approach works well for short-term studies. However, for long-term experiments it requires considerable analytic

	Calcium reactor	Balling method	Kalkwasser stirrer
Principle	Dissolution of aragonite or lime through CO_2 enrichment and low pH	Individual addition of $CaCl_2$, $MgCl_2$ and $NaHCO_3$	Dissolves $Ca(OH)_2$
Acquisition cost	Medium to high	Low	Medium
Operating cost	Medium	High	Low
Effecting pH	Decreases pH values	No effect	Increases pH values
Water volume	Applicable to tanks of various shapes and sizes of 200 to >20,000 L	Applicable to all tank sizes	Applicable to all tank sizes
Pros for microcosm experiments	Simultaneous increase of alkalinity and Ca^{2+} content, easy to handle	Individual adjustment of parameters, applicable to small water volumes	Efficient approach for increasing pH values
Cons for microcosm experiments	Risk of nutrient pollution (e.g. PO_4^{3-}), requires addition of CO_2	Requires high analytical efforts, risk of nutrient pollution, might cause ionic shifts that need to be compensated with NaCl-free sea salt	No alkalinity control, only marginal Ca^{2+} control

Table 4. Properties of popular systems/methods for controlling Ca^{2+} supply and alkalinity in coral-microcosm experiments.

efforts to avoid miscalculations and to prevent ionic shifts. The Kalkwasser method is an older approach for increasing the Ca^{2+} content by adding $Ca(OH)_2$ to the refill water [17, 28, 55]. Though its efficiency for Ca^{2+} control is relatively low, it may well be used for balancing daily fluctuations of pH values caused by photosynthetic activities [28].

Organic wastes and nutrients are typically removed from coral-reef tanks by protein skimmers (foam fractionators) and phosphate-binding agents. Protein skimmers are important for coral microcosms because they enable removing suspended particles and organic wastes before they enter the nutrient cycle [17, 28]. They are also of importance for increasing gas exchange, and constitute a good location for the application of ozone (see subsequent text). Besides removal of organic wastes, phosphate control is very important in coral-microcosm systems because phosphate enhances unwanted growth of algae and may inhibit calcification processes [54]. Though protein skimmers also help reduce phosphate concentrations in the water (particularly organic phosphate before it is converted into inorganic orthophosphate), phosphate-binding agents are more effective. However, the latter only help remove inorganic orthophosphate and not inorganic polyphosphate or organic phosphate. Thus, they may not mitigate an algae problem in the system as this is typically caused by organic phosphate. Various commercial phosphate-binding agents are available that are either based on aluminium oxide or on iron oxides and hydroxides [57], though the latter are preferred by most experts. It should be noted that, besides phosphate, these agents may also remove other chemical compounds such as heavy metals and silicate, which may or may not be desired.

Pollutants in reef aquariums, such as toxins, heavy metals, chlorine, ozone and drugs, are usually removed via chemical filtration with activated carbon. This popular filtration method may also eliminate water discoloration and plays an important role in the prevention of pollutant accumulation. Activated carbon is most efficiently used in a special flow-through filter arranged as bypass or equipped with an own pump. The direction of water flow is always upwards to reduce the risk of clogging. Filters with high water-flow-through (fluidized bed reactor) and low water-flow-through (slow flux filter) are in use. The former devices maximize reaction surface and respond rapidly. Disadvantages are high-carbon-abrasive rates and potentially abrupt changes in water parameters. The latter devices allow for an efficient usage of the substrate and enable a constant water quality but may increase sedimentation rates facilitated by detritus.

Reduction of parasite and pathogen loads may also help ensuring the health and vitality of the study organisms and the long-term stability of coral-microcosm systems. Moreover, depending on the goal of global-change experiments, it could be necessary to prevent the exchange of zooxanthellae across experimental tanks. This is most efficiently done through a combined ozone/ultraviolet (UV) radiation treatment—particularly in long-term experiments. UV irradiation is mainly effective in preventing infestations with pelagic microorganism. It does not introduce any harmful substances into the system, and high-quality products need relatively little maintenance. However, UV sterilizers will not be able to fully eradicate pathogens and have only little effect on benthic microbes. By contrast, an ozonizer uses the highly reactive ozone to efficiently kill pathogens in aquarium waters. Moreover, the gas helps transforming ammonia to nitrate, thus further increasing water quality. A disadvantage of ozone treatment is the high toxicity of the gas [58].

For global-change experiments in coral microcosms, a reliable temperature control is also crucial [52]. Depending on the tank size and goal of the experiment, several possibilities for temperature control exist, such as control via room-temperature regulation (heating/cooling), via a temperature-controlled water bath, or via heating rods. However, the internal temperature regulation of common heaters is often not reliable, typically requiring an independent sensor system.

Finally, for maintaining constant water conditions throughout the coral-microcosm system, an efficient water circulation is necessary. Pump selection depends on the capacity needed, the design of the delivery head, efficiency and the amount of excess heat produced [59].

Recommendations: Maintaining proper calcification conditions for long-term experiments is best achieved via a calcium reactor, whereas for smaller-water volumes (<200 L) and short-term experiments the 'Balling method' may be sufficient. For calculating the proper dosage, online calculators and apps are available (e.g. AquaCalculator; http://www.aquacalculator.com). Organic wastes and nutrients are efficiently removed from the system utilizing a combination of a protein skimmer and phosphate-binding agents. The latter could be used in a filter housing equipped with a slow-flux filter. To avoid unwanted side effects, phosphate-binding agents should be used cautiously. Moreover, phosphate levels in the system should be regularly monitored, and it is important to understand that these agents may also remove other chemical compounds.

Pollutants are typically eliminated from the system via chemical filtration with activated carbon. Good results can be obtained with a slow-flux filter and a daily-short time increase of the flow-through rate. However, since carbon loses its effectiveness when the surface pores close, its frequent replacement is important for optimal filtration. Efficient pathogen control is best done by combining an ozonizer and a UV sterilizer. As ozone is harmful to marine organisms and humans, excess gas must not enter the experimental tanks or the air and its application has to be monitored carefully.

Water temperature fluctuations should not exceed 1°C in 24 h. Temperature control can be best attained by controlling the lowest target water temperature in the system via room-temperature or water-bath temperature control. Higher temperature in individual tanks can then be achieved via heating rods. In the latter case, it is important to adjust the performance of the respective heating rods to tank size. Moreover, they need to be calibrated prior to the start of the experiment. More reliable, however, is the control of the water temperature through external, computer-based sensors. The internal temperature regulation of the heaters could then be used as a 'backup system' by adjusting it to 1°C above the target temperature.

Water circulation throughout the system can be achieved by using high-quality adjustable radial-flow pumps.

2.5. Safety and control systems

Coral-microcosm systems are often highly complex in terms of electrical and mechanical devices integrated, water parameters to be monitored and (dangerous) organisms to be maintained. This places high demands on the equipment used and the safety procedures implemented. Seawater, for example, is a good electrical conductor and also promotes corrosion. Thus, electrical hazards are of particular concern [28]. Moreover, minor failures such as a short-term deviation from the target temperature may endanger the success of the experiment and/or the health of the study organisms [17, 60].

Discussing all safety and control equipment required for coral-microcosm experiments is beyond the scope of this paper. However, important devices are listed in **Table 5** together with some basic recommendations.

Device	Function
Ground fault circuit interrupter (GFCI)	Reduces the risk of electric shock to humans and animals
Grounding probe	Reduces the risk of electrical shock to humans and animals
Uninterruptible power supply (UPS)	Buffers short-time power failures for the most important electrical devices (e.g. delivery pumps; measuring, monitoring and control systems) and prevents sensitive electronical devices from harmful power fluctuations; note that the capacity of most UPSs is too low to buffer all electrical devices in the system over a long period of time
Emergency power supply	Buffers power failures over an extended period of time; note that powering on emergency power generators might generate harmful spikes
Sensor system	Controls and monitors a wide range of parameters (e.g. water level, pH, temperature, O_2 content); systems are typically computer-controlled and linked to an alarm system
Alarm device	Triggers an alarm (acoustically, visually or via messaging) in case of malfunctions of devices or unusual readings
Webcam	Remotely monitors the system

Table 5. Important safety and control devices for coral-microcosm systems.

http://dx.doi.org/10.5772/intechopen.68770

Coral-microcosm experiments also require the implementation of a set of safety measures, including hazard assessments, safety-related labelling, emergency plans, staff training and regular security checks. A well-trained and experienced staff will not only help reduce the risk of accidents but also ensure a relatively problem-free operation of the experiment.

Recommendations: Depending on the goal, setup and duration of the experiment, several safety and control devices have to be installed. They include GFCIs as well as grounding probes for electrical safety purposes, a UPS (ideally in combination with an emergency power generator) for the continuous supply of electricity, sensor and webcam systems for monitoring proper operations and water parameters, as well as an alarm device to inform about malfunctions. It is highly recommended to spread electrical devices over several power circuits, each equipped with an own GFCI, to minimize the impact of failure of individual devices and to reduce electromagnetic interferences among electronical devices, respectively.

Prior to commissioning the installations, the responsible person should conduct a specific hazard assessment of the system together with the safety officer of the institution. This should include an evaluation of potential hazards through technical and electrical devices, irradiation (e.g. UV light), chemical substances (e.g. ozone or CO_2), as well as poisonous or otherwise dangerous marine organisms. This hazard assessment should also be used as a basis for the mandatory hazard-related labelling of devices and tanks, as well as for all staff-training measures to be conducted. Moreover, an emergency plan has to be developed and prominently displayed in the microcosm facility. Essential information should include, among others, the telephone numbers of the emergency poison centre and the institution's first-aiders. Finally, regular safety checks by a certified electrician and/or the safety officer of the institution should be conducted.

3. Study organisms

3.1. Selection of coral species

Selecting the proper study species for coral microcosms is a challenging task. Though the choice of species should largely be determined by the study question, other considerations such as availability, maintainability and legal aspects (e.g. CITES regulations [61]) also matter.

Scleractinian corals are a diverse and evolutionary old group that date back >250 million years ago [62]. However, many coral species are cryptic and (morphological) identification is not always straightforward (e.g. *Stylophora* spp.). Moreover, environmental parameters such as water temperature, water depth, water current, as well as light and nutrient availability not only effect the composition of species assemblages but also adaptations within species, leading to a variety of morpho- and ecotypes. Thus, different populations show different susceptibilities to changes in abiotic and biotic parameters [17].

Of relevance for microcosm experiments is also the fact that some species are more difficult to maintain than others. Moreover, there may be strong interspecific competition among species (e.g. *Galaxea* spp. have sweeper tentacles of up to 20 cm in length). In addition, branching coral species such as *Acropora formosa* can have growth rates of up to 2–3 cm per month, thus increasing space constraints in the tanks over time.

Another important practical consideration is the question whether wild or farmed corals should be used for the experiment. Corals taken from the wild are of particular interest if the natural composition of their endosymbionts is of concern. Moreover, often detailed information on the geographical origin and ecological setting is available. However, they typically need a long time of acclimatization to microcosm setups, growth rates are often lower and the susceptibility to diseases can be higher [17, 28]. Furthermore, permitting laws to collect and export/import specimens are typically stricter. By contrast, farmed corals are often healthier and more resistant in experimental systems than colonies taken from the wild [17, 28]. They are often readily available and some 'clonal lab strains' are being used across laboratories, enabling comparative analyses. However, their associated endosymbiont diversity may be depleted and/or altered, affecting coral growth and survivorship in experiments [17, 63]. Moreover, often limited information about their geographic origin is available. Finally, the selection of farmed coral species is much lower compared to wild-caught taxa.

Recommendations: Many coral species are cryptic and/or difficult to determine. Therefore, in some cases a molecular characterization of the study individuals might be necessary. Robust species such as *Pocillopora damicornis, A. formosa* and *Montipora digitata* are more appropriate for long-term studies than very sensitive ones. High growth rates and strong defence mechanisms of some species need to be considered for species and tank-size selection. As different populations show different susceptibilities to changes in abiotic and biotic parameters, precise information about the ecological and geographical origin of the study specimens might be important. Finally, the choice of wild versus farmed corals may have implications for questions concerning growth rates, disease susceptibility and endosymbiont composition.

3.2. Coral propagation

One of the biggest advantages of using stony corals for global-change microcosm experiments is the possibility of fragmenting larger individuals. Though some colonies may show intercolonial variation [37], individual fragments are typically considered to be 'clones' of the mother colony. This has benefits for the statistical design of the experiment as the same individual can be simultaneously exposed to different environmental parameters. Therefore, fragment propagation of scleractinian corals often forms the basis for coral-microcosm experiments.

However, fragmenting corals is not always straightforward. Whereas some species are relatively easy to handle (e.g. *M. digitata*), others need more care during fragmentation (e.g. *Catalaphyllia jardinei*) [28]. Moreover, the size of the fragments as well as the quality of maintenance will determine survival rates within the first weeks after fragmentation [64–66].

The minimum size of the fragment has been discussed in detail elsewhere [28, 64–66] and mainly depends on species, experiences with fragmenting, condition of the mother colony and maintenance conditions. Similar aspects apply to the actual fragmentation technique [17, 28]. Of concern are, for example, the size of the polyps (large-polyp vs. small-polyp stony corals)

http://dx.doi.org/10.5772/intechopen.68770

as well as shape (massive to fine-branched), hardness ('soft' to hard) and internal structure (dense to chambered) of the skeleton.

The most frequently used method involves a rotary tool equipped with a diamond-cutting disc (**Figure 3A**). It is applicable to most small-polyp stony coral species and works particularly well in species with a hard skeleton. For medium-hard and branched species, a coral clipper (bone cutter) is often used for the fragmentation of the mother colony. Finally, corals with a 'soft' skeleton such as *Alveopora* spp., *Goniopora* spp. and *Madracis* spp. are typically fragmented using a serrated knife.

All of the above methods may also cause injuries to humans through the tools used as well as through contact with toxic coral tissues or aerosols.

After fragmentation, different methods of treatment and rehabilitation can be realized. If the growth form of the corals is not of concern, the branched fragments are typically attached to a line hanging in the free water column. This approach reduces sedimentation and overgrowth by algae. If a more natural growth form is desired, the individual fragments are attached to a small pedestal using an adhesive [17, 28]. If the part of the coral that is to be connected to the pedestal is not covered by tissue, hot glue is used. Otherwise, cement, cyanoacrylate gel or epoxy adhesive provide good solutions. The latter two are particularly well suited for sensitive species and/or small fragments.

The choice of material for the pedestal depends on coral species and fragment size, as well as on the experimental design of the tanks. Common materials are unglazed tiles, specialized ceramic, plastic products ('reefplugs') or concrete. The latter can easily be used to produce custom-made structures with a range of labelling options (**Figure 3B**).

Figure 3. (A) Fragmentation of a *Porites rus* colony using a rotary tool. (B) Coral fragment attached to a concrete pedestal 4 weeks after propagation. Note: for the handling of some species or multiple colonies, the use of laboratory gloves is recommended to avoid human injuries or to minimize the risk of disease transmission among colonies, respectively.

Recommendations: The method to be used for fragmenting corals largely depends on the species, the size of the fragments as well as the treatment after propagation. All tools should be clean and sterile to avoid a potential transmission of coral diseases. The mother colony and fragments should be exposed as short as possible to the air, and high or low air temperatures must be avoided. Safety precautions have to be taken to prevent injuries to humans. Depending on method and species, this may include wearing laboratory gloves, safety glasses and a respirator. For most species, best fragmentation results are achieved by first superficially cutting the coral with a rotary tool and then carefully breaking off the fragment by hand, or using pliers or a small chisel. However, the heat generated by the cutting disc may harm the coral. If the fragments have to be mounted on individual pedestals, best results are obtained with cyanoacrylate gel and epoxy adhesive, though the former may dissolve in seawater after some time. A versatile and easy-to-handle material is low-pollutant Portland cement. To keep stress levels in the corals low, all fragments should remain within the origin water cycle for at least 1 week after fragmenting. Moreover, a slightly reduced light intensity and sufficient water movement might facilitate rehabilitation. To reduce sedimentation and to improve water circulation around the fragments, an elevated position within the tank might be helpful. This can be achieved, for example, via aquarium eggcrates ('lighting grids') (**Figure 1**).

3.3. Associated species

Semi-open and open coral microcosms often require the addition of associated species to ensure near-natural conditions, stabilize the system and facilitate the health of the corals [67, 68]. This applies in particular to long-term experiments. However, adding additional species also increases the complexity of the experiment and may affect the reproducibility of the data. A comprehensive discussion of individual species is beyond the scope of this paper. However, information for some of the most commonly associated organisms is listed in **Table 6**.

Taxon	Pros	Cons
Fishes		
Acanthurus spp.	Control macroalgae and periphyton	Large size, aggressive species
Chaetodon spp.	Control *Aiptasia* spp.	Feed on large-polyp stony corals
Chelmon rostratus	Controls *Aiptasia* spp.	Requires frozen food, may feed on *Tridacna* spp. and other invertebrates
Chromis spp.	Facilitate nutrient intake	–
Ctenochaetus spp.	Control periphyton	Potentially aggressive
Halichoeres spp.	Control some parasites (e.g. flatworms)	Require sand bed, feed on invertebrates
Pseudochelinus spp.	Control some parasites (e.g. flatworms)	Potentially aggressive, feed on invertebrates
Salarias spp.	Control periphyton	Need to be kept individually or in pairs
Siganus spp.	Control macroalgae	Large size, may be nervous, poisonous spines
Synchiropus spp.	Control some parasites (e.g. flatworms)	–
Zebrasoma spp.	Control macro algae and periphyton	Potentially aggressive

Taxon	Pros	Cons
Mollusks		
Aeolidiella stephanieae	Control *Aiptasia* spp.	Feed exclusively on *Aiptasia* spp.
Euplica spp.	Control macroalgae and periphyton	–
Nassarius spp.	Control carrion and detritus	Require sand or detritus
Stomatella spp.	Control periphyton	–
Tectus spp.	Control periphyton	May relocate corals because of size
Turbo spp.	Control periphyton	May relocate corals because of size
Crustaceans		
Hermit crabs	Control periphyton and detritus	Larger species may damage corals
Lysmata spp. (peppermint shrimps)	Control *Aiptasia* spp.	May stress corals while removing food residues and mucus
Mithrax spp.	Control macroalgae and periphyton	–
Percnon gibbesi	Controls macroalgae and periphyton	Large size
Stenopus spp.	Control flatworms and polychaetes	–
Trapezia spp.	Control parasites and reduce sedimentation in bushy corals	–
Echinoderms		
Sea urchins	Control periphyton and encrusting algae	May relocate and/or feed on corals, some species are poisonous
Macroalgae		
Halimeda spp.	Easy to maintain and better to control than *Caulerpa* spp.	High calcification rate has to be compensated
Chaetomorpha spp.	Control nutrient levels and pH in algae filter	Floating, not attached to substrate

Table 6. Advantages and disadvantages of common associated species in coral-microcosm experiments.

Whereas larger organisms are often deliberately placed into the tanks, essential microorganisms are typically introduced with substrates such as sand, (live) rock and mud. They play an important role for stabilizing the water system, especially the nutrient cycle [69, 70]. Furthermore, the use of live rocks may significantly increase the risk of introducing diseases (see also Section 4.2.).

Recommendations: Associated species for coral-microcosm experiments have to be carefully selected, and species that feed on, stress and/or move corals should be generally avoided. Moreover, as associated species may influence the water parameters in the experimental tanks, each tank should contain the same species in the same quantities and with similar sizes. In some cases, it might be advisable to rotate associated species among tanks. As water parameters affected by substrate-bound microorganisms are difficult to control, it may be advantageous to refrain from using substrate within the individual experimental tanks. Instead, a larger 'buffer tank' could be integrated in the water cycle, which contains a deep-sand bed as well as live rocks (**Figure 1**). The same may apply for some or all associated animal species discussed above.

4. Quality control and maintenance of seawater

Slightest unintended variations in water parameters can cause significant effects to reef organisms [71–74]. Therefore, high-quality seawater is an important prerequisite for ensuring meaningful and comparable results in experimental systems [20, 26]. Some water parameters are relatively straightforward to measure and control, such as water temperature, pH and salinity. Others are more challenging to assess, including alkalinity, Ca^{2+}, Mg^{2+}, NO_3^- and PO_4^{3-} [14, 18, 75], and yet others, such as the concentration of many trace elements and some metabolic-degradation products, cannot be determined with standard water tests.

Therefore, water parameters in coral-microcosm systems are usually controlled and maintained through a set of common measures. This comprises the routine measurement of key water parameters, maintaining water levels and salinity in the system, regular exchange of parts of the seawater and active control of selected water parameters. Each coral-microcosm experiment requires a detailed plan for water testing. Whereas such a plan is also mandatory for regular reef aquariums, coral microcosms are even more demanding as fluctuations in water parameters have to be kept within narrow limits. A detailed description of all test procedures is beyond the scope of this paper and there is an extensive literature on this subject [75–78]. However, essential information can be found in **Table 7**.

Maintaining the water level and salinity in microcosm systems is a first step towards assuring a high quality of water parameters and to reduce unintended fluctuations. Strong water movements in combination with high air and water temperatures lead to high evaporation rates. To maintain salinity and other water parameters, the water volume in the system has to be kept constant. In coral microcosms, this is typically achieved through an automatic refill with deionized or reverse-osmosis water.

Fluctuations in chemical water parameters can also be mitigated through a regular and partial exchange of seawater in the system [17, 28]. This measure will help to replenish essential trace elements and reduce accumulation of harmful substances. Depending on the experimental design (e.g. filtration measures, biomass volume, feeding strategies), the exchange rate may vary between 20% per day and 20% per month [17, 26].

Though moderate seawater exchange helps stabilizing some water parameters, other factors such as alkalinity, Ca^{2+}, Mg^{2+}, pH, PO_4^{3-} and NO_3^- often require an active control (see also Section 2.4.). Adjusting nutrient levels in coral-microcosm systems (e.g. PO_4^{3-} and NO_3^-) is even more difficult, particularly in long-term experiments. This is partly due to the fact that appropriate values for PO_4^{3-} and NO_3^- are close to the detection limit of most common water tests. Besides conducting partial water exchanges (see above), nutrient levels can also

Parameter	Target value*	Test frequency	Test method/equipment	Comments
Water temperature	24–28°C	Continuously or daily	Analogue or digital thermometer/sensors, data logging might be useful	Regular calibration required, at least two independent measurement systems required
Salinity	35 ppt	1 × per week	Analogue or digital refractometer	Regular calibration required
pH	8.1–8.3	Continuously or daily	Laboratory-grade pH meter, data logging might be useful	pH fluctuates during the day
Alkalinity	7–9 KH (2.5–3.2 $mmol \cdot L^{-1}$)	Daily to 2 × per week	Titration test	Use of standards is recommended, quality of commercial products differs considerably
Calcium (Ca^{2+})	380–440 $mg \cdot L^{-1}$	1–2 × per week	Titration test	Use of standards is recommended, quality of commercial products differs considerably
Magnesium (Mg^{2+})	1250–1350 $mg \cdot L^{-1}$	Biweekly	Titration test	Use of standards is recommended, quality of commercial products differs considerably
Phosphate (PO_4^{3-})	<0.03 $mg \cdot L^{-1}$	1 × per week	Photometric test, colorimetric test	Available aquarium-grade kits often insufficient
Nitrate (NO_3^-)	<0.5 $mg \cdot L^{-1}$	1 × per week	Photometric test, colorimetric test	Available aquarium-grade kits often insufficient, some tests are not applicable to seawater
Nitrite (NO_2^-)	<0.1 $mg \cdot L^{-1}$	0.5–2 × per week**	Photometric test, colorimetric test	Available aquarium-grade kits often insufficient
Ammonia (NH_3^+/ NH_4^+)	<0.1 $mg \cdot L^{-1}$	0.5–2 × per week**	Photometric test	Available aquarium-grade kits often insufficient
Redox potential	250–400 mV	Continuously	Laboratory-grade redox probe	Important for controlling ozone application

*Value also depends on coral species and geographic origin.

**Frequency of measurement depends on experimental design.

Table 7. Common water parameters to be monitored in coral-microcosm systems including typical target values, suggested test frequencies and test methods.

be reduced through a skimmer combined with an efficient water-flow-through system (see Section 2.4.). Other options include biological filters (see Section 2.4.) and chemical PO_4^{3-} adsorbers. Note that in some well-established coral tanks, nutrient values have to be increased and not decreased. This can be done through adding extra nutrients to the system [79] or by including associated animals, such as fish [80] (see Section 3.3.).

Recommendations: For maintaining salinity in larger microcosms, an automatic refill system for deionized or reverse-osmosis water equipped with double-protected sensors is recommended (see also Section 2.5.). The automatic refill system must be disabled during abstraction or exchange of saltwater.

Controlling alkalinity, Ca^{2+}, Mg^{2+} and pH requires a well-equipped laboratory (**Figure 4**; also see the technical recommendations provided in Section 2.4.). Particularly in systems with a high coral biomass, daily fluctuations of alkalinity need to be compensated. To buffer pH variations, either the addition of $Ca(OH)_2$ during the night or the use of an algae filter with an inverse lighting regime is suggested. All chemicals for controlling water parameters have to be administered individually and at places with a high water flow (e.g. outlet of pumps) to avoid precipitation. Moreover, if larger quantities of chemicals have to be added to the system, this should be done over a longer period of time. The use of special software tools (e.g. AquaCalculator; http://www.aquacalculator.com) is recommended.

If nutrients in the system are removed via a skimmer/water-flow-through system, skimmer size and water-circulation rates need to be adjusted carefully. As a rule of thumb, a 100-L tank requires a water-circulation rate of at least 300 $L{\cdot}h^{-1}$. If nutrient levels in the system need to be increased, particularly in long-term experiments, the use of carefully selected associated fish species might be less risky and achieves better results than the addition of extra nutrients. Special filter systems for nutrients, which are frequently used in fish aquacultures, are not recommended for coral-microcosm systems as accidentally released substances (e.g. H_2S or NO_2^-) may jeopardize the entire system.

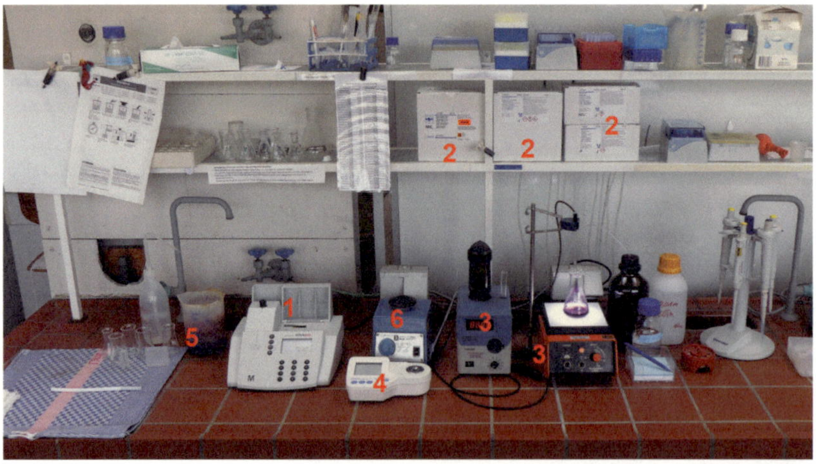

Figure 4. Laboratory workplace for seawater analyses. (1) Photometer, (2) test kits, (3) titration device with illuminated stirrer, (4) digital refractometer, (5) container for waste water and (6) lab shaker.

5. Pest and disease control

Maintaining stony corals and associated species in microcosms requires an effective pest and disease control. Particularly under experimental conditions, some usually inconspicuous organisms could become highly abundant [81], might compete with the study species for resources (e.g. light, nutrients or space) and may even prey on corals [28]. In addition, infectious diseases could be introduced into the system through animals, food and humans [17]. To minimize these risks, several procedures have to be implemented. They include a careful acclimation of study organisms, quarantine and prophylactic measures, a proper selection of associated species and the treatment of diseased corals.

When corals arrive at the facility, a slow acclimation to the new conditions is often suggested (but see the subsequent text). This is usually done via the drip method [17]. It eases acclimation stress both in the coral and its endosymbionts. After acclimation, the corals need to be inspected, unrelated organisms removed and the corals quarantined. Overlooked pest species or pest organisms introduced during the experiment should be eliminated through manual removal or chemical/biological treatment (see **Table 8**).

Moreover, a wide range of coral diseases is known [82–85] and the medication of diseased specimens is often problematic [81]. Typically, the infested tissue regions are removed, and the corals are treated with an iodine solution or other commercial products [17, 28]. Further spread of diseases may be reduced by covering infested parts with cyanoacrylate [17]. Often, the development of diseases is facilitated by inadequate water conditions. Therefore, an appropriate water exchange, filtration with activated carbon, or UV- and ozone treatments are typically used to reduce the risk of infection [86] (see also Section 2.4.). Note that UV- or ozone sterilization might unintentionally affect the exchange of symbionts among corals.

Recommendations: Experiences show that a slow-drip acclimation over ca. 30 min might be appropriate for most associated organisms. However, it might be better to directly transfer corals into fresh, temperature- and salinity-adjusted seawater (Schubert, unpublished data). Old transport water may contain high amounts of ammonium. If the pH rises during a slow acclimation process due to decreasing CO_2 levels, the proportion of toxic ammoniac will increase. This might harm the corals more than a sudden transfer into fresh seawater.

A common problem in newly arrived corals is flatworm infestations (e.g. *Amakusaplana* spp. [87]). Therefore, one treatment per week (10–20 min each) for a period of at least 2 weeks in an iodine bath is suggested. Particularly for *Acropora* spp., a prophylactic iodine treatment may be beneficial. Moreover, often it is useful to replace the complete base rock of the coral to reduce a potential parasite load. However, corals taken from the wild should be handled with particular care. At the beginning of quarantine, lighting should be dimmed to 50% and then gradually increased. After 3–4 weeks, the corals may be transferred to the experimental tanks.

Particularly in semi-closed and open microcosms, it is difficult to keep the coral tank free of pest species. They are best controlled through associated species such as fishes and invertebrates (**Table 8**). However, the former are often less suited for microcosm experiments due to their low abundances and more individual behaviour, potentially causing unintended differences among experimental tanks.

Pest species	Species/measures for pest control
Aiptasia spp. (Glass, rock or tube anemones)	*Aeolidiella stephanieae* (feeds exclusively on *Aiptasia* spp., high number of individuals needed for acute infestation or large tank, no preventive effect)
	Lysmata wurdemanni, L. rathbunae, L. seticaudata ('peppermint shrimps') (preventive effect, well suited for smaller tanks, shy species with limited radius of action, might stress corals while removing feed from tentacles)
	Chelmon rostratus (preventive effect, difficult to adapt to frozen food, may feed on other invertebrates)
Amakusaplana spp. (Acropora flatworms)	Treatment with iodine solution, fresh water or levamisol hydrochloride; removal of eggs (treatment might stress corals)
	Halichoeres cosmetus, H. marginatus, Thalassoma hardwicke, Pseudocheilinus hexataenia (effectiveness uncertain, may also prey on other invertebrates)
Convolutriloba spp. (Acoelomorph flatworms)	*Halichoeres cosmetus, Synchiropus marmoratus, S. stellatus* (effectiveness uncertain)
	Chelidonura varians (feeds exclusively on flatworms; specimens expensive—get easily sucked into pumps)
	Manual removal; treatment with freshwater or levamisole hydrochloride (dying flatworms may secrete toxic substances)
Embletonia spp. (Montipora-eating nudibranchs)	*Halichoeres cosmetus, Pseudocheilinus hexataenia* (usually effective but may also prey on other invertebrates)
Halofolliculina corallasia (Ciliate that causes the Skeletal Eroding Band syndrome)	Improvement of water conditions, freshwater or iodine treatment (effectiveness uncertain)
Heliocostoma spp. (Ciliates that may cause Rapid Tissue Necrosis—'brown jelly')	Treatment with iodine solution or fresh water; removal of infested areas; improvement of water conditions (treatment might stress corals)

Table 8. Overview of common pest species and respective species/measures for pest control.

6. Conclusions

Coral-microcosm systems offer an excellent opportunity for performing global-change simulation studies under controlled conditions. They may thus help to identify the individual and combined effects of stressors acting on reef-building corals, to better understand stress response and resilience and to identify policy implications. However, coral-microcosm experiments are a relatively new approach. To avoid a potential bias caused by unintended variations of system parameters, a broad spectrum of environmental factors has to be regulated within narrow limits. In fact, maintaining healthy conditions for corals over a long period of time remains challenging. Therefore, several problems have to be addressed during planning, setup and operation of coral microcosms, and the following key recommendations should be considered:

- The experimental setup has to be determined by the scientific question of interest, study species, associated fauna and flora, and the intended duration of the experiment; the longer the duration of the experiment, the more detailed planning is required.
- Unless natural seawater of high quality is readily available or the composition of its natural microbial community is of interest, synthetic seawater should be preferred in microcosm experiments.
- Closed coral-microcosm systems are typically only applicable for short-term experiments without the need for associated faunas, semi-closed systems are preferred for medium-term experiments without associated faunas, and open microcosm systems for long-term experiments or experiments with associated animals.
- For lighting, high-quality T5, LED or plasma lamps should be considered, though the latter will likely become more popular in the future.
- For ensuring proper water movement, water-oscillation systems should be preferred in most global-change studies as they produce a relatively homogeneous water movement, thus preventing a potential bias of the study results.
- Maintaining good calcification conditions for long-term experiments is best achieved via a calcium reactor; for smaller water volumes or short-term experiments, the 'Balling method' may be sufficient.
- Pollutants are best removed from the system via chemical filtration with activated carbon; efficient pathogen control is best done by combining an ozonizer and a UV sterilizer.
- Water-temperature regulation can be best achieved by controlling the lowest target water temperature in the system via room temperature; higher temperature in individual tanks can then be attained with heating rods; temperature values should be controlled by two independent systems.
- Several safety and control devices should be installed, including GFCIs, grounding probes, a UPS ideally in combination with an emergency power generator, sensor systems and a malfunction alarm device.
- A specific hazard assessment has to be conducted, all devices and tanks necessitate a hazard-related labelling, all staff requires safety training, an emergency plan has to be developed and regular safety checks should be performed.
- All coral study species require proper species identification, and the choice of wild versus farmed corals should take the study question into account.
- As associated species in microcosms may influence water parameters, each tank should contain the same species in the same quantities and with similar sizes; potential adverse interactions with the study species have to be considered.

- Water parameters should be usually controlled and maintained through adjusting water levels and salinity in the system, regular exchange of parts of the seawater and active control of selected water parameters.
- All microcosms require an effective pest and disease control, including a careful acclimation of study organisms, quarantine and prophylactic measures, a proper selection of associated species, and the treatment of diseased corals.

In the years to come, we expect significant advances in coral-microcosm setups. They will likely involve improved lighting and water-circulation equipment, as well as sophisticated sensor systems for the continuous control of essential water parameters. Moreover, we might see important improvements in chemical water testing, aiming at quantifying essential trace elements and some metabolic-degradation products. This, in turn, may open new possibilities for closed microcosm setups and will likely further promote the use of synthetic seawater.

However, despite all technical improvements we may see in the future, the key factors for the success of global-change microcosm experiments are well-trained and dedicated people planning, setting up and operating the system. Therefore, the authors hope that this book chapter not only helps to better understand the advantages and pitfalls of coral-microcosm experiments, and the excellent opportunities such systems provide, but also encourages the reader to utilize this fascinating methodology for answering some of the key questions mankind faces relative to global-change processes in our 'rainforests of the oceans'.

Acknowledgements

This work was supported by the CEMarin (Bogota, Colombia). We would like to thank Nina Paul (Leibniz-Zentrum für Marine Tropenforschung, Bremen, Germany), Kai Chepa (Vivarium, Staatliches Museum für Naturkunde, Karlsruhe, Germany), André Billion (pro marin, Giessen, Germany) and Thomas Tikatsch (Frankfurt Zoo, Frankfurt am Main, Germany) for the interesting discussions. Further thanks go to Ronny Schöpke (Bad Geiesbach, Germany) and Immo Gerber (Neuhausen ob Eck, Germany) for the information provided regarding LED and plasma lighting.

Author details

Patrick Schubert* and Thomas Wilke

*Address all correspondence to: Patrick.Schubert@fg.bio.uni-giessen.de

Institute of Animal Ecology and Systematics, Justus Liebig University Giessen, Giessen, Germany

References

[1] Bouchet P. The magnitude of marine biodiversity. In: Duarte C, editor. The Exploration of Marine Biodiversity: Scientific and Technological Challenges. Bilbao: FundaciÓn BBVA; 2006. pp. 32-64

[2] Reaka-Kudla ML. The global biodiversity of coral reefs: A comparison with rain forests. In: Reaka-Kudla ML, Wilson DE, Wilson EO, editors. Biodiversity II: Understanding and Protecting Our Biological Resources. Washington DC: Joseph Henry Press; 1997. pp. 83-108. Available from: https://books.google.de/books?id=-X5OAgAAQBAJ&printsec=copyright&hl=de&source=gbs_pub_info_r#v=onepage&q&f=false

[3] Knowlton N, Brainard R, Fisher R, Moews M, Plaisance L, Caley J. Coral reef biodiversity. In: McIntyre AD, editor. Life in the World's Oceans. Oxford: Blackwell Publishing Ltd.; 2010

[4] Rocha J, Peixe L, Gomes NCM, Calado R. Cnidarians as a source of new marine bioactive compounds—An overview of the last decade and future steps for bioprospecting. Marine Drugs. 2011;**9**:1860-1886. DOI: 10.3390/md9101860

[5] Lalli CM, Parsons TR. Chapter 8 – Benthic Communities. Biological Oceanography: An Introduction. 2nd ed. Oxford: Elsevier; 1997. pp. 196-246. DOI: 10.1016/B978-075063384-0/50064-5

[6] Wilkinson C, editor. Status of Coral Reefs of the World. Townsville: Australian Institute of Marine Science (AIMS); 2004

[7] Hoegh-Guldberg O, Mumby PJ, Hooten AJ, Steneck RS, Greenfield P, Gomez E, et al. Coral reefs under rapid climate change and ocean acidification. Science. 2007;**318**:1737-1742. DOI:10.3390/md9101860

[8] Hughes TP, Baird AH, Bellwood DR, Card M, Connolly SR, Folke C, et al. Climate change, human impacts, and the resilience of coral reefs. Science. 2003;**301**:929-933. DOI: 10.1126/science.1085046

[9] Baker AC, Glynn PW, Riegl B. Climate change and coral reef bleaching: An ecological assessment of long-term impacts, recovery trends and future outlook. Estuarine, Coastal and Shelf Science. 2008;**80**:435-471. DOI: 10.1016/j.ecss.2008.09.003

[10] McNeil BI, Matear RJ, Barnes DJ. Coral reef calcification and climate change: The effect of ocean warming. Geophysical Research Letters. 2004;**31**:1-4. DOI: 10.1029/2004GL021541

[11] Tambutté E, Venn AA, Holcomb M, Segonds N, Techer N, Zoccola D, et al. Morphological plasticity of the coral skeleton under CO_2-driven seawater acidification. Nature Communications. 2015;**6**:7368. DOI: 10.1038/ncomms8368

[12] Pandolfi JM, Connolly SR, Marshall DJ, Cohen AL. Projecting coral reef futures under global warming and ocean acidification. Science. 2011;**333**:418-422. DOI: 10.1126/science.1204794

[13] Breitberg D, Salisbury J, Bernhard J, Cai W-J, Dupont S, Doney S, et al. And on top of all that … coping with ocean acidification in the midst of many stressors. Oceanography. 2015;**28**:48-61 DOI: 10.5670/oceanog.2015.31

[14] Crain CM, Kroeker K, Halpern BS. Interactive and cumulative effects of multiple human stressors in marine systems. Ecology Letters. 2008;**11**:1304-1315. DOI: 10.1111/j.1461-0248.2008.01253.x

[15] Riebesell U, Fabry VJ, Hansson L, Gattuso J-P. Guide to Best Practices for Ocean Acidification Research and Data Reporting [Internet]. Luxembourg: Publications Office of the European Union; 2010 DOI: 10.2777/58454

[16] Benton TG, Solan M, Travis JMJ, Sait SM. Microcosm experiments can inform global ecological problems. Trends in Ecology & Evolution. 2007;**22**: 516-521. DOI: 10.1016/j.tree.2007.08.003

[17] Bartlett TC. Small scale experimental systems for coral research: Considerations, planning, and recommendations. NOAA Technical Memorandum NOS NCCOS 165 and CRCP 18. 2013;68 pp

[18] Cornwall CE, Hurd CL. Experimental design in ocean acidification research: Problems and solutions. International Council for the Exploration of the Sea Journal of Marine Science. 2015;**73**:572-581. DOI: 10.1093/icesjms/fsv118

[19] Adey WH, Loveland K. Chapter 1: Introduction. Dynamic Aquaria. 3rd ed. London: Academic Press. 2007. pp. 1-10. DOI: 10.1016/B978-0-12-370641-6.50010-8

[20] Wernberg T, Smale DA, Thomsen MS. A decade of climate change experiments on marine organisms: Procedures, patterns and problems. Global Change Biology. 2012;**18**:1491-1498. DOI: 10.1111/j.1365-2486.2012.02656.x

[21] Sommer U. Experimental Systems in Aquatic Ecology. eLS. Chichester, UK: John Wiley & Sons, Ltd.; 2012. DOI: 10.1002/9780470015902.a0003180.pub2

[22] Hovanec TA, Toy PDEL, Westerlund J, Coshland JL. The toxicity of synthetic sea salts and natural seawater to the development of white sea urchin (*Lytichinus pictus*) larvae. In: Advanced Aquarist [Internet]. 2005 [cited 6 Feb 2017]. Available from: http://www.advancedaquarist.com/2005/3/aafeature

[23] Gallagher ML, Brown WD. Comparison of artificial and natural sea water as culture media for juvenile lobsters. Aquaculture. 1976;**9**:87-90. DOI: 10.1016/0044-8486(76)90051-X

[24] Hemdal J. Micro-ecosystems. In: Advanced Aquarist [Internet]. 2008 [cited 6 Feb 2017]. Available from: http://www.advancedaquarist.com/2008/9/aafeature3

[25] Huguenin JE, Colt J. Chapter 4: Seawater Sources. Design and Operating Guide for Aquaculture Seawater Systems. Amsterdam: Elsevier B.V.; 2002. pp. 59-71. DOI: 10.1016/S0167-9309(02)80004-7

[26] Adey WH, Loveland K. Chapter 4: Water Composition: Management of Salinity, Hardness, and Evaporation. Dynamic Aquaria. 3rd ed. London: Academic Press. 2007. pp. 61-74. DOI: 10.1016/B978-0-12-3 70641-6.50013-3

[27] Adey WH, Loveland K. Chapter 20: Models of Coral Reef Ecosystems. Dynamic Aquaria. 3rd ed. London: Academic Press. 2007. pp. 343-369. DOI: 10.1016/B978-0-12-370641-6. 50029-7

[28] Calfo A. Book of Coral Propagation. 2nd ed. Williams C, editor. Monroeville: Reading Tree; 2007. 416 pp

[29] Kühl M, Cohen Y, Tage D, Jorgensen BB, Revsbech NP. Microenvironment and photosynthesis of zooxanthellae in scleractinian corals studied pH and light with microsensors for O_2, pH and light. Marine Ecology Progress Series. 1995;**117**:159-172

[30] Cohen I, Dubinsky Z, Erez J. Light enhanced calcification in hermatypic corals: New insights from light spectral responses. Frontiers in Marine Science. 2016;**2**:122. DOI: 10.3389/fmars.2015.00122

[31] Boeuf G, Le Bail P-Y. Does light have an influence on fish growth? Aquaculture. 1999;**177**:129-152. DOI: 10.1016/S0044-8486(99)00074-5

[32] Hunt P. Marine aquarium basics. In: TropicalFishHobbyistMagazine [Internet]. 2013 [cited 14 Feb 2017]. Available from: http://www.tfhmagazine.com/details/articles/marine-aquarium-basics-full-article.htm

[33] Riddle D, Olaizola M. Lighting the reef aquarium - spectrum or Intensity? In: Advanced Aquarist [Internet]. 2002 [cited 30 Jan 2017]. Available from: http://www.advancedaquarist.com/2002/2/aafeature

[34] Venn AA, Loram JE, Douglas AE. Photosynthetic symbioses in animals. Journal of Experimental Botany. 2008;**59**:1069-1080. DOI: 10.1093/jxb/erm328

[35] Chang S, Iaccarino G, Ham F, Elkins C, Monismith S. Local shear and mass transfer on individual coral colonies: Computations in unidirectional and wave-driven flows. Journal of Geophysical Research: Oceans. 2014;**119**:2599-2619. DOI: 10.1002/2013JC009751

[36] Wijgerde T, Henkemans P, Osinga R. Effects of irradiance and light spectrum on growth of the scleractinian coral *Galaxea fascicularis* — Applicability of LEP and LED lighting to coral aquaculture. Aquaculture. 2012;**344**:188-193. DOI: 10.1016/j.aquaculture.2012.03.025

[37] Ulstrup KE, Ralph PJ, Larkum AWD, Kühl M. Intra-colonial variability in light acclimation of zooxanthellae in coral tissues of *Pocillopora damicornis*. Marine Biology. Springer-Verlag. 2006;**149**:1325-1335. DOI: 10.1007/s00227-006-0286-4

[38] Brown BE. Coral bleaching: Causes and consequences. Coral Reefs. Springer-Verlag. 1997;**16**:129-138. DOI: 10.1007/s003380050249

[39] Schlacher TA, Stark J, Fischer ABP. Evaluation of artificial light regimes and substrate types for aquaria propagation of the staghorn coral *Acropora solitaryensis*. Aquaculture. 2007;**269**:278-289. DOI: 10.1016/j.aquaculture.2007.04.085

[40] Rocha RJM, Pimentel T, Serôdio J, Rosa R, Calado R. Comparative performance of light emitting plasma (LEP) and light emitting diode (LED) in ex situ aquaculture of scleractinian corals. Aquaculture. 2013;**402**:38-45. DOI: 10.1016/j.aquaculture.2013.03.028

[41] Adams J. Feature article: Water flow is more important for corals than light. Part II: The science of corals and water flow. In: Advanced Aquarist [Internet]. 2006 [cited 30 Jan 2017]. Available from: http://www.advancedaquarist.com/2006/8/aafeature

[42] Finelli CM, Helmuth BST, Pentcheff ND, Wethey DS. Water flow influences oxygen transport and photosynthetic efficiency in corals. Coral Reefs. Springer-Verlag. 2006;**25**:47-57. DOI: 10.1007/s00338-005-0055-8

[43] Nakamura T. Importance of water-flow on the physiological responses of reef-building corals. Galaxea, Journal of Coral Reef Studies. The Japanese Coral Reef Society. 2010;**12**:1-14. DOI: 10.3755/galaxea.12.1

[44] Reidenbach MA, Koseff JR, Monismith SG, Steinbuckc JV, Genin A. The effects of waves and morphology on mass transfer within branched reef corals. Limnology and Oceanography. 2006;**51**:1134-1141. DOI: 10.4319/lo.2006.51.2.1134

[45] Nakamura T, van Woesik R. Water-flow rates and passive diffusion partially explain differential survival of corals during the 1998 bleaching event. Marine Ecology Progress Series. 2001;**212**:301-304. DOI: 10.3354/meps212301

[46] Brown B, Bythell J. Perspectives on mucus secretion in reef corals. Marine Ecology Progress Series. 2005;**296**:291-309. DOI: 10.3354/meps296291

[47] Brander RW, Kench PS, Hart D. Spatial and temporal variations in wave characteristics across a reef platform, Warraber Island, Torres Strait, Australia. Marine Geology. 2004;**207**:169-184. DOI: 10.1016/j.margeo.2004.03.014

[48] Monismith SG. Hydrodynamics of coral reefs. Annual Review of Fluid Mechanics. 2006;**39**:37-55. DOI: 10.1146/annurev.fluid.38.050304.092125

[49] Warrick J. Advanced beginnings: The basics of water movement in the reef aquarium. In: Advanced Aquarist [Internet]. 2012 [cited 30 Jan 2017]. Available from: http://www.advancedaquarist.com/2003/4/beginner

[50] Huguenin JE, Colt J. Design and operating guide for aquaculture seawater systems. In: Developments in Aquaculture and Fisheries Science. Amsterdam: Elsevier B.V.; 2002. DOI: 10.1016/S0167-9309(13)70002-4

[51] Riddle D. Feature article: Tunze Wavebox, Turbelle stream pump, and wave controller. In: Advanced Aquarist [Internet]. 2010 [cited 30 Jan 2017]. Available from: http://www.advancedaquarist.com/2010/9/aafeature

[52] Putnam HM, Edmunds PJ. The physiological response of reef corals to diel fluctuations in seawater temperature. Journal of Experimental Marine Biology and Ecology. 2011;**396**:216-223. DOI: 10.1016/j.jembe.2010.10.026

[53] Langdon C, Takahashi T, Sweeney C, Chipman D, Goddard J, Marubini F, et al. Effect of calcium carbonate saturation state on the calcification rate of an experimental coral reef. Global Biogeochemical Cycles. 2000;**14**:639-654. DOI: 10.1029/1999GB001195

[54] Holmes-Farley R. Aquarium chemistry: The chemical and biochemical mechanisms of calcification. In: Advanced Aquarist [Internet]. 2002 [cited 14 Feb 2017]. Available from: http://www.advancedaquarist.com/2002/4/chemistry

[55] Aspinall R. Calcium reactors: Understanding the basics. In: UltraMarine Magazine [Internet]. 2016 [cited 14 Feb 2017]. Available from: http://ultramarinemagazine.co.uk/features/calcium-reactors-understanding-basics/

[56] Brockmann D, Janse M. Calcium and carbonate in closed marine aquarium systems. In: Leewis RJ, Janse M, editors. Advances in Coral Husbandry in Public Aquariums. Arnhem: Burgers' Zoo; 2008. pp. 133-142. Available from: https://www.researchgate.net/publication/228361538_Calcium_and_carbonate_in_closed_marine_aquarium_systems

[57] Holmes-Farley R. Chemistry and the aquarium: Phosphorus: algae's best friend. In: Advanced Aquarist [Internet]. 2002 [cited 14 Feb 2017]. Available from: http://www.advancedaquarist.com/2002/9/chemistry

[58] Schröder J. Potential and limitations of ozone in marine recirculating aquaculture systems—Guidelines and thresholds for a safe application [Doctoral thesis/PhD]. Kiel, Germany; 2011

[59] Huguenin JE, Colt J. Chapter 7: Pump Selection. Design and Operating Guide for Aquaculture Seawater Systems. Amsterdam: Elsevier B.V.; 2002. pp. 101-116. DOI: 10.1016/S0167-9309(02)80007-2

[60] Huguenin JE, Colt J. Chapter 14: Alarms, Monitoring and Automatic Control Systems. Design and Operating Guide for Aquaculture Seawater Systems. Amsterdam: Elsevier B.V.; 2002. pp. 193-199. DOI: 10.1016/S0167-9309(02)80014-X

[61] CITES. Checklist of CITES species [Internet]. 2014 [cited 5 Feb 2017]. Available from: http://checklist.cites.org/#/en

[62] Simpson C, Kiessling W, Mewis H, Baron-Szabo RC, Müller J. Evolutionary diversification of reef corals: A comparison of the molecular and fossil records. Evolution. 2011;**65**:3274-3284. DOI: 10.1111/j.1558-5646.2011.01365.x

[63] McIlroy SE, Gillette P, Cunning R, Klueter A, Capo T, Baker AC, et al. The effects of *Symbiodinium* (Pyrrhophyta) identity on growth, survivorship, and thermal tolerance of newly settled coral recruits. Journal of Phycology. 2016;**52**:1114-1124. DOI: 10.1111/jpy.12471

[64] Smith LD, Hughes TP. An experimental assessment of survival, re-attachment and fecundity of coral fragments. Journal of Experimental Marine Biology and Ecology. 1999;**235**:147-164. DOI: 10.1016/S0022-0981(98)00178-6

[65] Bowden-Kerby A. Low-tech coral reef restoration methods modeled after natural fragmentation processes. Bulletin of Marine Science. University of Miami – Rosenstiel School of Marine and Atmospheric Science. 2001;**69**:915-931

[66] Forsman ZH, Rinkevich B, Hunter CL. Investigating fragment size for culturing reef-building corals (*Porites lobata* and *P. compressa*) in ex situ nurseries. Aquaculture. 2006;**261**:89-97. DOI: 10.1016/j.aquaculture.2006.06.040

[67] Fossa SA, Nilsen AJ. The Modern Coral Reef Aquarium. Vol. 3. 1st ed. Bornheim: Schmettkamp; 2006. 449 pp

[68] Fossa SA, Nilsen AJ. The Modern Coral Reef Aquarium. Vol. 4. 1st ed. Bornheim: Schmettkamp; 2006. 480 pp

[69] Adey WH, Loveland K. Chapter 3 – Substrate: The Active Role of Rock, Mud, and Sand. Dynamic Aquaria. 3rd ed. London: Academic Press. 2007. pp. 43-60. DOI: 10.1016/B978-0-12-370641-6.50012-1

[70] Fossa SA, Nilsen AJ. The Modern Coral Reef Aquarium. Vol. 1. 1st ed. Bornheim: Schmettkamp; 2006. 368 pp

[71] Langdon C, Atkinson MJ. Effect of elevated pCO_2 on photosynthesis and calcification of corals and interactions with seasonal change in temperature/irradiance and nutrient enrichment. Journal of Geophysical Research. 2005;**110**:C09S07. DOI: 10.1029/2004JC002576

[72] Inoue M, Suwa R, Suzuki A, Sakai K, Kawahata H. Effects of seawater pH on growth and skeletal U/Ca ratios of *Acropora digitifera* coral polyps. Geophysical Research Letters. 2011;**38**:L12809. DOI: 10.1029/2011GL047786

[73] Ferrier-Pagès C, Godinot C, D'Angelo C, Wiedenmann J, Grover R. Phosphorus metabolism of reef organisms with algal symbionts. Ecological Monographs. 2016;**86**:262-277. DOI: 10.1002/ecm.1217

[74] Pretet C, Reynaud S, Ferrier-Pagès C, Gattuso J-P, Kamber BS, Samankassou E. Effect of salinity on the skeletal chemistry of cultured scleractinian zooxanthellate corals: Cd/Ca ratio as a potential proxy for salinity reconstruction. Coral Reefs. Springer Berlin Heidelberg. 2014;**33**:169-180. DOI: 10.1007/s00338-013-1098-x

[75] Dickson AG, Sabine CL, Christian JR. Guide to best practices for ocean CO_2 measurements. PICES Special Publication. 2007;**3**:191. DOI: 10.1159/000331784

[76] Dickson AG. Part 1: Seawater carbonate chemistry. The Carbon Dioxide System in Seawater: Equilibrium Chemistry and Measurements. Guide to Best Practices for Ocean Acidification Research and Data Reporting. Publications Office of the European Union. 2010. pp. 17-40. DOI: 10.2777/66906

[77] Bockmon EE, Frieder CA, Navarro MO, White-Kershek LA, Dickson AG. Technical note: Controlled experimental aquarium system for multi-stressor investigation of carbonate chemistry, oxygen saturation, and temperature. Biogeosciences. Copernicus GmbH. 2013;**10**:5967-5975. DOI: 10.5194/bg-10-5967-2013

[78] Fangue NA, O'Donnell MJ, Sewell MA, Matson PG, MacPherson AC, Hofmann GE. A laboratory-based, experimental system for the study of ocean acidification effects on marine invertebrate larvae. Limnology and Oceanography: Methods. 2010;**8**:441-452. DOI: 10.4319/lom.2010.8.441

[79] Adey WH, Loveland K. Chapter 9: The Primary Nutrients – Nitrogen, Phosphorus, and Silica: Limitation and Eutrophication. Dynamic Aquaria. 3rd ed. London: Academic Press. 2007. pp. 131-140. DOI: 10.1016/B978-0-12-370641-6.50018-2

[80] Adey WH, Loveland K. Chapter 12: Community Structure: Biodiversity in Model Ecosystems. Dynamic Aquaria. 3rd ed. London: Academic Press. 2007. pp. 173-189. DOI: 10.1016/B978-0-12-370641-6.50 021-2

[81] Borneman E. Introduction to the husbandry of corals in aquariums: A review. In: Leewis RJ, Janse M, editors. Advances in coral husbandry in public aquariums. Arnhem: Burgers' Zoo; 2008. pp. 3-14

[82] Cróquer A, Bastidas C, Lipscomp D, Rodríguez-Martínez RE, Jordan-Dahlgren E, Guzman HM. First report of folliculinid ciliates affecting Caribbean scleractinian corals. Coral Reefs. Springer-Verlag. 2006;**25**:187-191. DOI: 10.1007/s00338-005-0068-3

[83] Luna GM, Bongiorni L, Gili C, Biavasco F, Danovaro R. *Vibrio harveyi* as a causative agent of the White Syndrome in tropical stony corals. Environmental Microbiology Reports. Blackwell Publishing Ltd. 2009;**2**:120-127. DOI: 10.1111/j.1758-2229.2009.00114.x

[84] Sweet M, Jones R, Bythell J. Coral diseases in aquaria and in nature. Journal of the Marine Biological Association of the United Kingdom. 2011;**92**:791-801. DOI: 10.1017/S0025315411001688

[85] Thurber RLV, Correa AMS. Viruses of reef-building scleractinian corals. Journal of Experimental Marine Biology and Ecology. 2011;**408**:102-113. DOI: 10.1016/j.jembe.2011.07.030

[86] Huguenin JE, Colt J. Chapter 13: Disinfection. Design and Operating Guide for Aquaculture Seawater Systems. Amsterdam: Elsevier B.V.; 2002. pp. 183-192. DOI: 10.1016/S0167-9309(02)80013-8

[87] Hume BCC, D'Angelo C, Cunnington A, Smith EG, Wiedenmann J. The corallivorous flatworm *Amakusaplana acroporae*: An invasive species threat to coral reefs? Coral Reefs. Springer Berlin Heidelberg. 2014;**33**:267-272. DOI: 10.1007/s00338-013-1101-6

Chapter 9

Coral Reef Resilience Index for Novel Ecosystems: A Spatial Planning Tool for Managers and Decision Makers - A Case Study from Puerto Rico

Edwin A. Hernández-Delgado, Sonia Barba-Herrera, Angel Torres-Valcárcel, Carmen M. González-Ramos, Jeiger L. Medina-Muñiz, Alfredo A. Montañez-Acuña, Abimarie Otaño-Cruz, Bernard J. Rosado-Matías and Gerardo Cabrera-Beauchamp

Additional information is available at the end of the chapter

http://dx.doi.org/10.5772/intechopen.71605

Abstract

Timely information is critical for coral reef managers and decision-makers to implement sustainable management measures. A Coral Reef Resilience Index (CRRI) was developed with a GIS-coupled decision-making tool applicable for Caribbean coral reef ecosystems. The CRRI is based on a five-point scale parameterized from the quantitative characterization of benthic assemblages. Separate subindices such as the Coral Index, the Threatened Species Index, and the Algal Index also provide specific information regarding targeted benthic components. This case study was based on assessments conducted in 2014 on 11 reef sites located across 3 geographic zones and 3 depth zones along the southwestern shelf of the island of Puerto Rico, Caribbean Sea. There was a significant spatial and bathymetric gradient ($p < 0.05$) in the distribution of CRRI values indicating higher degradation of inshore reefs. Mean global CRRI ranged from 2.78 to 3.17 across the shelf, ranking them as "fair." The Coral Index ranged from 2.60 to 3.76, ranking reefs from "poor" to "good," showing a general cross-shelf trend of improving conditions with increasing distance from pollution sources. Turbidity and ammonia were significantly correlated to CRRI scores. Multiple recommendations are provided based on coral reef conditions according to observed CRRI rankings.

Keywords: benthic community structure, coral reefs, Coral Reef Resilience Index (CRRI), Caribbean Sea, Puerto Rico, ecosystem health, marine management, marine biodiversity, novel ecosystems, conservation, coral bleaching, tropical ecosystems

1. Introduction

1.1. The emergence of novel ecosystems

Coral reefs across regional to global scales are showing unequivocal signs of decline. The long-term combined impacts of local human-driven factors, such as land-based source pollution (LBSP), water quality decline and overfishing, as well as large-scale climate change-related factors, such as massive coral bleaching, coral disease outbreaks, and mass coral mortalities, have resulted in a large-scale alteration of coral reef community dynamics and in the irreversible demise of coral assemblages [1–4]. These have resulted in a net coral reef decline and in often irreversible benthic community regime shifts [5–9], with significant impacts on multiple coral and fish functional groups [10]. These alterations might impair considerable coral reef ecosystem functions. Three massive coral bleaching events occurred across the northeastern Caribbean region in 1987, 1998, and 2005. But the 2005 sea surface warming episode and massive coral reef bleaching event caused an unprecedented coral mortality episode across the northeastern Caribbean region, including P.R., that mostly impacted large reef-building taxa [11–14]. More than a decade later, there is still no net recovery among many of the impacted coral species, and reef communities have followed a significantly different trajectory resulting in the emergence of novel ecosystems largely dominated by ephemeral coral species [15] and macroalgal growth [16–18]. Although such impacts have been well documented, long-term impacts associated to the emergence of novel benthic assemblages on reef functions, values, and benefits still remain largely unknown. Such rapidly changing reefs have been deemed as unhealthy. However, there are still no clear definitions of what exactly is a healthy reef.

Large-scale declines in Caribbean coral reef fish communities have also been documented across fishery target species, mostly resulting from long-term fishing effects [19, 20], but also across multiple nontarget taxa resulting from large-scale, long-term coral reef habitat decline and flattening [21, 22]. Coral cover and topographic complexity are critical components of habitat structure for supporting diverse fish assemblages and must be managed accordingly [23–25]. Evidence from a multiplicity of fish assemblage data sets across the Caribbean suggests that specialist reef fish species have largely declined across very large spatial scales, implying the large-scale nature of reef decline and its negative consequences on multiple fish taxa [22, 25]. Highly altered novel ecosystems have emerged from largely declining benthic communities. Novel ecosystems can be defined as: "*ecosystems containing new combinations of species that arise through human action, environmental change, and the impacts of the deliberate and inadvertent introduction of species from other regions. Novel ecosystems (also termed 'emerging ecosystems') result when species occur in combinations and relative abundances that have not occurred previously within a given biome. Key characteristics are novelty, in the form of new species combinations and the potential for changes in ecosystem functioning, and human agency, in that these ecosystems are the result of deliberate or inadvertent human action*" [26]. Novel coral reef ecosystems have emerged out of the dramatic changes in benthic community trajectory that have followed long-term reef decline and slowly evolving regime shifts, favoring macroalgal and nonreef building taxa dominance [27]. Coral reefs across regional and global scales are showing unequivocal signs of distress, with the emergence of novel assemblages of multiple taxa, including corals, algae, sponges, fish, and seagrasses. Such significant regime shifts have pushed out many coral reefs beyond the point of recovery. Hobbs et al. [28] suggested that *these novel systems will require significant revision of conservation and restoration norms and practices away from the traditional place-based focus on existing or historical assemblages*. But how much have such changes impacted ecosystem functions, resilience, benefits, and values is still poorly understood due to the lack of appropriate indicators of reef condition. This information is essential for reef managers and decision-makers.

1.2. The concept of "coral reef health" in the context of novel ecosystems

One fundamental challenge is still the need to develop an operational/functional definition of "coral reef health," particularly in the context of novel ecosystems. According to McField and Kramer [29], a healthy reef would be "*the presence of indicator species*," "*maintaining key processes like herbivory*," "*having higher fishing catches/landings*," or even "*just looking like it did in years past*." These seem to be obvious indicators of reef health. But there is not an exact definition relying on a single indicator species, taxa, or group due to the highly variable nature of coral reefs. For instance, a coral reef with high fish species richness, abundance, or biomass may appear to be healthy, but if its living coral cover is very low, then it may not, depending on which indicator we use. Therefore, the definition of reef health must incorporate a suite of indicator variables and then combine and weight them in such a way that a more holistic index can be defined to rank a coral reef as healthy, fair, or unhealthy. A more holistic definition of a healthy reef was provided by McField and Kramer [29]: "*A reef is healthy if it maintains its structure and function and allows for the fulfillment of reasonable human needs*." Alternatively, we suggest a broader definition: *A reef is healthy if it maintains its structure, function, and self-replenishing capacity, if it can naturally recover from disturbance, and if it can maintain its natural connectivity with other ecosystems and allows for the fulfillment of reasonable human needs*. In this sense, the interaction of six factors can influence reef health (**Figure 1**). These include (1) ecosystem structure, (2) ecosystem processes, (3) connectivity, (4) human well-being, (5) governance, and (6) drivers of change.

The interaction of multiple processes is fundamental for maintaining reef health, including maintaining biodiversity, community structure, habitat extent, and abiotic factors (e.g., low sediment inputs, water quality, and sea surface temperature). Also, coral condition, reproduction, and recruitment success, high reef accretion:bioerosion rates (a positive carbon budget balance), and herbivory are important. Maintaining functional terrestrial-marine, genetic, ecological, and energetic connectivity is vital to support high productivity. In addition, a healthy reef should contribute to support human health (e.g., through food protein), local economy and livelihoods (e.g., fisheries, tourism-based businesses, coastal protection, and pharmacological products), and culture (e.g., traditional artisanal fisheries and other uses). Governance is a critical factor for sustaining healthy reefs, particularly if appropriate and operational public policies are fully implemented and supported by a strong legal framework and enforcement. However, the lack of available human resources (e.g., natural resource managers, scientific staff, enforcement officers) is central for governance efficiency. Finally, a combination of local, regional, and global drivers of change will determine reef health, including

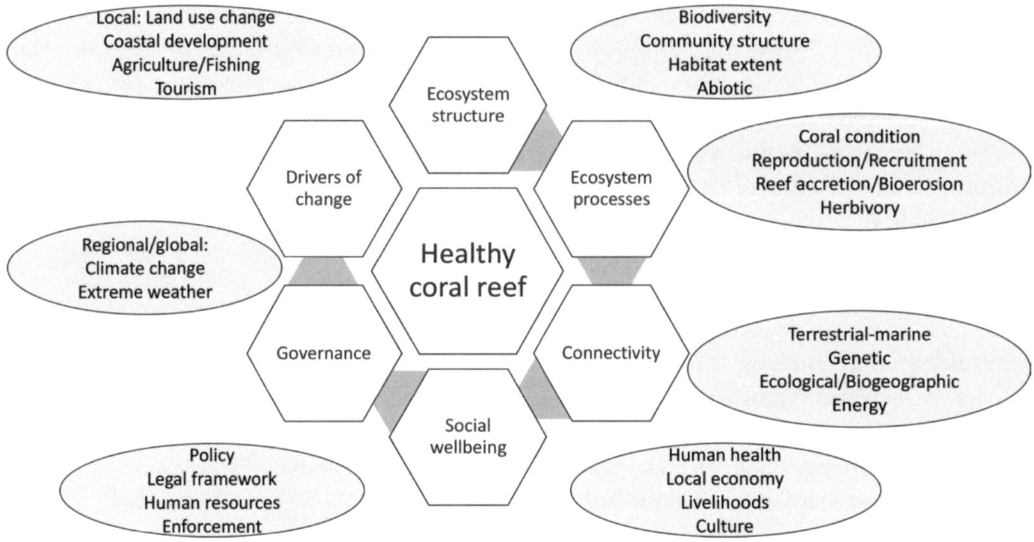

Figure 1. Conceptual model of factors affecting coral reef health.

factors that operate on different spatiotemporal scales. This may include local factors such as land use changes, tourism, agriculture and fishing, and regional/global factors such as climate change and extreme weather events.

As more of the Earth becomes transformed by human actions, novel ecosystems increase in importance, but these still remain barely studied. In the particular case of emergent novel coral reefs, their impact on fish assemblages or whether these new systems are persistent over large spatial and temporal scales still remains largely unknown. Also, how such alteration can affect ecosystem functions, resilience, benefits, and values remains poorly understood. There is also limited information with regard to novel reef ecosystem's health and how reef health responds to gradients of human pressure. It might be difficult or costly to return such systems to their previous state, and hence consideration needs to be given to developing appropriate *real-time* metrics applied to develop, modify, or adapt management goals and conservation approaches through the fine-tuning and implementation of coral reef health indices. This would provide rapid and effective tools for managers and decision-makers, information that would be critical to adapt management plans to face increasing climate change–related threats.

1.3. The development of coral reef health indices

There are multiple known attempts to implement indices to address reef health [30, 31]. Most classical examples of indicator parameters are based on single indicators such as percent live coral cover [32], the Mortality Index [33], the ratio between living and dead corals [34], or the size–frequency distribution of corals, with emphasis on estimating the proportion of small corals, which may indicate recruitment [35, 36]. There is also the Deterioration Index, which is based on the ratio between mortality and recruitment rates of branching corals [37]. Crosby and Reese [38] proposed an index for Pacific coral reefs using butterflyfish diversity as a bioindicator of reef condition. Edinger et al. [39] proposed the use of coral growth rates as indicators of eutrophication impacts. Holmes et al. [40] proposed the use of branching coral rubble bioerosion as indicators of reef trophic condition. Lirman et al. [41] suggested the use of percent recent mortality as indicators of reef adverse conditions. Edinger and Risk [42] also suggested the pattern of coral morphotypes as indicators of Pacific coral reef condition. Jameson et al. [43] developed a Coral Damage Index (CDI) based on the abundance of broken coral and coral rubble to address SCUBA diving impacts on reefs. Hawkins et al. [44] also developed a method to assess coral fragmentation and overall reef condition across reefs impacted by SCUBA diving. Swain et al. [45] developed a coral taxon–specific bleaching response index (taxon-BRI) by averaging taxon-specific response over all sites where a taxon was present. Nonetheless, the most significant limitation of methods based on a single or few bioindicators is that many of them can have significant variability due to factors that may not necessarily reflect changes in reef health. This suggests the need to use a combination of parameters to improve the accuracy of reef condition assessments.

Jokiel and Rodgers [46] used reef fish biomass, reef fish endemicity, total living coral cover, population of the endangered Hawaiian monk seal (*Monachus schauinslandi*), and the number of female green sea turtles (*Chelonia mydas*) nesting annually on each Hawaiian island as bioindicators, developing a simple integrated, composite scoring and ranking system. Rodgers et al. [47] further expanded this approach by integrating 46 different indicators, developing a reference site model and an ecological gradient model to assess impacts on coral reefs. Kaufman et al. [48] also developed the Coral Health Index aimed at assessing the condition of benthic fish and microbial communities. Lasagna et al. [49] developed the Coral Condition Index, which was based on the proportional abundance of coral colonies belonging to six categories: recently dead, bleached, smothered, upturned, broken, and healthy. This index ranges from 0 (100% of dead corals) to 1 (100% of healthy corals), with low values suggesting large scale disturbances (e.g., climate impacts) and high values suggesting disturbances acting on a small scale. Jameson et al. [50], Fore et al. [51], and Bradley et al. [52] suggested the development of a multiparameter Coral Reef Biocriteria Index for addressing coral reef ecological condition. Fabricius et al. [53] tested the use of 38 indicators, where 33 of them (including coral physiology, benthic composition, coral recruitment, macrobioeroder densities, and a foraminifera index) showed significant relationships with a composite index of 13 water quality variables. However, many of these methods based on multiple parameters, although scientifically robust, can be significantly complex and difficult to implement by non-academic personnel (e.g., managers, NGOs, and base communities). Thus, there is still a need to develop robust yet simple methods with multiple potential applications and which can be used by a wide range of users.

Risk et al. [54] suggested the use by coastal communities of simple techniques that have been shown to identify stress on reefs including coral mortality indices, benthic bioindicators (e.g., stomatopods, foraminifera, and amphipods), coral associate counts, and coral rubble bioerosion. McField and Kramer [29, 55] developed the Coral Reef Health Index (CRHI) in the Mesoamerican Barrier System based on assessing several parameters of benthic and fish assemblages. This method has been successfully used across the Caribbean [56–59]. McField and Kramer [60] summarized a set of multiple simple criteria to be used by coastal communities. In a comparative study between two reef health indices and different metrics of biological, ecological, and functional diversity of fish and corals, Díaz-Pérez et al. [61] found out that health indices should be complemented with classic community indices to improve the accuracy of the estimated health status of Caribbean coral reefs. This brings in the idea that coral reef health indices must be made more robust by complementing them with a suite of biological and water quality parameters often easily obtained from standard reef characterization and long-term monitoring data sets.

According to Ben-Tzvi et al. [37], any broad-based reef health index monitoring should *(1) enable reliable comparison between different reef types (e.g., reefs of different live cover); (2) be simple to apply, including by nonscientific personnel (e.g., recreational divers); (3) provide an indication of the trend in reef health rather than only the current state of the reef; (4) provide a quantitative, or at least semiquantitative, indication of the reef state, to enable comparisons between distinct reefs of different characters; and (5) not require repeated serial surveys, but be able to provide some indication of the state of the health of the reef from a single survey event*. An easy-to-implement rapid assessment method of novel coral reef assemblages was tested, in combination with a rapid diagnostic tool of reef condition useful for managers and decision-makers for both small- and large-scale assessments, which could also be implemented in standard long-term monitoring programs.

1.4. Goals and objectives

The goal of this chapter is to test an easy-to-implement rapid assessment, reef characterization, and decision-making tool for coral reef managers. Many countries, particularly, small island nations, with limited socioeconomic resources, lack appropriate governance infrastructure, human resources, and economic and technological tools to incorporate scientific information into decision-making regarding the management of coral reefs and fishery resources. The lack of appropriate management is a critical concern in the face of current and forecasted climate change–impacts. Coral reefs are often the first line of defense against storm swells and sea level rise, besides their importance as a source of food protein, for sustaining biodiversity, as a sinkhole of ATM CO_2, as a source of natural products of biomedical importance, and as a source of revenue for multiple local economies. Coral reef conservation becomes particularly important in novel coastal ecosystems adjacent to large urban centers, subjected to significant local sources of human stressors. We propose the application of a Coral Reef Resilience Index (CRRI) focused on scoring the ecological condition of coral reef benthic and fish communities, based on actual quantitative data obtained from ecological characterization surveys or from long-term monitoring efforts. Complex quantitative data, difficult to analyze and interpret, are changed into a five-point scale scoring system, similar to the one developed by McField and Kramer [29], and also converted into GIS-based format to produce a set of indicator maps. This will provide managers with easy-to-interpret tools for decision-making regarding conservation- and restoration-oriented management strategies. A step-by-step guide for the implementation of the tool is discussed. This chapter also provides a case study from coral reefs across a water quality stress gradient from the Southwestern Puerto Rico shelf and provides a basic guide for management recommendations based on different scores of the CRRI with application across multiple coral reef ecosystems on a global scale.

2. Methods

2.1. Study sites

Field data used to parameterize the CRRI were obtained from a study of coral reef condition across a water quality stress gradient through the southwestern Puerto Rico insular shelf during the month of July 2014 [62]. Sampling was conducted at 11 locations along a water quality stress gradient and a distance gradient from the coast (**Figure 2**). Coral reefs were subdivided into three different geographic zones: (1) inshore reefs [<4 km] (Punta Ostiones [OST], Punta Lamela [LAM], Punta Guaniquilla [GUA], Cayo Ratones [RAT], Bajo Enmedio [EME]), (2) mid-shelf reefs [4–8 km] (Arrecife Resuello [RES], Corona del Norte [CON], Arrecife El Ron [RON]), and (3) outer-shelf reefs [8–20 km] (Escollo El Negro [NEG], Arrecife Papa San [PPS], Arrecife Gallardo [GAL]). A total of 55% of the studied reefs were located within natural reserves managed by the Puerto Rico Department of Natural and Environmental Resources (DNER), including inshore location RAT (Isla Ratones Natural Reserve), OST (Finca Belvedere Natural Reserve Marine Extension), and GUA (Punta Guaniquilla Natural Reserve Marine Extension). Mid-shelf locations RON and CON, and outer-shelf location PPS

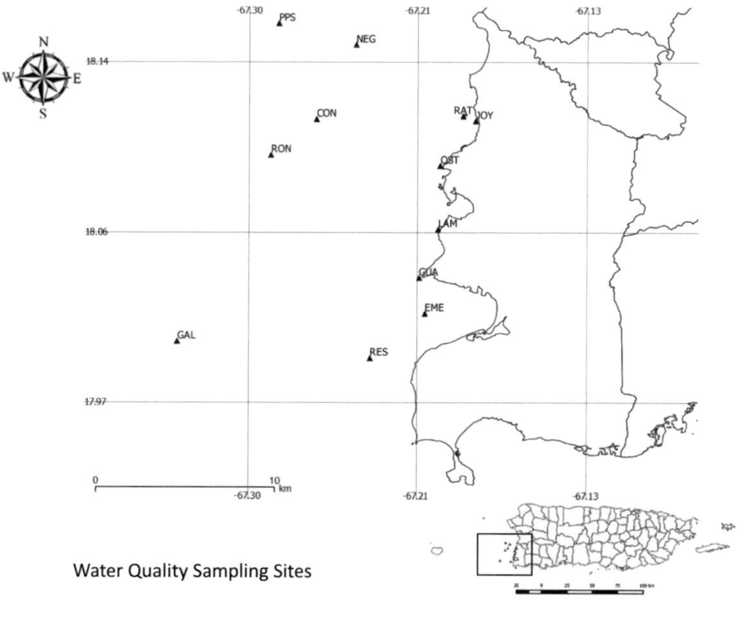

Figure 2. Locations of study sites through the southwestern Puerto Rico insular platform. These were divided into three geographic areas: inshore reefs (<4 km)—Cayo Ratones (RAT), Punta Ostiones (OST), Punta Lamela (LAM), Punta Guaniquilla (GUA), Bajo Enmedio (EME); mid-shelf reefs (4–8 km)—Arrecife Resuello (RES), Corona del Norte (CON), El Ron (RON); and outer-shelf reefs (8–20 km)—Escollo El Negro (NEG), Arrecife Papa San (PPS), Bajo Gallardo (GAL). Acronyms of protected areas: BEB = Bosque Estatal de Boquerón; CRNWR = Cabo Rojo National Wildlife Refuge; EMRNFB = Extensión Marina Reserva Natural Finca Belvedere; EXRNPG = Extensión Marina Reserva Natural Punta Guaniquilla; EMBEB = Extensión Marina Bosque Estatal Boquerón; RVSIAB = Refugio de Vida Silvestre y de Aves de Boquerón; RNAT = Reserva Natural Arrecifes Tourmaline; RNCR = Reserva Natural Cayo Ratones; RNFB = Reserva Natural Finca Belvedere; RNLJ = Reserva Natural Laguna Joyuda; RNPG = Reserva Natural Punta Guaniquilla. Gray-shaded areas in the left image represent coral reefs.

were located within Arrecife's Tourmaline Natural Reserve, which has a six-month seasonal fishing closure (December 1–May 31). Other studied reserve and nonreserve locations are open to fishing.

2.2. Sampling design

With the exception of inshore locations OST, LAM, GUA, and RAT, characterized only by shallow areas, each remaining locality was subdivided into three depth zones: depth 1 (<5 m), depth 2 (5–10 m), and depth 3 (10–20 m). Only depths 1 and 2 were studied in EME, and depth 3 and depth 4 (20–30 m) were studied in PPS. In each of these depths, from 5 to 15 random belt phototransects (10 × 1 m) were studied by taking 5 high-resolution, nonoverlapping, digital images of 1.0 × 0.7 m per transect at fixed intervals, obtaining a total of 25–75 images per depth zone from each location. A 48-point dot grid was digitally projected over each image and benthic components under each point were identified to the lowest taxon possible (e.g., Scleractinian corals, hydrocorals, octocorals, sponges, algal functional groups, cyanobacteria, and open substrate [sand, rubble, and pavement]). The relative number of points per category was counted and divided by the total number of points to obtain the percentage of coverage of the benthic components.

2.3. Coral Reef Resilience Index (CRRI)

A modification and expansion of McField and Kramer [60] and NEPA [63] was used to define CRRI's parameters. An average index score for each indicator listed in **Table 1** was calculated for each individual transect, depth zone, and location and compared to threshold value ranges listed in the table. CRRI rankings were similar to those defined by McField and Kramer [60], with a scale of 1–5 points as follows: 5 = very good, 4 = good, 3 = fair, 2 = poor, and 1 = critical. Four different indices were calculated: (1) Global Index = an average of all the parameters; (2) Coral Index = an average of all coral parameters; (3) Threatened Species Index = an average of all threatened coral parameters; and (4) Algal Index = an average of all algal parameters. Mean scores were calculated for all four indices, for each geographic zone and location and for each depth zone. The final mean value of each index is deemed as very good (4.2–5), good (3.4–4.2), fair (2.6–3.4), poor (1.8–2.6), and critical (1–1.8).

Fifteen indicators were selected to calculate the benthic index (**Table 1**). In the coral index, percentage of living tissue coverage, species richness, coral recruit density (diameter < 5 cm), and percentage of bleaching frequency were used. In the Threatened Species Index, based on the International Union for the Conservation of Nature (IUCN) Red List and on the U.S. Endangered Species Act listed coral species, the following species were used: Staghorn coral (*Acropora cervicornis*), Elkhorn coral (*A. palmata*), Columnar star columnar coral (*Orbicella annularis*), and Laminar star coral (*O. faveolata*). Of the seven threatened species in the Caribbean, these were the most common species throughout the study areas [62]. In the Algal Index, macroalgae, turf, crustose coralline algae (CCA), *Halimeda* spp., *Dictyota* spp., *Lobophora*

http://dx.doi.org/10.5772/intechopen.71605

Indices	Very good(5)	Good(4)	Fair(3)	Poor(2)	Critical(1)
Coral Index					
% Coral cover	>40%	20–39.9%	10–19.9%	5–9.9%	<5%
Species richness	>10	7–9.9	5–6.9	3–4.9	<2.9
Recruitment density (#/m^2)	>10	5–9.9	3–4.9	2–2.9	<2
% Bleaching	0%	<2%	2–9.9%	10–50%	>50%
Threatened Species Index					
Acropora cervicornis	>20%	10–19.9%	5–9.9%	2–4.9%	<2%
Acropora palmata	>20%	10–19.9%	5–9.9%	2–4.9%	<2%
Orbicella annularis	>40%	20–39.9%	10–19.9%	5–9.9%	<5%
Orbicella faveolata	>40%	20–39.9%	10–19.9%	5–9.9%	<5%
Algal Index					
Macroalgae	<10%	10–19.9%	20–39.9%	40–59.9%	>60%
Turf	<10%	10–19.9%	20–39.9%	40–59.9%	>60%
Crustose coralline algae	>30%	20–29.9%	10–19.9%	5–9.9%	<5%
Halimeda spp.	<5%	5–9.9%	10–19.9%	20–29.9%	>30%
Dictyota spp.	<5%	5–9.9%	10–19.9%	20–29.9%	>30%
Lobophora variegata	<5%	5–9.9%	10–19.9%	20–29.9%	>30%
Ramicrusta/Peyssonnelia	<5%	5–9.9%	10–19.9%	20–29.9%	>30%

Table 1. Benthic community indicators, with their corresponding CRRI scores.

variegata, and red encrusting algae *Ramicrusta* spp./*Peyssonnelia* spp. (species that can overgrow living corals) were used.

2.4. Statistical testing

A three-way permutational analysis of variance (PERMANOVA) was used to test the null hypothesis of no significant difference in CRRI scores among geographic zones, locations, and depth zones [64]. Multivariate tests were carried out in statistical package.

PRIMER v7 + PERMANOVA 1.06 (PRIMER-e, Auckland, New Zealand). Scores were $\log_{10}$-transformed and Bay-Curtis similarity resemblance matrices were calculated for each individual index. Nonmetric multidimensional scaling (nMDS) was used to illustrate spatial pattern of mean scores of each index [65]. A 'linkage tree' of coral reef benthic community structure based on the BIOENV routine to environmental variables was also carried out to determine the influence of environmental variables on the spatial patterns of benthic community structure and thus on the CRRI.

3. Results

3.1. Water quality stress gradients

Water turbidity showed a highly significant decline with increasing distance from the shoreline ($r^2 = 0.7119$; $p = 0.0006$), suggesting a strong cross-shelf spatial gradient. Turbidity was significantly different among geographic zones ($p < 0.0001$) and among locations ($p < 0.0001$). The zone × location interactions were also significant ($p < 0.0001$). Higher mean values across inshore locations showed a range from 1.0 to 3.8 NTU (**Figure 3**). Mid-shelf locations averaged 0.9–1.0 NTU, and outer-shelf locations averaged 0.4–0.9 NTU. Turbidity patterns show often complex spatial and temporal variability across the western shelf due to complex circulation patterns.

There was also a highly significant ($r^2 = 0.4961$; $p = 0.0458$) nonlinear decline in ammonia (NH_3^+) and increasing distance from the shore (**Figure 4**), suggesting a similar strong cross-shelf spatial

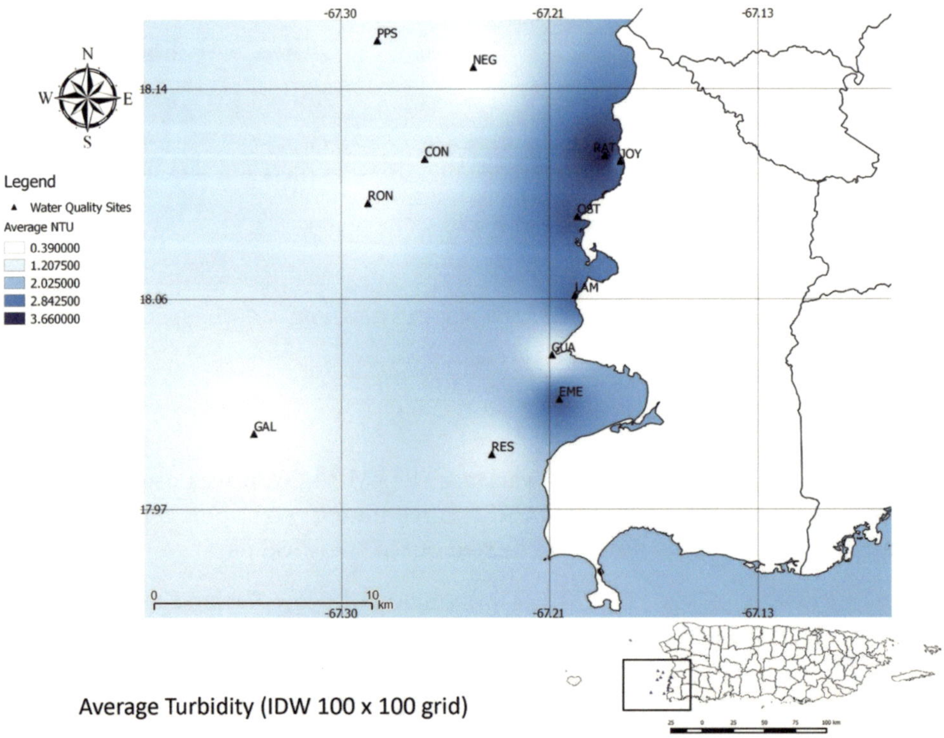

Figure 3. GIS-based inverse distance weighting (IDW) interpolation showing water turbidity spatial patterns. For location acronyms refer to **Figure 2**.

http://dx.doi.org/10.5772/intechopen.71605

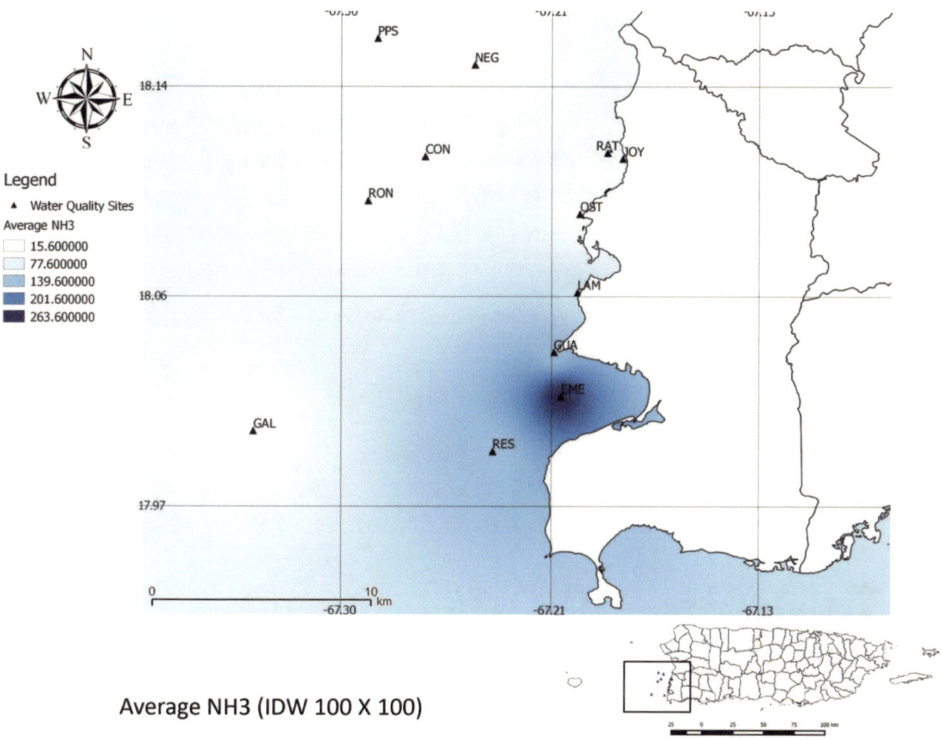

Figure 4. GIS-based inverse distance weighting (IDW) interpolation showing ammonia (NH_3^+) concentration spatial patterns. For site acronyms refer to **Figure 2**.

gradient. NH_3^+ was significantly different among geographic zones ($p < 0.0001$) and among locations ($p < 0.0001$). The geographic zone × location interaction was also significant ($p < 0.0001$). NH_3^+ concentrations showed large spatial variability, with inshore locations ranging from 25 to 264 μM. Mid-shelf locations ranged from 22 to 133 μM, and outer-shelf sites ranged from 15 to 16 μM. EME (264 μM), GUA (136 μM), and RES (133 μM), which are located just off Boquerón Bay and are known to receive recurrent raw sewage illegal discharges and poorly treated sewage effluents from a malfunctioning treatment facility from Boquerón Bay, showed the highest NH_3^+ concentrations. NH_3^+ concentration at nearby, sewage-polluted LAM, located just off Puerto Real, showed a concentration of 94 μM, which is also considered very high.

3.2. Global Coral Reef Resilience Index (CRRI)

A significant cross-shelf increase ($p < 0.0001$) was observed in the mean global CRRI score in coral reefs (**Figure 5a**, **Table 2**). Mean global CRRI across inshore sites was 2.83, with a range of 2.79–2.90 (**Table 3**). The average on the mid-shelf reefs was 3.04 with a range of 2.88–3.20. Meanwhile, the average reef at the outer shelf was 3.12, with a range of 3.00–3.26. The global CRRI spatial gradient was evident (**Figure 6**). Differences among geographic zones, locations, and depth zones were highly significant ($p < 0.0001$). All possible interaction combinations were also significant. However, cross-shelf mean values of global CRRI ranked all locations as "fair."

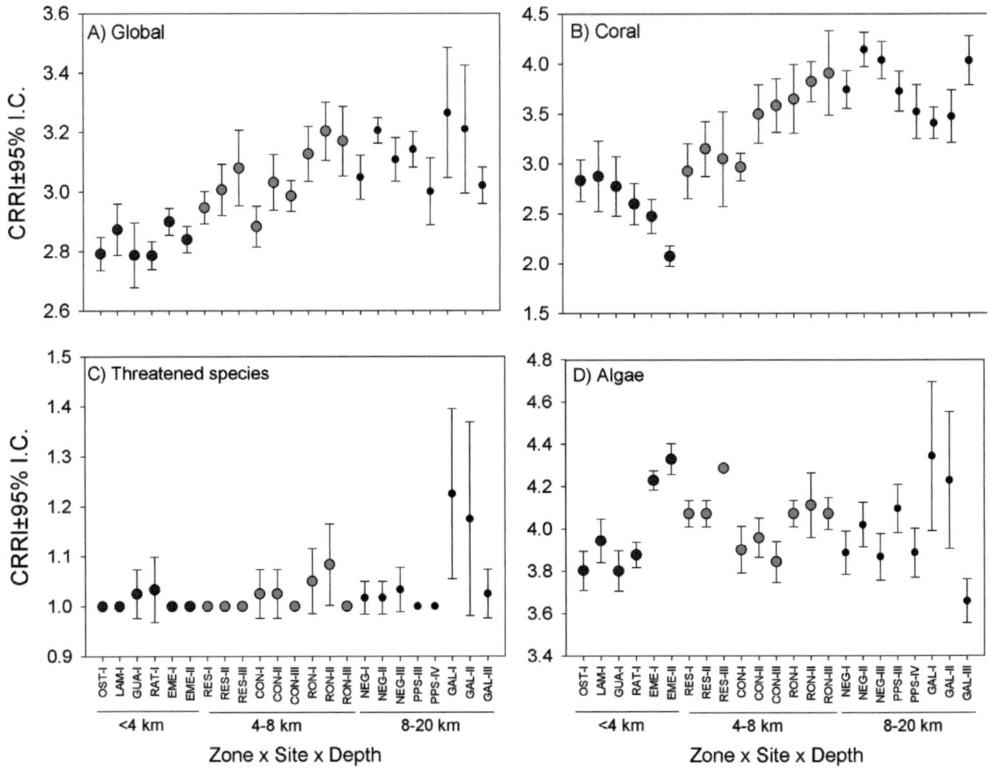

Figure 5. Coral Reef Resilience Index: (A) Global Index; (B) Coral Index; (C) Threatened Coral Species Index; and (D) Algal Index. Mean ± 95% confidence intervals. For site acronyms refer to **Figure 2**.

Variable	d.f.	Global CRRI	Coral Index	Threatened Species Index	Algal Index
Geographic zone (Z)	2254	41.85<0.0001	115.5<0.0001	2.980.0469	1.010.3651
Location (L)	10,246	10.96<0.0001	35.31<0.0001	4.290.0006	7.01<0.0001
Depth (D)	3253	9.73<0.0001	22.30<0.0001	1.530.1910	5.690.0014
Z × L	10,246	10.96<0.0001	35.31<0.0001	4.290.0009	7.01<0.0001
Z × D	8248	13.49<0.0001	36.53<0.0001	2.660.0124	4.75<0.0001
L × D	22,234	7.42<0.0001	19.97<0.0001	3.160.0003	7.00<0.0001
Z × L × D	22,234	7.42<0.0001	19.97<0.0001	3.160.0005	7.00<0.0001

Table 2. Summary of a three-way PERMANOVA on global CRRI. Pseudo-F value and statistical probability.

The nMDS analysis showed a spatial pattern confirming a significant cross-shelf gradient of global CRRI (stress = 0.01) (**Figure 7**). Three clustering patterns were observed. The first cluster was dominated by locations across the inshore geographic zone. The second cluster was a mixed group of some inshore and mid-shelf reefs. The third mixed group was composed of some mid-shelf and outer-shelf reefs. The location with the highest global CRRI value was GAL (depth I)

Zone	Global CRRI	Coral Index	Threatened Species Index	Algal Index
Entire shelf	3.02 (fair)	3.32 (fair)	1.03 (critical)	4.01 (good)
Inshore	2.83 (fair)	2.60 (poor)	1.01 (critical)	4.00 (good)
Mid-shelf	3.05 (fair)	3.40 (fair)	1.02 (critical)	4.04 (good)
Outer shelf	3.13 (fair)	3.76 (good)	1.06 (critical)	4.00 (good)

Table 3. Mean CRRI values across the western Puerto Rican shelf.

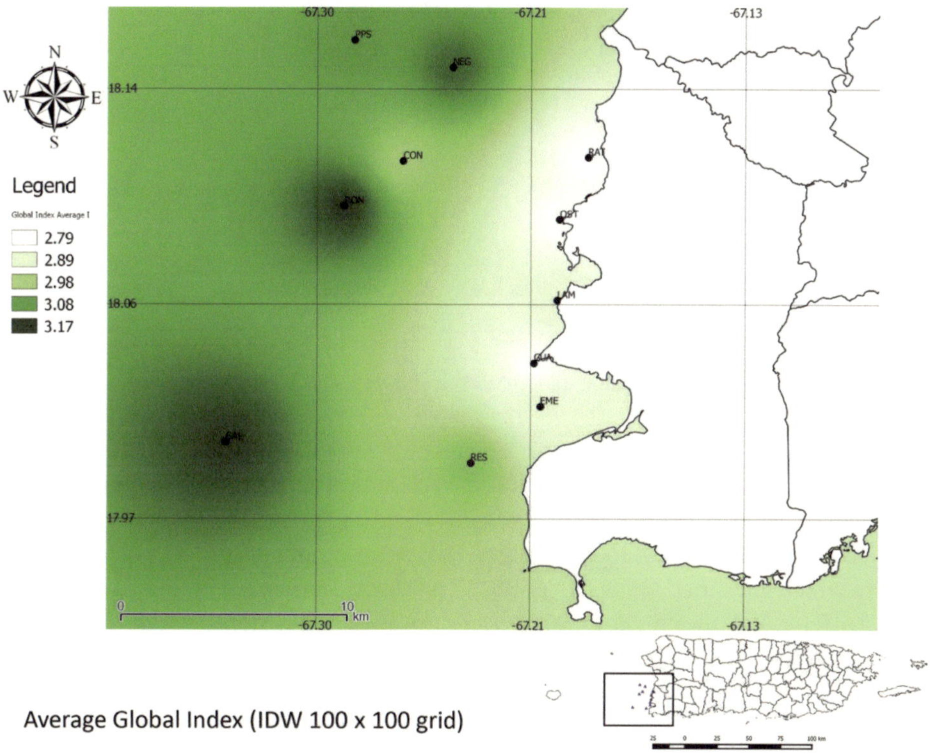

Figure 6. GIS-based inverse distance weighting (IDW) interpolation showing mean global CRRI spatial patterns. For site acronyms refer to **Figure 2**.

with 3.27. The locality with the lowest overall CRRI value was RAT (depth I) with 2.79. In general, depth zones II and III showed global CRRI values greater than those documented in zone I. Variation in depth was related to geographic patterns of variation.

3.3. Coral Index

A significant cross-shelf increase ($p < 0.0001$) was also observed in the mean Coral Index score in coral reefs (**Figure 5b**, **Table 2**). Mean Coral Index across inshore sites was 2.60, with a range of 2.07–2.87 (**Table 3**). On average, inshore coral reefs were classified as "poor," although three of them were classified as "fair." Mid-shelf reef Coral Index averaged 3.40, with a range

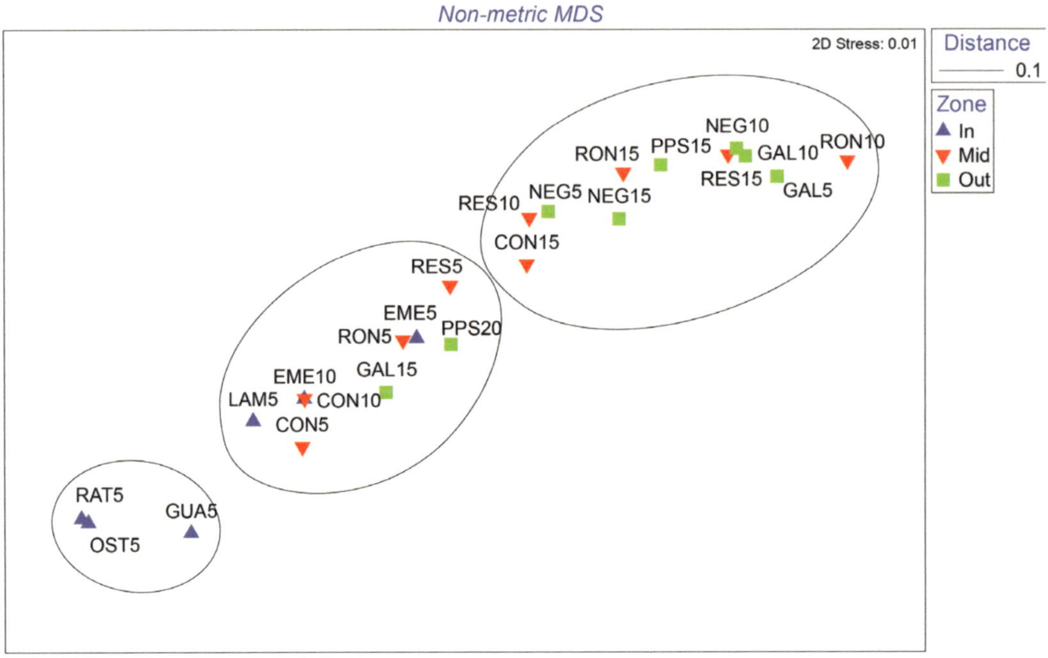

Figure 7. Nonmetric multidimensional scaling plot (nMDS) based on global CRRI scores across geographic zones × location × depth.

of 2.92–3.90. Of these, all depth areas of RES were classified as "fair," the flat area of CON was classified as "fair," but its deeper zones were classified as "good." RON reef was categorized as "good." Coral Index mean values averaged 3.76 across outer-shelf locations, ranging from 3.41 to 4.14, which classified reefs as "good." The Coral Index spatial gradient was evident (**Figure 8**). Differences among geographic zones, locations, and depth zones were highly significant ($p < 0.0001$). All possible interaction combinations were also significant.

The nMDS analysis showed a nearly similar spatial pattern confirming a significant cross-shelf gradient of the Coral Index (stress = 0.01) (**Figure 9**). Clustering patterns were nearly similar as those documented for global CRRI. The first cluster was dominated by locations across the inshore geographic zone. The second cluster was a mixed group of some inshore and mid-shelf reefs. The third mixed group was composed of some mid-shelf and outer-shelf reefs. The location with the highest Coral Index value was NEG (depth II) with 4.14. The locality with the lowest overall Coral Index value was EME (depth II) with 2.08. In general, depth zones II and III showed Coral Index values greater than those documented in zone I. Variation in depth was related to geographic patterns of variation.

3.4. Threatened Coral Index

A significant cross-shelf increase ($p = 0.0469$) was also observed in the mean Threatened Coral Index score in coral reefs (**Figure 5c**, **Table 2**). Mean Threatened Coral Index across inshore sites was 1.00, with a range of 1.00–1.03 (**Table 3**). On average, inshore coral reefs were classified as

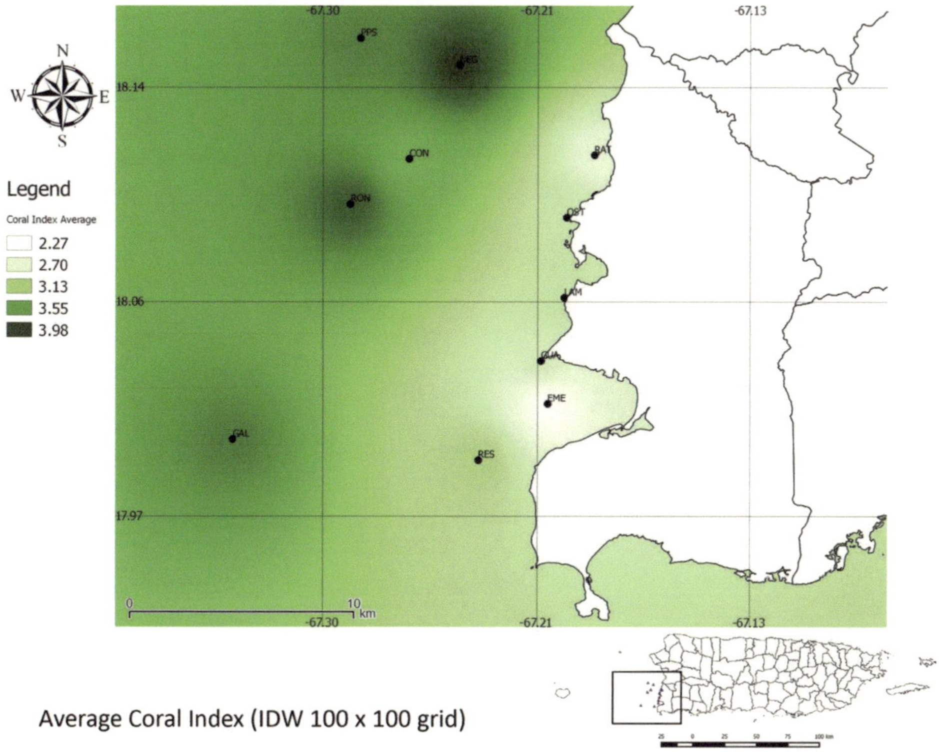

Figure 8. GIS-based inverse distance weighting (IDW) interpolation showing mean Coral Index spatial patterns. For site acronyms refer to **Figure 2**.

"critical." Mid-shelf reef Coral Index averaged 1.02, with a range of 1.00–1.08. Mid-shelf reefs were also classified as "critical." Threatened Coral Index mean values averaged 1.06 across outer-shelf locations, ranging from 1.00 to 1.22, which also classified outer-shelf reefs as "critical." However, the Threatened Coral Index spatial gradient was also evident (**Figure 10**). Differences among geographic zones ($p = 0.0469$) and locations ($p = 0.0006$) were significant, but not among depth zones ($p = 0.1910$). All possible interaction combinations were also significant.

The nMDS analysis confirmed a significant cross-shelf gradient of the Threatened Coral Index (stress <0.01) (**Figure 11**). The first cluster was dominated by two depth zones of outer-shelf GAL location. The second cluster was a mixed group of some inshore and mid-shelf reefs, which had sporadic colonies of threatened species. The third mixed group was composed of some inshore and mid-shelf reefs, which lacked threatened species. The location with the highest Threatened Coral Index value was GAL (depth I) with 1.23. Multiple locations shared the lowest overall Threatened Coral Index value, with 1.00.

3.5. Algal Index

A significant cross-shelf increase was observed in the mean Algal Index score among locations ($p < 0.0001$) and among depth zones ($p = 0.0014$), but not among geographic zones (**Figure 5d**,

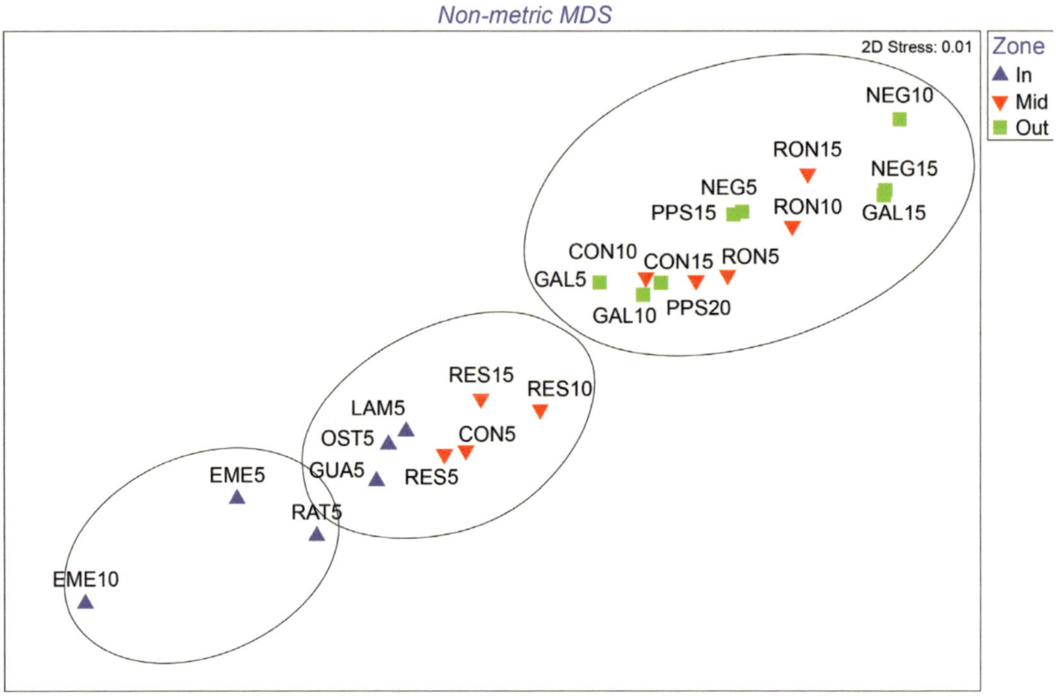

Figure 9. Nonmetric multidimensional scaling plot (nMDS) based on Coral Index scores across geographic zones × location × depth.

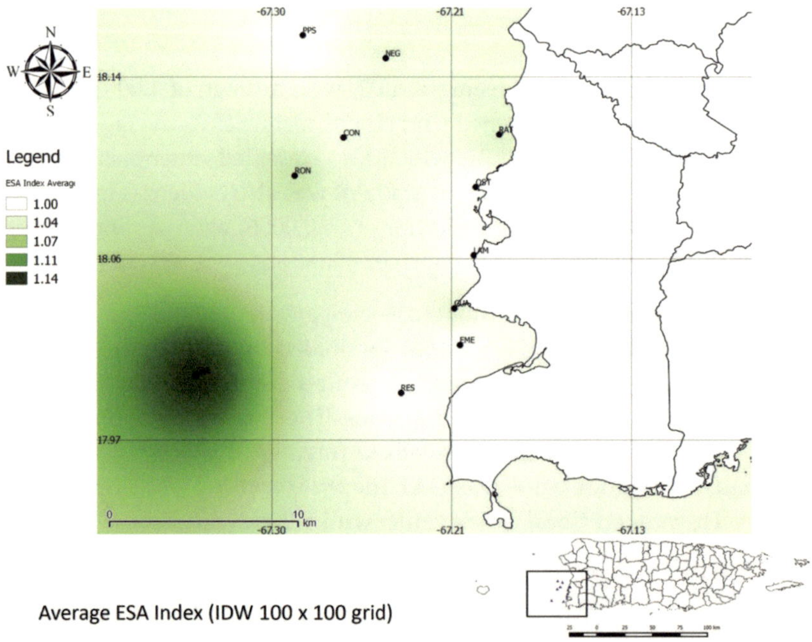

Figure 10. GIS-based inverse distance weighting (IDW) interpolation showing average Threatened Coral Index spatial patterns. For site acronyms refer to Figure 2.

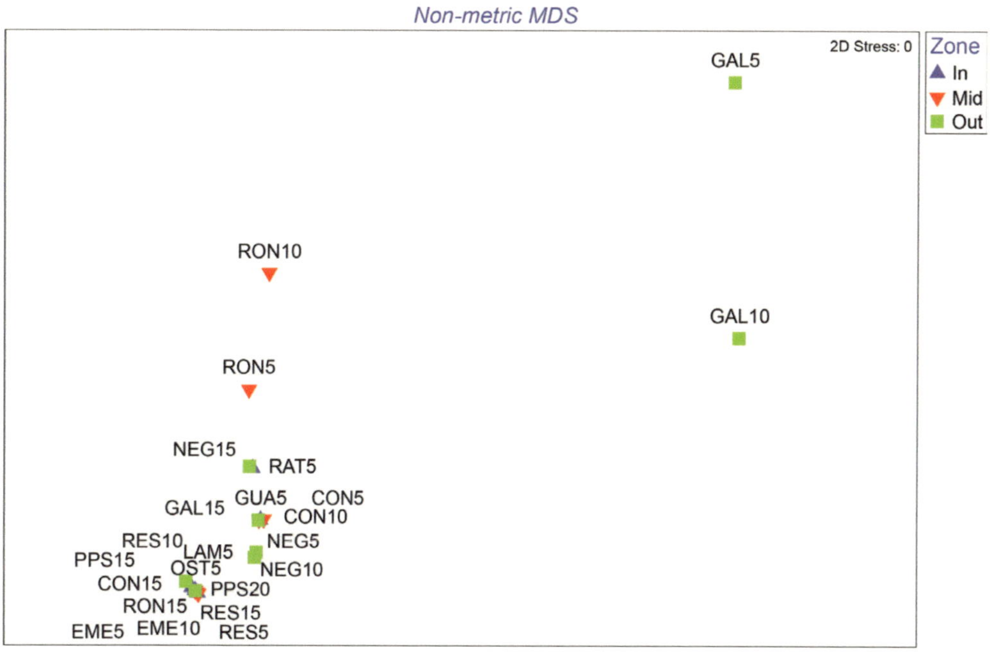

Figure 11. Nonmetric multidimensional scaling plot (nMDS) based on Threatened Coral Index scores across geographic zones × location × depth.

Table 2). All possible interaction combinations were also significant. Mean Algal Index across inshore sites was 4.00, with a range of 3.80 to 4.33 (**Table 3**). On average, inshore coral reefs were classified as "good." Mid-shelf reef Algal Index averaged 4.04, with a range of 3.84 to 4.11. Mid-shelf reefs were also classified as "good." Algal Index mean values averaged 4.00 across outer-shelf locations, ranging from 3.87 to 4.34, which also classified outer-shelf reefs as "good." The Algal Index spatial gradient was also evident (**Figure 12**).

The nMDS analysis confirmed a significant cross-shelf gradient of the Algal Index (stress = 0.01) (**Figure 13**). The first cluster was dominated by two depth zones of outer shelf GAL location. The second cluster was a mixed group of some inshore and mid-shelf reefs. The third mixed group was composed of some inshore and mid-shelf reefs. Spatial patterns of algal assemblages varied depending on the location and reef's trophic state, as well as on the cross-shelf complex water circulation pattern. The locality with the highest Algal Index value was GAL (depth I) with 4.34, and it was classified as "very good." The locality with a lower Algal Index was found on the same reef (GAL) but at depth III, with 3.66, with a category of "good."

3.6. Impacts of water quality stress gradient on CRRI

A 'linkage tree' of coral reef benthic community structure based on the BIOENV routine to environmental variables was carried out and a binary split on the basis of the best single environmental variable was thresholded to maximize the analysis of similitude (ANOSIM) R statistic for the two groups formed. This observed ANOSIM of R = 0.57 and B = 85.9%, which suggests that most of the observed variation can be explained by this solution (**Figure 14**). The

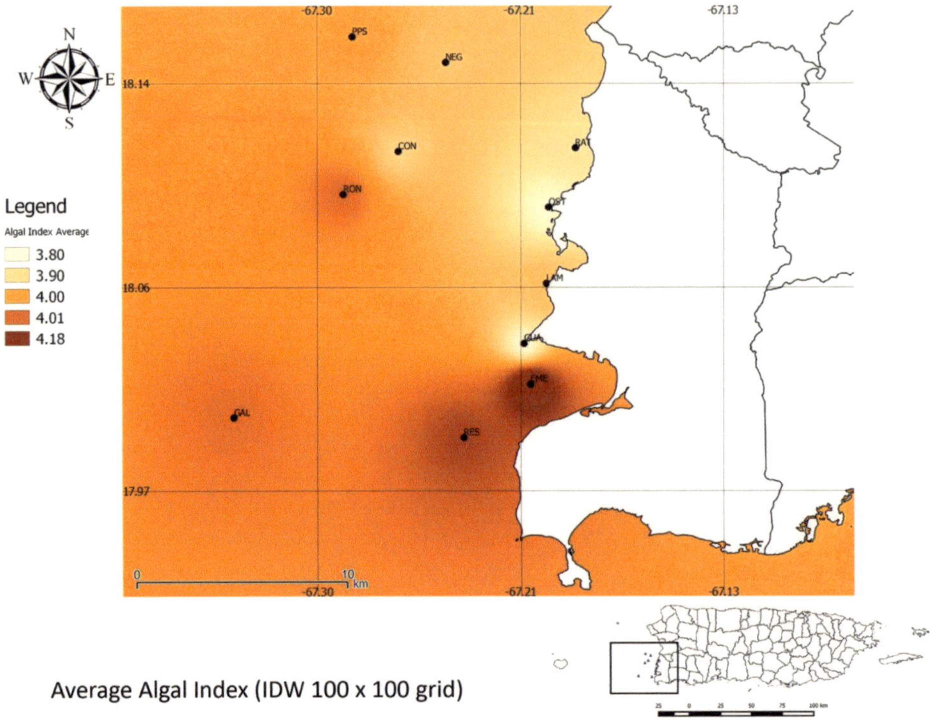

Figure 12. GIS-based inverse distance weighting (IDW) interpolation showing average Algal Index spatial patterns. For site acronyms refer to **Figure 2**.

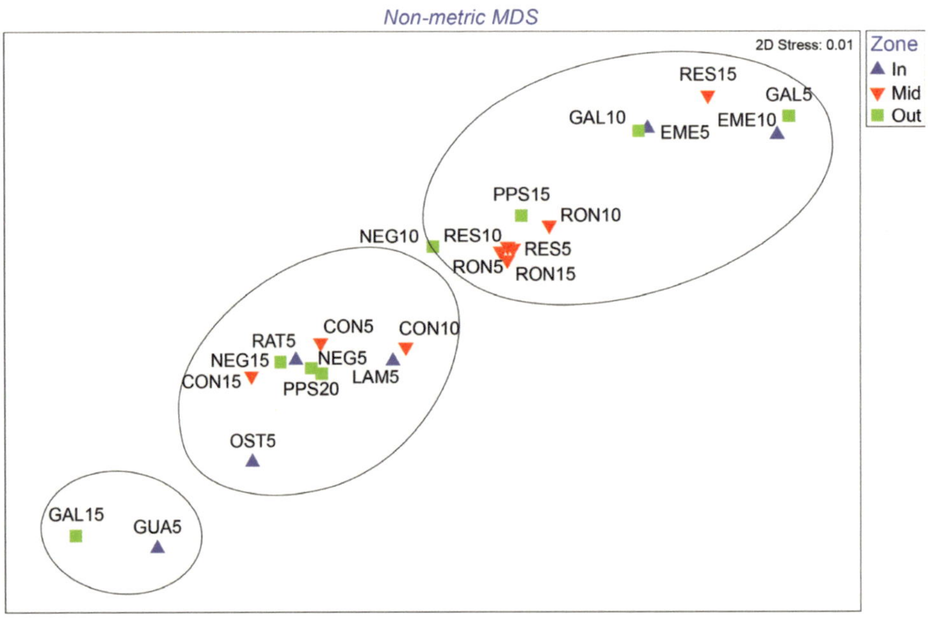

Figure 13. Nonmetric multidimensional scaling plot (nMDS) based on the Algal Index scores across geographic zones × location × depth.

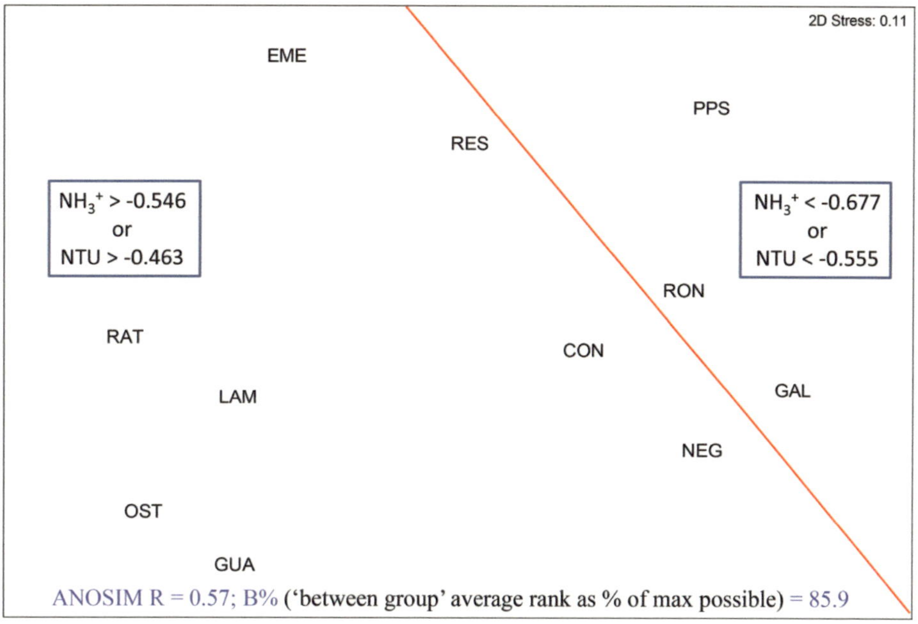

Figure 14. Multidimensional scaling (MDS) plot of the first stage in a 'linkage tree' of coral reef benthic community structure to environmental variables. Binary split on the basis of the best single environmental variable, thresholded to maximize the analysis of similitude (ANOSIM) R statistic for the two groups formed.

pattern was characterized by lower NH_3^+ to the right side of the plot (NH_3^+ Euclidean distance < −0.677) at outer-shelf sites PPS and GAL and at mid-shelf site RON and by higher values (NH_3^+ Euclidean distance > −0.546) to the left side of the plot across the remaining inshore and mid-shelf sites. Alternatively, the same split of sites was obtained by choosing lower turbidity to the right side of the plot (Turbidity Euclidean distance < −0.555) at outer-shelf sites PPS and GAL and at mid-shelf site RON and high turbidity (Turbidity Euclidean distance > −0.463) to the right side of the plot. ANOSIM R was the same whichever of the two variables was used as they gave the same split of biotic data. LINKTREE analysis showed that variation in NH_3^+ and turbidity explained most of the spatial variation observed in coral reef benthic community structure, therefore, in the CRRI spatial distribution.

4. Discussion

4.1. Spatial variation patterns of water quality conditions

This study showed important evidence of an LBSP gradient across the western Puerto Rico shelf and that chronic water quality decline has significantly affected the face of coral reef benthic communities, which was reflected on the mean CRRI scores. A snapshot view of LBSP showed that particularly turbidity and NH_3^+ concentrations increased along inshore locations. It is particularly concerning that EME reef site and to some extent GUA, LAM, and OST are being exposed to recurrent pulses of sewage effluents from malfunctioning sewage treatment facilities at Boquerón Bay and from multiple nonpoint sewage sources. Elevated NH_3^+ concentrations at EME suggest that tidal cycles may continuously expose coral reefs adjacent to Boquerón Bay to recurrent sewage pollution and eutrophication impacts. Turbidity was also higher at inshore locations such as JOY, RAT, and OST. Their proximity to Joyuda Bay and Puerto Real Bay continuously expose these sampling sites to recurrent polluted, turbid runoff pulses. A particular concern was degraded water quality pulses even across outer-shelf sites, where NH_3^+ concentrations exceeded recommended levels for healthy coral reefs. Pollution across outer-shelf sites may come from other significant sources such as the Río Guanajibo, Río Yagüez, and the Mayagüez Bay.

Documented turbidity spatial patterns were highly consistent with findings of cross-shelf scale pollution patterns documented by Bonkosky et al. [66]. Turbidity patterns were also consistent with previous unpublished observations from year 2000 (Hernández-Delgado, unpub. Data). Therefore, it is reasonable to assume that observed spatial patterns of water quality conditions in this study were highly consistent with chronic large-scale degradation at least over the last two decades and that the observed LBSP stress gradient in the form of chronic turbidity and eutrophication, mostly associated to sewage pollution, represent a nearly permanent state. Observed NH_3^+ concentrations in this study also reflected an evident cross-shelf gradient with increasing distance from known sewage pollution sources. Lapointe and Clark [67] suggested that NH_3^+ concentrations for coral reefs should not exceed 0.1 μM and that any concentration above 24 μM were deemed as too high. Our findings are highly concerning as observed NH_3^+ concentrations were from 150 to 2600 times higher than recommended limits for healthy coral reefs. Eight out the twelve sampled sites (75%) showed NH_3^+ concentrations exceeding dangerous concentrations for coral reefs as much as 10.8 times.

Regression analyses have previously shown that several water quality indicator parameters reflected significant gradients with increasing distance from LBSP [62]. These authors found a significant relationship among turbidity, phosphate (PO_4), chlorophyll-a, and dissolved oxygen concentration, implying that increasing chronic water quality degradation can significantly affect multiple parameters, adversely impacting coral reefs. Although this study just provided a snapshot view of water quality across the western Puerto Rico shelf, results were concerning as critical water quality parameters resulted significantly higher than recommended limits for sustaining coral reef health. These results suggest that human-driven LBSP across the western Puerto Rico shelf is highly significant; it is a large-scale, chronic phenomenon and deserves full long-term monitoring across large spatial and temporal scales. It also suggests the need to rapidly implement best management practices (BMPs) to reduce LBSP impacts across the shelf.

4.2. Spatial variation patterns of the benthic CRRI

The observed spatial pattern in CRRI values was significantly influenced by an LBSP stress gradient across the entire western Puerto Rican shelf. Overall, the global CRRI averaged 3.02 ("fair") across the entire shelf, the Coral Index averaged 3.32 ("fair"), the Threatened Species Index 1.03 ("critical"), and the Algal Index 4.01 ("good"). Based on the spatial distribution of the global CRRI mean values, coral reefs across the western Puerto Rican shelf can be classified as "fair." But based on the spatial patterns of the Coral Index, reefs showed a more consistent cross-shelf gradient of conditions, ranging from "poor" to "fair" across inshore locations, from "fair" to "good" along mid-shelf locations, and "good" across outer-shelf locations. There was also an evident depth-related gradient, with deeper reef zones showing higher CRRI and higher Coral Index values, in comparison to shallower zones. Based on the global CRRI, 100% of the surveyed reefs in this study were classified as "fair." But based on the Coral Index, 45% of the surveyed reefs across the western Puerto Rican shelf were classified as "good," 36% as "fair," and 19% as "poor," reflecting a strong inshore-offshore environmental stress gradient. This implies that a potential combination of human and natural factors can be influencing reef condition and CRRI values in Puerto Rico. The cross-shelf spatial gradient can be the result of chronic water quality degradation along inshore zones, which are located adjacent to known pollution sources. But the bathymetric gradient in reef conditions and CRRI values can be the potential combined result of variation in water turbidity, and the combined long-term impacts of postbleaching coral mortality, coral disease outbreaks, and impacts from hurricanes and north-western winter swells.

In comparison, previous studies using a nearly similar Coral Reef Health Index in Jamaica showed a mean value of 2.1 ("poor"), with ranges from 1.6 to 2.6 [63]. A similar study from 326 locations across four countries of the Mesoamerican Barrier Reef System (Belize, Guatemala, Honduras, and México) showed that 47% of the reefs were in "poor" condition in 2008, 6% were "critical," 41% "fair," 6% "good," and none were classified as "very good" [57]. A survey of 130 locations across the same region in 2010 showed that 40% of the reefs were in "poor" condition, 30% were "critical," 21% "fair," 8% "good," and only 1% "very good" [57]. A similar study from 193 locations across the same region in 2012 showed that 40% of the reefs were in "poor" condition, 24% were "critical," 25% "fair," 9% "good," and only 2% "very good" [58]. A similar study from 149 locations across the same region in 2015 showed that 40% of the reefs were still in "poor" condition, 17% were "critical," 34% "fair," 8% "good," and only 1% "very good" [59]. From this comparison, it is evident that multiple reef locations across the wider Caribbean region are significantly degraded by a multiplicity of factors, including a combination of overfishing [19, 21, 68], LBSP [7], and climate change [11]. Many of these locations are not showing signs of recovery [16, 17, 68].

Findings in this study of a strong cross-shelf stress gradient on coral reefs is also consistent with the literature that suggests significant impacts of LBSP [69], eutrophication [70, 71], sewage pollution [72], turbidity [73, 74], sedimentation [75–77], and bioerosion [78] on coral reefs adjacent to sources of stress.

4.3. Implications for coral reef conservation

Coral reef benthic assemblages in this study were showing signs of a cross-shelf environmental stress (e.g., turbidity, sewage pollution, eutrophication, sedimentation, and sediment bedload), therefore potentially compromising coral reefs' long-term reef accretion sustainability and ecosystem resilience. Coral reefs across the southwestern shelf of Puerto Rico have shown evidence of significant environmental degradation over the last four decades. Loya [79] and Goenaga and Cintrón [80] documented signs of degradation across inshore and mid-shelf reefs from chronic sedimentation. Many of these have suffered damage over time due to high terrigenous sediment loads [81, 82] and massive coral bleaching [83]. Schärer et al. (2010). High percent cover of threatened Elkhorn coral, *Acropora palmata*, were documented across offshore western mid-shelf reefs, but populations were largely declining in reefs adjacent to the coast due to water quality degradation [72]. Other studies have shown further reef degradation associated to LBSP, including the combination of sedimentation and turbidity [84, 85] and sewage and eutrophication [66, 72, 86]. Declining environmental conditions across the shelf have resulted in declining coral growth rates [81] and in significant declines of *A. palmata* populations across inshore reefs adjacent to areas impacted by LBSP [72, 84, 87–90]. Chronic decline in water quality could also have significant negative impacts on fish assemblages as several fish taxa can be sensitive to environmental degradation [91].

Findings in this study imply potential LBSP impacts across very large temporal and spatial scales, with very wide and persistent implications on coral reef benthic communities and on reef-associated fauna. LBSP impacts (i.e., sewage pollution from human and animal sources) were documented across the entire southwestern shelf in Puerto Rico, even in waters complying with existing microbiological quality standards [66]. This points out at the increasing spatial scale of chronic LBSP impacts across multiple coral reef systems and at the potentially increasing turnover rates of reef communities. The lack of adequate controls of LBSP across the region constitutes one of the most significant concerns regarding the conservation and recovery of declining coral reef ecosystems.

Efforts are being currently developed to implement erosion and sedimentation controls across watershed scales in southwestern Puerto Rico. But so far, these efforts have completely missed a long-term ecological monitoring component to determine if current land-based efforts have had any meaningful impacts on improving adjacent coral reef ecosystems. Therefore, the use of rapid assessment approaches, such as the one implemented in this study, could provide a meaningful approach to address the spatial patterns of coral reef conditions, understand its potential causes of stress, and identify alternative strategies to implement BMPs to reduce stressors.

4.4. Management recommendations for decision-making

A summary of management recommendations for decision-making has been included in **Table 4**. These are based on the CRRI score rankings. Suggested actions were subdivided by sector following the suggestions of HRI [56] into government, NGOs, private sector, and the academia. Recommendations included a combination of broad and targeted management actions aimed at improving governance by regulatory agencies, including improving enforcement capacity of water quality regulations and land use plan and fostering the implementation of BMPs of erosion control. They are also aimed at supporting NGOs and academic research to strengthen ecosystem-based management of coral reefs and other coastal resources. The government should also provide economic incentives for conservation and sustainable business, implement a green tax system to support these initiatives, and establish a functional network of no-take marine protected areas (MPAs).

Recommendations are also aimed to empower base communities to undertake management actions and engage into citizen science programs, including coral farming and reef rehabilitation through community-based NGO efforts. Also, base communities should strengthen their advocacy for coral reef conservation and fully support government initiatives, which promote community-based participation in management. The private sector should also become more active in supporting government efforts to manage MPAs, as well supporting coral farming and reef restoration efforts led by government, NGOs, or other sectors. The academia needs also to develop management-oriented research aimed at responding to multiple questions by natural resource and MPA managers. Applied research should also aim to understand the long-term dynamics of change of novel coral reef ecosystems. Multidisciplinary research should also be implemented to address the impacts of potential sources of stress on coral reefs. Communications and outreach need also to be improved between the academia and other sectors.

Based on the observed global CRRI and on the Coral Index scores in this study, the government should focus their efforts on implementing many of the above-mentioned suggestions, but in particular, strengthening the implementation of BMPs for erosion and runoff control, and support the ecological restoration of depleted coral reefs. NGOs should also strengthen community-based coral farming and reef restoration efforts. The private sector should also implement/support "adopt a reef" programs to promote reef conservation and restoration, and/or fully support NGO efforts. The academia should also strengthen long-term ecological monitoring programs to address sources of stress and should engage in research to understand the dynamics of emergent, novel coral reef ecosystems.

Nevertheless, the successful implementation of coral reef conservation will largely depend on the effective implementation of a coastal zone management plan, in the successful networking and effective communication among multiple stakeholders, in the implementation of effective communication among and in translating scientific information to managers, decision-makers, government leaders, and base communities, and in building trust and transparency among different sectors of society. It would also be critical to reduce pollution sources across watersheds (e.g., raw sewage discharges, agricultural, livestock, urban, and industrial runoff) through the implementation of sustainable BMPs and strict enforcement of existing regulations. Effective enforcement of fishery regulations and improved no-take MPA governance are also fundamental for achieving sustainable coral reef resilience. Further, there is a need to comply with internationally recommended protection of 20% of territorial sea as no-take MPAs. There are Caribbean islands that comply with that recommended goal, such as the U.S. Virgin Islands, where 15% of the area within their MPA boundaries had no-take regulations, in contrast to Puerto Rico, which only had 3% [92].

It would also be critical to implement sustainable development practices, particularly for small tropical island nations [88], including establishing setbacks from vulnerable areas along the shoreline and measures to protect local community livelihoods. A climate change adaptation program must also be implemented focused on the sustainable adaptability of coupled social-ecological systems, on the sustainability of the ecosystem services provided by the first line of defense against storm swells (e.g., coral reefs and mangroves) and on fishery sustainable adaptability [93]. The implementation of alternative livelihood programs for displaced fishers and an improved effectiveness in the management of no-take MPAs through consistent enforcement, sustainable funding, and technical capacity building is also paramount.

Category	Sectors			
	Government	NGOs	Private sector	Academic researchers
Very good	Provide economic incentives for conservation and sustainable businessDesignate no-take MPAs to maintain resilient reef fish assemblagesFully support citizen science programsFully support long-term ecological monitoring led by NGOs and academiaEnforce existing water quality regulations	Support efforts to fully protect more reefs (MPAs)Increase public participation in managementDevelop management-oriented citizen science programs	Sustain local MPAs through financial, staff, or technical assistanceCollaborate and support government, academic, and NGO efforts for reef conservation and restorationImprove the implementation of BMPs, sustainable codes of conduct, and other strategies to reduce environmental impacts	Engage in research to respond questions by natural resource and MPA managersDevelop long-term ecological monitoring programs to address ecological change and climate change impactsPromote integration of citizen science programsEstablish communication and outreach programs with other sectors
Good	As in "very good" +Implement coral farming and reef restoration to maintain healthy coral populationsImplement a green tax system to support coral reef conservation and restoration initiative	As in "very good" +Implement community-based coral farming and reef restoration	As in "very good" +Promote partnerships with other sectors to support coral farming and reef restoration	As in "very good" +Promote partnerships with other sectors to support coral farming and reef restorationDevelop multidisciplinary research integrating social sciences and economy
Fair	As in "good" +Implement BMPs for erosion and runoff controlRestore depleted coral reef	As in "good" +Strengthen community-based coral farming and reef restoration	As in "good" +Implement/support "adopt a reef" programs to promote reef conservation and restoration	As in "good" +Strengthen long-term ecological monitoring programs to address sources of stress
Poor	As in "fair" +Strengthen the implementation of the coastal zone management plan and the land use planAggressive implementation of BMPs for erosion and runoff controlStrengthen enforcement of fisheries regulations to enhance herbivorous fish populationsImprove land use, management of soil erosion, wastewater, and urban runoffImplement local moratoriums on coastal development projects	As in "fair" +Strengthen community-based advocacy in coral reef conservationStrengthen community-based coral farming and reef restoration	As in "fair" +Strengthen partnerships and support of coral reef management efforts by governmentStrengthen partnerships and support of coral farming and reef restoration	As in "fair" +Strengthen collaborations and communication with natural resource and MPA managersConduct management-oriented research on novel reef ecosystemsAssist government and other sectors in developing or strengthening management plans

Category	Sectors			
	Government	**NGOs**	**Private sector**	**Academic researchers**
Critical	As in "poor" +Establish emergency measures to reduce environmental stressors to reefsEstablish priority mechanisms to implement BMPs to reduce sediment delivery to coastal waters and to improve efficiency of wastewater and urban runoff management	As in "poor" +Promote effective enforcement of fishery regulations to enhance herbivorous fish populationsImplement community-based reef restoration	As in "poor" +Strengthen partnerships and fully support efforts led by government, NGOs, and the academia for coping critical declining coral reefs	As in "poor" +Strengthen multidisciplinary approaches to reef management to understand the role of human uses of reef ecosystems

Table 4. Summary of recommended management actions.

Government agencies also need to establish effective partnerships with the academia, NGOs, and the private sector to promote applied research aimed at responding to management-oriented research questions regarding emergent novel coral reef ecosystems, which are characterized by altered benthic and fish assemblages as a result of multiple human impacts. Also, in a moment of complex and profound socioeconomic crisis, it is pivotal that governments need to promote and adopt sustainable consumption guidelines for marine resources; protect vulnerable coastal habitats, watersheds, and water sources; and secure food security and sovereignty [93]. Local governments should establish effective mechanisms, such as green taxes, to enhance available funding to support MPA management, coral farming, reef rehabilitation, and sustainable natural resource-based recreation. The private sector should contribute significantly to MPA and coral reef conservation and restoration through financial assistance and through supporting human and technical resources. Moreover, there is a critical need to reduce impacts by massive tourism activities [88], to reduce carbon emissions [94], and to adopt and expand a reward system for carbon sequestration, with the reduction of hydrocarbon dependency [56]. Approximately 85% of the energy produced in Puerto Rico is derived from hydrocarbon burning. There is a need to promote the use of alternative renewable energy sources.

4.5. Other potential applications of the modified CRRI

Multiple coral reef health indices have been successfully implemented across the globe to address a multiplicity of management-oriented questions. Some of them are very specific, while others can be applied to a variety of questions. The proposed CRRI is a very useful method to address coral reef conditions under a variety of scenarios. With the proper sampling design, the method can provide rapid, robust data to address spatial and temporal variability in coral reef conditions across multiple environmental conditions and across a variety of reef morphotypes and depth zones. It can also be implemented across leeward (protected) habitats, as well as across windward (exposed) sites. The CRRI can be used to address the long-term environmental impacts of any coastal development project, such as dredging, the construction of seawalls, marinas, beach renourishment, and other activities. With the proper sampling design, it can even be used following a before-after-control-impact (BACI) approach to simultaneously address multiple research questions. The proposed CRRI can also be implemented to address impacts by acute factors such as vessel groundings. In addition, it can address impacts of large-scale phenomena such as hurricanes, winter swells, coral mortality events, and massive bleaching. The CRRI can even be applied during assessments of the effectiveness of coral outplanting and reef restoration.

With minimal training, the CRRI can be fully adapted and implemented through a combination of academic, government, or community-based NGO and private-led citizen science programs. It can further be easily combined with other standard long-term monitoring efforts (e.g., Atlantic and Gulf Rapid Reef Assessment [AGRRA]). Therefore, its implementation can become a paramount tool to facilitate the interpretation of large data sets by the scientific community, politicians, government decision-makers, natural resource managers, economists, private stakeholders, base communities, fishermen, and other interested sectors. This element of cross-participation, integration, and understanding of science is fundamental for helping planning and decision-making processes.

5. Conclusions

Coral reefs across the western Puerto Rican platform are showing signs of environmental stress. This was reflected on a cross-shelf spatial gradient of water turbidity and NH_3^+ that is affecting coral reef ecosystems across the entire shelf. CRRI mean values reflected this trend and pointed out at a gradient of reef conditions from inshore, highly degraded locations, to mid-shelf moderately degraded reefs, to less degraded outer-shelf locations. This suggests the need to implement a suite of management strategies by multiple societal sectors, from government, to NGOs, the private sector, and the academia. When coupled with a long-term permanent monitoring program or any reef rapid assessment method, the proposed CRRI can become a useful tool for all sectors, in particular for natural resource and MPA managers, and for community-based, NGO-led citizen science programs in support of government management efforts and of academic research. The successful implementation of the CRRI would provide the basic framework for wide participation of stakeholder networks, which would provide baseline information for improving coral reef management. However, successful and effective coral reef conservation can be achieved only if such efforts are multidisciplinary and are broadly participatory (fair and meaningful engagement of multiple sectors) and if science is translated into easy-to-understand information for all sectors of society, including decision-makers. A key benefit of the proposed CRRI method is that, with proper training, it can be implemented by any members of any sector and that complex quantitative information generated can be rapidly translated into easy-to-interpret formats. This is critical for the timely implementation of adaptive management actions, particularly in the context of rapidly shifting ecosystems by climate change–related impacts and by other ecological surprises.

Coastal ecosystem resilience and sustainability are fundamental goals for many small island nations. The implementation of long-term ecological monitoring programs is important to address management effectiveness. However, it could be difficult for many small islands and developing countries to implement such programs due to economic constraints and/or lack of trained personnel or appropriate resources. Therefore, easy-to-implement, economic, reliable, rapid assessment methods such as the CRRI can become valuable tools for achieving such goals, particularly in a time of socioeconomic crisis and accelerating climate change.

Nevertheless, Sammarco et al. [95] found that a key problem regarding coral reef assessment and monitoring strategies was that differences in objectives can create communication and information gaps. These may even prevent direct comparisons among studies. There is a need to improve communications among government agencies, managers, academia, and groups engaged in reef assessment and monitoring activities and to promote community-based participation through fully supported citizen science programs. Only improved science and communication will lead to improved decision-making on both local and Caribbean-wide regional scales [96]. It is also important to understand the ultimate requirements of local, state, and national governments and understand their staff and funding limitations and management needs. These will help identify clear management questions and goals and design hypothesis-driven research, which will ultimately determine which specific indicators would be required. As a final thought, given the continuously declining conditions of multiple coral reefs around the Caribbean region, promoting community-based efforts of coral farming and reef restoration, coupled with continuous monitoring, must become a top priority. There are important published success stories of community-based coral reef restoration in Puerto Rico (e.g., [97, 98]). The take-home message is that planning and selection of bioindicators for coral reef assessment and monitoring need to start from the end in mind in order to achieve the common ultimate goal of coral reef conservation and the sustainability of ecosystem productivity, resilience, functions, benefits, and services. This will require strengthening networking among different stakeholders and promoting stronger community-based participation in planning, decision-making, and management-oriented science.

Acknowledgements

This study was possible thanks to funding provided by NGOs Protectores de Cuencas, Inc. and Ridge to Reefs, Inc. to Sociedad Ambiente Marino under a Coral Reef Conservation grant from the National Fish and Wildlife Foundation (NFWF). Partial support was also provided by the National Science Foundation (HRD #0734826) through the Center for Applied Tropical Ecology and Conservation (CATEC) and the University of Puerto Rico's Vice-Presidency of Research and Technology to E.A. Hernández-Delgado. Our appreciation goes to the logistical field support provided by the crew of M/V *Tourmarine* and Capt. Elick Hernández. This publication is a contribution from CATEC's Coral Reef Research Group and SAM's collaborative *Coral Reefs Conservation and Rehabilitation Project*.

Author details

Edwin A. Hernández-Delgado[1,2,3]*, Sonia Barba-Herrera[1,5], Angel Torres-Valcárcel[3], Carmen M. González-Ramos[4,6], Jeiger L. Medina-Muñiz[7], Alfredo A. Montañez-Acuña[1,4,6], Abimarie Otaño-Cruz[1,3,4], Bernard J. Rosado-Matías[4] and Gerardo Cabrera-Beauchamp[4]

*Address all correspondence to: edwin.hernandezdelgado@gmail.com

1 Center for Applied Tropical Ecology and Conservation, University of Puerto Rico, San Juan, Puerto Rico, USA

2 University of Puerto Rico, College of Natural Sciences, Interdisciplinary Program, San Juan, Puerto Rico, USA

3 Sociedad Ambiente Marino, San Juan, Puerto Rico, USA

4 Department of Environmental Sciences, University of Puerto Rico, San Juan, Puerto Rico

5 Departamento de Biología, Universidad de Málaga, Málaga, Spain

6 Department of Biology, University of Puerto Rico, San Juan, Puerto Rico

7 Protectores de Cuencas, Yauco, Puerto Rico

References

[1] Hoegh-Guldberg O. Climate change, coral bleaching and the future of the world's coral reefs. Marine and Freshwater Research. 1999;**50**:839-866

[2] Hoegh-Guldberg O, Mumby PJ, Hooten AJ, Steneck RS, Greenfield P, Gómez E, Harvell CD, Sale PF, Edwards AJ, Caldeira K, Knowlton N, Eakin CM, Iglesias-Prieto R, Muthiga N, Bradbury RH, Dubi A, Hatziolos ME. Coral reefs under rapid climate change and ocean acidification. Science. 2007;**318**:1737-1742

[3] Hoegh-Guldberg O, Bruno JF. The impact of climate change on the world's marine ecosystems. Science. 2010;**328**:1523-1528

[4] Veron JEN, Hoegh-Guldberg O, Lenton TM, Lough JM, Obura DO, Pearce-Kelly P, Sheppard CRC, Spalding M, Stafford-Smith MG, Rogers AD. The coral reef crisis: The critical importance of <350 ppm CO_2. Marine Pollution Bulletin. 2009;**58**:1428-1436

[5] Bruno JF, Selig ER. Regional decline of coral cover in the Indo-Pacific: Timing, extent, and subregional comparisons. PLoS One. 2007;**2**(8):e711

[6] Elmhirst T, Connolly SR, Hughes TP. Connectivity, regime shifts and the resilience of coral reefs. Coral Reefs. 2009;**28**(4):949-957

[7] Hughes TP. Catastrophes, phase shifts, and large-scale degradation of a Caribbean coral reef. Science. 1994;**265**:1547-1551

[8] McClanahan T, Polunin N, Done T. Ecological states and the resilience of coral reefs. Conservation Ecology. 2002;**6**(2):1-27, 18

[9] McClanahan T, Cinner JE. Adapting to a Changing Environment: Confronting the Consequences of Climate Change. New York, NY: Oxford University Press; 2012. 193 pp

[10] Bellwood DR, Hughes TP, Folke C, Nyström M. Confronting the coral reef crisis. Nature. 2004;**429**:827-833

[11] Eakin CM, Morgan JA, Smith TB, Liu G, Alvarez-Filip L, Baca B, Bouchon C, Brandt M, Bruckner A, Cameron A, Carr L, Chiappone M, James M, Crabbe C, Day O, de la Guardia-Llanso E, DiResta D, Gilliam D, Ginsburg R, Gore S, Guzmán H, Hernández-Delgado EA, Husain E, Jeffrey C, Jones R, Jordán-Dahlgren E, Kramer P, Lang J, Lirman D, Mallela J, Manfrino C, Maréchal JP, Mihaly J, Miller J, Mueller E, Muller E, Noordeloos M, Oxenford H, Ponce-Taylor D, Quinn N, Ritchie K, Rodríguez S, Rodríguez-Ramírez A, Romano S, Samhouri J, Schmahl G, Steiner S, Taylor M, Walsh S, Weil E, Williams E. Caribbean corals in crisis: Record thermal stress, bleaching and mortality in 2005. Plos One. 2010;**5**(11):e13969, pp. 1-10

[12] Hernández-Pacheco R, Hernández-Delgado EA, Sabat AM. Demographics of bleaching in the Caribbean reef-building coral *Montastraea annularis*. Ecosphere. 2011;**2**(1):1-13, art9

[13] Miller J, Waara R, Muller E, Rogers C. Coral bleaching and disease combine to cause extensive mortality on reefs in US Virgin Islands. Coral Reefs. 2006;**25**:418

[14] Miller J, Muller E, Rogers C, Waara R, Atkinson A, Whelan KRT, Patterson M, Witcher B.Coral disease following massive bleaching in 2005 causes 60% decline in coral cover on reefs in the US Virgin Islands. Coral Reefs. 2009;**28**:925-937

[15] Hernández-Delgado EA, González-Ramos CM, Alejandro-Camis PJ. Large-scale coral recruitment patterns in Mona Island, Puerto Rico: Evidence of shifting coral community trajectory after massive bleaching and mortality. Revista de Biología Tropical. 2014;**62**(Suppl. 3):49-64

[16] Gardner TA, Côté IM, Gill JA, Grant A, Watkinson AR. Long-term region-wide declines in Caribbean corals. Science. 2003;**301**:958-960

[17] Gardner TA, Côté IM, Gill JA, Grant A, Watkinson AR. Hurricanes and Caribbean coral reefs: Impacts, recovery patterns, and role in long-term decline. Ecology. 2005;**86**:174-184

[18] Williams SM, Sánchez-Godínez C, Newman SP, Cortés J. Ecological assessments of the coral reef communities in the Eastern Caribbean and the effects of herbivory in influencing coral juvenile density and algal cover. Marine Ecology. 2017;**38**(2):1-11

[19] Hawkins JP, Roberts CM. Effects of artisanal fishing on Caribbean coral reefs. Conservation Biology. 2004;**18**:215-226

[20] Roberts CM. Effects of fishing on the ecosystem structure of coral reefs. Conservation Biology. 1995;**9**:988-995

[21] Pratchett MS, Munday P, Wilson SK, Graham NAJ, Cinner JE, Bellwood DR, Jones GP, Polunin NVC, McClanahan TR. Effects of climate-induced coral bleaching on coral-reef fishes. Ecological and economic consequences. Oceanogaphy and Marine Biology: Annual Review. 2008;**46**:251-296

[22] Paddack MJ, Reynolds JD, Aguilar C, Appeldoorn RS, Beets J, Burkett EW, Chittaro PM, Clarke K, Esteves R, Fonseca AC, Forrester GE. Recent region-wide declines in Caribbean reef fish abundance. Current Biology. 2009;**19**:590-595

[23] Alvarez-Filip L,Côté IM, Gill JA, Watkinson AR, Dulvy NK. Region-wide temporal and spatial variation in Caribbean reef architecture: Is coral cover the whole story? Global Change Biology. 2011;**17**:2470-2477

[24] Alvarez-Filip L, Gill JA, Dulvy NK, Perry AL, Watkinson AR, Côté IM. Drivers of region-wide declines in architectural complexity on Caribbean reefs. Coral Reefs. 2011;**30**:1051-1060

[25] Alvarez-Filip L, Paddack MJ, Collen B, Robertson DR, Côté IM. Simplification of Caribbean reef-fish assemblages over decades of coral reef degradation. PLoS One. 2015;**10**(4):e0126004. DOI: 10.1371/journal.pone.0126004

[26] Hobbs RJ, Arico S, Aronson J, Baron JS, Bridgewater P, Cramer VA, Epstein PR, Ewel JJ, Klink CA, Lugo AE, Norton D. Novel ecosystems: theoretical and management aspects of the new ecological world order. Global Ecology and Biogeography. 2006;**15**:1-7

[27] Hughes TP, Linares C, Dakos V, van de Leemput IA, van Nes EH. Living dangerously on borrowed time during slow, unrecognized regime shifts. Trends in Ecology and Evolution. 2013;**28**:149-155

[28] Hobbs RJ, Higgs E, Harris JA. Novel ecosystems: implications for conservation and restoration. Trends in Ecology and Evolution. 2009;**24**(11):599-605

[29] Mcfield MD, Kramer PR. Healthy Reefs for Healthy People: A guide to indicators of reef health and social well-being in the Mesoamerican Reef Region. Belize: Healthy Reefs Initiative; 2007. pp. 1-208

[30] Cooper TF, Gilmour JP, Fabricius KE. Bioindicators of changes in water quality on coral reefs: review and recommendations for monitoring programmes. Coral Reefs. 2009;**28**(3):589-606

[31] Jameson SC, Erdmann MV. Charting a course toward diagnostic monitoring: a continuing review of coral reef attributes and a research strategy for creating coral reef indexes of biotic integrity. Bulletin of Marine Science. 2001;**69**(2):701-744

[32] Loya Y. Community structure and species diversity of hermatypic corals at Eilat, Red Sea. Marine Biology. 1972;**13**(2):100-123

[33] Gómez ED, Alino PM, Yap HT, Licuanan WY. A review of the status of Philippine reefs. Marine Pollution Bulletin. 1994;**29**(1-3):62-68

[34] Yap HT. Bioindication in coral reef ecosystems. Acta Biologica Hungarica. 1986;**37**(1): 55-58

[35] Bak RPM, Meesters EH. Coral population structure: the hidden information of colony size-frequency distribution. Marine Ecology Progress Series. 1998;**162**:301-306

[36] Meesters EH, Hilterman M, Kardinall E, Keetman M, de Vries M, Bak RPM. Colony size-frequency distributions of scleractinian coral populations: spatial and interspecific variation. Marine Ecology Progress Series. 2001;**209**:43-54

[37] Ben-Tzvi O, Loya Y, Abelson A. Deterioration Index (DI): A suggested criterion for assessing the health of coral communities. Marine Pollution Bulletin. 2004;**48**(9):954-960

[38] Crosby MP, Reese ES. A Manual for Monitoring Coral Reefs with Indicator Species: Butterflyfishes as Indicators of Change on Indo Pacific Reefs. Silver Spring, MD: Office of Ocean and Coastal Resource Management, National Oceanic and Atmospheric Administration; 1996. pp. 1-45

[39] Edinger EN, Limmon GV, Jompa J, Widjatmoko W, Heikoop JM, Risk MJ. Normal coral growth rates on dying reefs: Are coral growth rates good indicators of reef health? Marine Pollution Bulletin. 2000;**40**(5):404-425

[40] Holmes KE, Edinger EN, Limmon GV, Risk MJ. Bioerosion of live massive corals and branching coral rubble on Indonesian coral reefs. Marine Pollution Bulletin. 2000;**40**(7):606-617

[41] Lirman D, Formel N, Schopmeyer S, Ault JS, Smith SG, Gilliam D, Riegl B. Percent recent mortality (PRM) of stony corals as an ecological indicator of coral reef condition. Ecological Indicators. 2014;**44**:120-127

[42] Edinger EN, Risk MJ. Reef classification by coral morphology predicts coral reef conservation value. Biological Conservation. 2000;**92**(1):1-13

[43] Jameson SC, Ammar MSA, Saadalla E, Mostafa HM, Riegl B. A coral damage index and its application to diving sites in the Egyptian Red Sea. Coral Reefs. 1999;**18**(4):333-339

[44] Hawkins JP, Roberts CM, Van'T Hof T, De Meyer K, Tratalos J, Aldam C. Effects of recreational scuba diving on Caribbean coral and fish communities. Conservation Biology. 1999;**13**(4):888-897

[45] Swain TD, Vega-Perkins JB, Oestreich WK, Triebold C, DuBois E, Henss J, Baird A, Siple M, Backman V, Marcelino L. Coral bleaching response index: A new tool to standardize and compare susceptibility to thermal bleaching. Global Change Biology. 2016;**22**(7):2475-2488

[46] Jokiel PL, Rodgers KS. Ranking Coral Ecosystem'Health and Value' for the Islands of the Hawaiian Archipelago. Pacific Conservation Biology. 2007;**13**(1):60-76

[47] Rodgers KUS, Jokiel PL, Bird CE, Brown EK. Quantifying the condition of Hawaiian coral reefs. Aquatic Conservation: Marine and Freshwater Ecosystems. 2010;**20**(1): 93-105

[48] Kaufman L, Sandin S, Sala E, Obura D, Rohwer F, Tschirky T. Coral Health Index (CHI): measuring coral community health. Conservation International, Arlington, VA, USA: Science and Knowledge Division; 2011

[49] Lasagna R, Gnone G, Taruffi M, Morri C, Bianchi CN, Parravicini V, Lavorano S. A new synthetic index to evaluate reef coral condition. Ecological Indicators. 2014;**40**:1-9

[50] Jameson SC, Erdmann MV, Gibson GR Jr, Potts KW. Development of biological criteria for coral reef ecosystem assessment. Washington, DC: USEPA, Office of Science and Technology, Health and Ecological Criteria Division; 1998. pp. 1-96

[51] Fore LS, Fisher WS, Davis WS. Bioassessment Tools for Stony Corals: Monitoring Approaches and Proposed Sampling Plan for the US Virgin Islands. Washington, DC, USA: United States Environmental Protection Agency; 2006, Office of Environmental Information EPA-260-R-06-004

[52] Bradley P, Fisher WS, Bell H, Davis W, Chan V, LoBlue C, Wiltse W. Development and implementation of coral reef biocriteria in U.S. jurisdictions. Environmental Monitoring and Assessment. 2009;**150**:43-51

[53] Fabricius KE, Cooper TF, Humphrey C, Uthicke S, De'ath G, Davidson J, LeGrand H, Thompson A, Schaffelke B. A bioindicator system for water quality on inshore coral reefs of the Great Barrier Reef. Marine Pollution Bulletin. 2012;**65**(4):320-332

[54] Risk MJ, Heikoop JM, Edinger EN, Erdmann MV. The assessment 'toolbox': Community-based reef evaluation methods coupled with geochemical techniques to identify sources of stress. Bulletin of Marine Science. 2001;**69**(2):443-458

[55] Mcfield MD, Kramer PR. The Healthy Mesoamerican Reef Ecosystem Initiative: A conceptual framework for evaluating reef ecosystem health. Proceedings of the 10th International Coral Reef Symposium. 2006. pp. 1118-1124

[56] HRI. Report Card for the Mesoamerican Reef. Belize: Healthy Reefs Initiative; 2008. pp. 1-15

[57] HRI. Report Cad for the Mesoamerican Reef. Belize: Healthy Reefs Initiative; 2010. pp. 1-22

[58] HRI. Report Card for the Mesoamerican Reef. Belize: Healthy Reefs Initiative; 2012. pp. 1-22

[59] HRI. Report Card for the Mesoamerican Reef. Belize: Healthy Reefs Initiative; 2015. pp. 1-29

[60] McField M, Kramer P. Quick Reference Guide: 2008 A Companion to A Guide to Indicators of Reef Health and Social Well-Being in the Mesoamerican Reef Region. Belize: Healthy Reefs Initiative; 2008. pp. 1-26

[61] Díaz-Pérez L, Rodríguez-Zaragoza FA, Ortiz M, Cupul-Magaña AL, Carriquiry JD, Ríos-Jara E, Rodríguez-Troncoso AP, del Carmen García-Rivas M. Coral reef health indices versus the biological, ecological and functional diversity of fish and coral assemblages in the Caribbean Sea. PloS One. 2016;**11**(8):e0161812

[62] Hernández-Delgado EA, González-Ramos CM, Medina-Muñiz JL, Montañez-Acuña AA,Otaño-Cruz A, Rosado-Matías BJ, Cabrera-Beauchamp G. Widespread Impacts of Land-Based Source Pollution on Southwestern Puerto Rican Coral Reefs. Final Report submitted to Protectores de Cuencas, Inc., and Ridge to Reef, Inc. Yauco, Puerto Rico. 2014. pp. 1-134

[63] NEPA. An Evaluation of Ecosystem Health: 2013 – A Report Card for Reefs. Kingston, Jamaica: National Environment and Planning Agency; 2014. pp. 1-15

[64] Anderson M, Gorley R, Clarke K. PERMANOVA+ for PRIMER: Guide to Software and Statistical Methods. Plymouth: PRIMER-E; 2008

[65] Clarke K, Gorley R, Somerfield P, Warwick R. Change in Marine Communities: An Approach to Statistical Analysis and Interpretation. 3rd ed. Plymouth: PRIMER-E; 2014

[66] Bonkosky M, Hernández-Delgado EA, Sandoz B, Robledo IE, Norat-Ramírez J, Mattei H.Detection of spatial fluctuations of non-point source fecal pollution in coral reef surrounding waters in southwestern Puerto Rico using PCR-based assays. Marine Pollution Bulletin. 2009;**58**(1):45-54

[67] Lapointe BE, Clark MW. Nutrient inputs from the watershed and coastal eutrophication in the Florida Keys. Estuaries and Coasts. 1992;**15**(4):465-476

[68] Jackson J, Donovan M, Cramer K, Lam V. Status and trends of Caribbean coral reefs: 1970-2012. Global Coral Reef Monitoring Network. International Union for the Conservation of Nature, Gland, Switzerland. 2014:1-304

[69] Fisher WS, Fore LS, Hutchins A, Quarles RL, Campbell JG, LoBue C, Davis WS. Evaluation of stony coral indicators for coral reef management. Marine Pollution Bulletin. 2008;**56**(10):1737-1745

[70] Díaz-Ortega G, Hernández-Delgado EA. Land-based source pollution in a climate of change: A roadblock to the conservation and recovery of Elkhorn coral *Acropora palmata* (Lamarck 1816). Natural Resources. 2014;**5**(10):561-581

[71] Ennis RS, Brandt ME, Grimes KRW, Smith TB. Coral reef health response to chronic and acute changes in water quality in St. Thomas, United States Virgin Islands. Marine Pollution Bulletin. 2016;**111**(1):418-427

[72] Hernández-Delgado EA, Sandoz B, Bonkosky M, Mattei H, Norat J. Impacts of non-point source sewage pollution in Elkhorn coral, *Acropora palmata* (Lamarck), assemblages of the southwestern Puerto Rico shelf. In: Proceedings of the 11th International Coral Reefs Symposium. 2010. pp. 747-751

[73] Fabricius KE. Effects of terrestrial runoff on the ecology of corals and coral reefs: Review and synthesis. Marine Pollution Bulletin. 2005;**50**(2):125-146

[74] Te FT. Turbidity and its effects on corals: A model using the extinction coefficient (k) of photosynthetic active radiance (PAR). Proceedings of the 8th International Coral Reef Symposium. 1997;**2**:1899-1904

[75] Nowlis JS, Roberts CM, Smith AH, Siirila E. Human-enhanced impacts of a tropical storm on nearshore coral reefs. Ambio. 1997;**26**(8):515-521

[76] Ramos-Scharrón CE, Amador JM, Hernández-Delgado EA. (2012). An interdisciplinary erosion mitigation approach for coral reef protection – A Case Study from the Eastern Caribbean. 127-160. In: A. Cruzado (Ed.), Marine Ecosystems. InTech. ISBN: 978-953-51-0176-5

[77] Ramos-Scharrón C, Torres-Pulliza D, Hernández-Delgado EA. Watershed- and island-scale land cover changes in Puerto Rico (1930s–2004) and their potential effects on coral reef ecosystems. Science of the Total Environment. 2015;**506-507**:241-251

[78] Chazottes V, Le Campion-Alsumard T, Peyrot-Clausade M, Cuet P. The effects of eutrophication-related alterations to coral reef communities on agents and rates of bioerosion (Reunion Island, Indian Ocean). Coral Reefs. 2002;**21**(4):375-390

[79] Loya Y. Effects of water turbidity and sedimentation on the community structure of Puerto Rican corals. Bulletin of Marine Science. 1976;**26**(4):450-466

[80] Goenaga C, Cintrón G. Inventory of the Puerto Rican coral reefs. Commonwealth of Puerto Rico: Report submitted to the Coastal Zone Management, Department of Natural Resources; 1979. pp. 1-190

[81] Goenaga C. The distribution and growth of *Montastraea annularis* (Ellis and Solander) in Puerto Rican platform reefs. [PhD dissertation], Mayagüez: University of Puerto Rico, Dept. Marine Sciences; 1988. pp. 1-215

[82] Goenaga C, Boulon Jr RH. The State of Puerto Rican and U.S. Virgin Islands Corals: An Aid to Managers. Report submitted to the Caribbean Fishery Management Council, Hato Rey, PR. 1992. pp. 1-66

[83] Goenaga C, Canals M. Island-wide coral bleaching in Puerto Rico. Caribbean Journal of Science. 1990;**26**:171-175

[84] Hernández-Delgado EA. Historia natural, caracterización, distribución y estado actual de los arrecifes de coral Puerto Rico. 281-356. In: Joglar RL, editor. Biodiversidad de Puerto Rico: Vertebrados Terrestres y Ecosistemas. Serie Historia Natural. San Juan, PR: Editorial Instituto de Cultura Puertorriqueña; 2005. pp. 1-563

[85] Morelock J, Ramírez WR, Bruckner AW, Carlo M. Status of coral reefs, southwest Puerto Rico. Caribean Journal of Science, Special Publication. 2001;**4**:1-57

[86] Hernández-Delgado EA, Sandoz-Vera B. Impactos antropogénicos en los arrecifes de coral. 62-72. In: Seguinot-Barbosa J, editor. Islas en Extinción: Impactos Ambientales en las Islas de Puerto Rico. Cataño, PR: Ediciones SM; 2011. pp. 1-255

[87] Hernández-Delgado EA. Effects of anthropogenic stress gradients in the structure of coral reef epibenthic and fish communities. [Ph.D. dissertation]. San Juan, PR: Department of Biology, University of Puerto Rico; 2000. pp. 1-330

[88] Hernández-Delgado EA, Ramos-Scharrón CE, Guerrero C, Lucking MA, Laureano R, Méndez-Lázaro PA, Meléndez-Díaz JO. Long-term impacts of tourism and urban development in tropical coastal habitats in a changing climate: Lessons learned from Puerto Rico. 357-398. In: Kasimoglu M, editor. Visions from Global Tourism Industry-Creating and Sustaining Competitive Strategies. Prague, Czech Republic: Intech Publications; 2012

[89] Norat-Ramírez J, Méndez-Lázaro P, Hernández-Delgado EA, Cordero-Rivera L. El impacto de aguas usadas de fuentes dispersas en el litoral costero de Ia Reserva Marina

Tres Palmas (Rincón-Puerto Rico). Revista Asociación Venezolana de Ingeniería Sanitaria y Ambiental. 2013;**Vll**(015):27-30

[90] Weil E, Hernández-Delgado EA, Bruckner AW, Ortiz AL, Nemeth M, Ruiz H. Distribution and status of Acroporid (Scleractinia) populations in Puerto Rico. 71-98. In: Bruckner AW, editor. Proceedings of the Caribbean *Acropora* Workshop: Potential Application of the U.S. Endangered Species Act as a Conservation Strategy. Memorandum NMFS-OPR-24, Silver Spring, MD: NOAA Tech; 2003. pp. 1-199

[91] Bejarano I, Appeldoorn RS. Seawater turbidity and fish communities on coral reefs of Puerto Rico. Marine Ecology Progress Series. 2013;**474**:217-226

[92] Schärer-Umpierre M, Mateos-Molina D, Appeldoorn R, Bejarano I, Hernández-Delgado EA, Nemeth R, Nemeth M, Valdés-Pizzini M, Smith T. Marine managed areas and associated fisheries in the US Caribbean. Advances in Marine Biology. 2014;**69**:129-152

[93] Hernández-Delgado EA. The emerging threats of climate change on tropical coastal ecosystem services, public health, local economies and livelihood sustainability of small islands: Cumulative impacts and synergies. Marine Pollution Bulletin. 2015;**101**(1):5-28

[94] Côté IM, Darling ES. Rethinking ecosystem resilience in the face of climate change. PLoS Biology. 2010;**8**(7):e1000438

[95] Sammarco PW, Hallock P, Lang JC, LeGore RS. Roundtable discussion groups summary papers: environmental bio-indicators in coral reef ecosystems: the need to align research, monitoring, and environmental Regulation. Environmental Bioindicators. 2007;**2**(1):35-46

[96] Ramos-Scharrón CE, Rogers C, Hernández-Delgado EA, Restrepo J, Botero F, Coldren S, Garza-Pérez JR, Sánchez-Navarro P, Dokken Q, Ferguson R, Koss J, Martindale R, Vandiver L, Viqueira-Ríos RA. Caribbean coral reefs at risk: Improved decision making through better science and communication. Reef Encounter. 2016;**31**:61-66

[97] Hernández Delgado EA, Montañez-Acuña A, Otaño-Cruz A, Suleimán-Ramos SE. Bomb-cratered coral reefs in Puerto Rico, the untold story about a novel habitat: From reef destruction to community-based ecological rehabilitation. Revista de Biología Tropical. 2014;**62**(Suppl. 3):183-200

[98] Hernández-Delgado EA, Mercado-Molina AE, Alejandro-Camis PJ, Candelas-Sánchez F, Fonseca-Miranda JS, González-Ramos CM, Guzmán-Rodríguez R, Mège P, Montañez-Acuña AA, Olivo-Maldonado I, Otaño-Cruz A, Suleimán-Ramos SE. Community-based coral reef rehabilitation in a changing climate: Lessons learned from hurricanes, extreme rainfall, and changing land use impacts. Open Journal of Ecology. 2014;**4**(14):918-944

Chapter 10

Multi-Disciplinary Lessons Learned from Low-Tech Coral Farming and Reef Rehabilitation: I. Best Management Practices

Edwin A. Hernández-Delgado,
Alex E. Mercado-Molina and
Samuel E. Suleimán-Ramos

Additional information is available at the end of the chapter

http://dx.doi.org/10.5772/intechopen.73151

Abstract

Low-tech coral farming and reef rehabilitation have become important community-based coral reef management tools. At least in the wider Caribbean region, these strategies have been successfully implemented to recover depleted populations of staghorn (*Acropora cervicornis*) and elkhorn coral (*A. palmata*). They have also been used with relative success to recover depleted fish assemblages. Indirectly, coral reef rehabilitation has also resulted in enhanced benthic spatial heterogeneity, in providing multiple new microhabitats for fish and invertebrate species; have contributed to the recovery of coastal resilience, increasing the protection of shorelines against erosion; and have fostered an increased interest of the tourism sector as an enhanced attraction for visitors and recreationists. Nevertheless, there is still a need to implement best management practices to improve the success of these strategies. In this chapter, lessons learned from the *Community-Based Coral Aquaculture and Reef Rehabilitation Program* in Culebra Island, Puerto Rico, are shared from a multi-disciplinary standpoint. Learning from past experiences is a critical process to improve science. In a time of significant projected climate change impacts and sea level rise, improving the scale of coral farming and reef rehabilitation has become a critical tool for coral reef conservation. But multiple roadblocks must still be overcome.

Keywords: coral farming, coral reefs, ecological rehabilitation, lessons learned, Puerto Rico, Caribbean Sea, reef fish communities, threatened coral species

1. Introduction

1.1. Coral reef decline and the emergent role of ecological restoration as a management tool

Coral reefs have largely declined across regional and global scales over the last four to five decades as a combined result of multiple local, regional, and global human stressors. Local stressors are factors that affect ecological processes which occur within reef communities and often include water quality degradation [1], eutrophication [2], sedimentation [3], turbidity [4, 5], fishing [6, 7], blast fishing [8], vessel groundings [9], military training activities [10], and recreational abuse [11, 12], among many others. Regional-to-global scale stressors are climate change related, including sea surface warming [13, 14], massive coral bleaching [15, 16], disease outbreaks [17, 18], mass coral mortalities [19, 20], and its concomitant effect in reef fish assemblages [21]. Ocean acidification [22], in combination with eutrophication [23], has been shown to accelerate coral diseases and erode overall reef resilience [24]. Bioerosion has also become a key widespread process that, in combination with rapidly declining reefs, eutrophication, and ocean acidification, represents an increasing threat, though data on actual bioerosion rates are limited to a few well-studied cases [25]. In addition, hurricanes have become a significant concern in the Atlantic due to their apparent link to increasing climate change impacts [26–28]. Such multiple combined interactions have contributed to the large-scale demise of coral reef ecosystem resilience, compromising their capacity to sustain ecosystem services; threatening the sustainability of reef fisheries, its productivity, and its ability to produce food protein to feed hundreds of millions of people; and protecting shorelines from erosion, unless trends are otherwise reversed [29].

Indo-Pacific coral reefs have shown significant declines following past disturbances [30, 31], but have shown rapid signs of recovery, often within decadal time scales [32–34]. However, Caribbean coral reef ecosystems have also shown significant impacts [20, 21], but very limited natural recovery [7, 35]. Instead, Caribbean coral reef benthic communities have shown a rapidly shifting trajectory from coral dominance to dominance by non-reef-building taxa, mostly macroalgae [36–38]. Recent coral recruitment trends also point out at a shifting trajectory and dominance by small-sized, ephemeral species [39]. Limited natural recovery ability and shifting benthic community trajectories are the direct result of limited functional redundancy of Caribbean reef ecosystems, in comparison to their Indo-Pacific counterparts [40]. Such declining trends may imply the onset of permanent alterations on ecosystem resilience and persistence, ecological functions, values, and benefits. Therefore, reef ecosystems are shifting into what has been designated as novel ecosystems [41–45], with often significantly altered biological assemblages and ecological functions, and yet unknown long-term effects on ecosystem composition, functions, and productivity. This makes necessary to examine the role of assisted recovery of depleted coral diversity, restoring coral functional groups, and the rehabilitation of coral reefs at the reefscape, functional level, as a new strategy to buffer and restore present declining trends.

Multiple low-tech coral farming and reef rehabilitation efforts have proliferated across a global scale over the past 10–15 years with the aim to foster a rapid recovery of depleted coral species [46–51] and to recover reef structure, function, and ecosystem services [52]. However, most of these experiments have been of very limited spatial scales (often <100 m^{-2}) and often of very short duration (<1 y) [53–61]. Therefore, impacts have been of very limited ecological significance. Nevertheless, multiple methods have shown to be promising as future tools for depleted coral species restoration, for the rehabilitation of reef's ecosystem functions and services, and for recovering coastal resilience. But there is still a general lack of published systematized information regarding lessons learned from such activities that could serve as a step-by-step guide for coral reef managers to restore depleted coral reefs. Reviews on coral farming strategies to replenish degraded coral reefs are still scarce and have focused on very limited questions and geographical scales [47, 49, 50], on the role of integrating habitat enhancement to aquaculture and fisheries management [62], and on the potential benefits to habitat conservation [51, 63]. However, there is still a general lack of published systematized information regarding lessons learned from such activities that could serve as a step-by-step guide for coral reef managers to restore depleted coral reefs.

1.2. Goals and objectives

The goal of this chapter is to briefly update the state of knowledge regarding low-tech coral farming efforts around the globe and address a wide range of multi-disciplinary lessons learned through the 15-year-old *Community-Based Coral Aquaculture and Reef Rehabilitation Program* led in Puerto Rico by NGO Sociedad Ambiente Marino, with the collaboration of the Center for Applied Tropical Ecology and Conservation of the University of Puerto Rico. Lessons learned cover topics regarding: coral biology; the science of coral collection, handling, transporting, and out-planting to farming units; maintenance; long-term monitoring of corals in farms; out-planting site selection and methods; and the long-term monitoring of coral out-plants. Finally, the chapter also included a discussion on general recommendations and needs for implementing best management practices.

2. State of knowledge in the development of low-tech coral farming and reef rehabilitation

The state of knowledge regarding the development of low-tech coral farming and reef rehabilitation has largely expanded across the globe during the recent two decades. **Table 1** summarizes some of the recent efforts across different geographical areas. Pioneering work commenced across the Caribbean due to its limited natural recovery ability and the need to implement low-tech reef restoration efforts. But a suite of different methods has been developed across the globe involving multiple benthic coral culture units, floating units, rope nurseries, and combinations of these. Also, different methods have been implemented for

Location	Method	Survival	Time (m)	Reference
Israel	Cement tiles; plastic net	25–83% fragments *Stylophora pistillata*, 6 m, cement tiles, site-specific 83%, 6 m; 61%, 18 m; plastic net	6–18 m	[64]
Egypt	Plastic mesh	14–48% without epoxy, species-specific	6 m	[65]
		86–91% with epoxy, species-specific	12 m	
		8–11% without epoxy, species-specific		
		11–21% with epoxy, species-specific		
Japan	Concrete armor blocks	15–20% species-specific	20 m	[66]
Philippines	Concrete blocks	ND	12 m	[67]
Singapore	Table nursery; plastic mesh	34%	14 m	[68]
Palau	Pushmounts for coral spat	73–80%; asexual fragments *Acropora digitifera*, *A. hyacinthus*	18 m	[69]
	Plastic ties and pushmounts for fragments	14–24%; after sexual larval settlement	12 m	
Indonesia	Cathode and electric field	68%; *Acropora yongei* in cathode wire; 99% in electric field and in control	4 m	[70]
		83%; *A. pulchra* in cathode wire; 91% in electric field, and 87% in control		
Puerto Rico and Pohnpei, Micronesia	Lose fragments Fragments attached to a fishing line in the bottom	Survival in *A. cervicornis* and *A. prolifera* strongly treatment and size dependent. 0% in fragments 8–12 cm; 95% in fragments >30 cm	3	[74]
Puerto Rico		Survival treatment, size and location dependent: *Acropora cervicornis*	6	[73]
		3–5 cm 70%; 8–12 cm 80%; 15–22 cm 95% in backreef areas;3–5 cm 50%; 8–12 cm 90%; 15–22 cm 95% in reef front areas; *A. prolifera*—3–5 cm 80%; 8–12 cm 70%; 15–22 cm 70%		
Puerto Rico	Wire mesh "A frames"; Horizontal line nurseries (HLN)	Survival rate in *A. cervicornis* "A frame" units strongly dependent on method and exposure to extreme rainfall and runoff; "A frames" 73% in 2011–2012; 81% in 2012–2013; 97% in HLN in 2012–2013	24	[80]

ND, no data.

Table 1. Global-scale variation in coral farming and reef restoration methods.

out-planting activities, involving the use of masonry nails and plastic ties, as well as the use of different artificial substrates used for compensatory mitigations of environmental impacts or for habitat enhancement. However, coral colony survival rates either during coral farming or after out-planting to natural reefs have been largely variable and often species-, size-, and site-specific. Nevertheless, low-tech methods have been successfully implemented with the participation of base communities across multiple locations.

The use of wild and/or captive-bred coral larvae is also a promising alternative for coral propagation without compromising source population fitness [71]. However, such methods involve high tech, more expensive methods, equipment, and laboratory facilities. They also require highly trained scientific personnel, with limited possibilities of involving base communities in the process, unless extensive technical training has been provided to participants. Recent advances integrating population genetic structure of corals and holosymbionts have also pointed out the importance of collecting a genetically diverse nursery stock and of maintaining poor-performing holobionts in culture to avoid selecting only nursery-fit genets [72]. Alternatively, nurseries may be established in multiple habitat types to maintain a wide range of holobiont types acclimatized to different environmental conditions. Probably, the most significant advantage of high-tech propagation strategies is the ability to propagate *ex situ* significant amounts of coral spat during each reproductive cycle. Further, *ex situ* propagation in flow tanks can have the potential to produce 100–1000 of small coral fragments (e.g., sizes of only a few polyps) several times per year. The combination of low-tech and high-tech, genetic-based propagation strategies can lead to enhance coral propagation and out-planting success across multiple coral reef locations. With improved capabilities of long-distance transportation, these methods can also improve the ability to restore multiple locations within shorter time scales. Long-distance transportation may imply hours to several days, depending on distance of source coral reefs and on logistics of transportation. It may require developing simple to sophisticated methods of keeping corals wet, aerated, and protected from direct sunlight and high temperature. Providing mechanisms such as plastic buckets or coolers provided with a battery-powered water pump and a small PVC pipe system with multiple small holes drilled on them to allow water to sprinkle corals will allow to keep them wet during prolonged transportation. But using a Z-shaped 2″ PVC pipe in a vessel can allow natural pumping of oxygenated seawater to sprinkle corals during the ride. If a cooler is used, it can also be provided with a battery-powered chiller and an air pump. Previous experience in Puerto Rico using such battery-powered systems has allowed transportation of corals of up to 11 h from source to farming site and involving multiple transportation systems (e.g., small boat, vehicle, ferry, and another small boat). This effort resulted in a 100% survival rate, with no stress to out-planted staghorn coral (*Acropora cervicornis*) and to fused staghorn coral (*Acropora prolifera*).

2.1. A quick glimpse of previous successful experiences in Puerto Rico

Coral farming and reef rehabilitation science in Puerto Rico evolved since year 1980 with low-scale pioneering experiments by Carlos Goenaga and Vance Vicente in Cayo Enrique reef, La Parguera. However, that experiment, though successful, generated no publications. Then, by 1993, Austin Bowden-Kerby developed low-tech coral farming and reef restoration work involving staghorn coral (*A. cervicornis*) and fused staghorn coral (*A. prolifera*) [73, 74]. Ortiz-Prosper et al. [75] in 1998 out-planted corals to reef ball artificial units and to dead coral surfaces. Sociedad Ambiente Marino (SAM), in collaboration with Culebra Fishers Association and Correlations, established in 2003 the *Community-Based Coral Aquaculture and Reef Rehabilitation*

Program in Culebra Island, with over 60,000 staghorn coral colonies out-planted in 15 years [76–80]. Elkhorn coral (*Acropora palmata*) out-planting under high wave energy conditions has been successfully conducted at Vega Baja since 2008 by Vegabajeños Impulsando Desarrollo Ambiental Sustentable (VIDAS) and SAM [76]. Additional work with extensive out-planting of staghorn coral has been conducted in southwestern Puerto Rico and more recently in northeastern Puerto Rico by the National Oceanic and Atmospheric Administration—Restoration Center (NOAA-RC) and by Sea Ventures [81, 82]. Additional coral farming and out-planting has been carried out since 2010 in southwestern Puerto Rico by HRJ Reefscaping in collaboration with NOAA-RC. Another important coral restoration effort was conducted by NOAA at Mona Island following a major vessel grounding in 1997 [83, 84]. Recent smaller efforts have also been developed across the northern coast of Puerto Rico by NOAA-RC and VIDAS.

In summary, a suite of methods is currently being successfully implemented in Puerto Rico involving multiple locations (**Figure 1**) and a variety of methods (**Figures 2** and **3**). Similarly, a combination of low-tech approaches has been implemented during coral out-planting on natural depauperate reef substrates. The longest continuous coral farming and reef restoration project is led by SAM in Culebra Island (since 2003), with the close collaboration of the University of Puerto Rico's Center for Applied Tropical Ecology and Conservation (CATEC). VIDAS has led the Vega Baja restoration project since 2008. HJR Reefscaping has led projects along southwestern Puerto Rico. NOAA-RC has led and/or collaborated with basically all other initiatives. Also, the agency has led multiple reef restoration efforts across the U.S. Virgin Islands and Florida, USA. All of the above-listed efforts have placed Puerto Rico at the top leading role of coral reef restoration and rehabilitation efforts across the northeastern Caribbean region (**Figure 4**).

Figure 1. Active coral farming and reef restoration sites in Puerto Rico. From northwest to northeast: Isabela; Arecibo; Vega Baja-Manatí; Arrecifes La Cordillera Natural Reserve, Fajardo (Cayo Diablo, Palominos Island); Canal Luis Peña Natural Reserve, Culebra (Bahía Tamarindo, Punta Tamarindo Chico); and Punta Soldado, Culebra. From southwest to east: La Parguera Natural Reserve, La Parguera (San Cristobal, El Mario); Guánica Biosphere Reserve, Guayanilla; and Caja de Muerto Natural Reserve, Ponce.

Figure 2. Examples of low-tech coral farming methods used in Puerto Rico. From top left: (A) plastic-covered wire mesh; (B) large wire mesh; (C) COndominium pvc unit; (D) horizontal line nursery; (E) floating underwater coral array (FUCA); and (F) benthic underwater coral array (BUCA). The first five models have been used with staghorn coral (*Acropora cervicornis*) and fused staghorn coral (*A. prolifera*). The BUCA has been used for elkhorn coral (*A. palmata*).

Figure 3. Examples of additional low-tech coral farming methods used in Puerto Rico. From top left: (A) pvc plastic grid; (B) "cathedral" line nursery; (C) tree unit; (D) modified benthic underwater coral array (m-BUCA); (E) concrete cookies; and (F) tree unit. Models A and B have been used with staghorn coral (*Acropora cervicornis*). Models C–E have been used with elkhorn coral (*A. palmata*). Model F has also been used with pillar coral (*Dendrogyra cylindrus*) and star coral (*Orbicella faveolata*).

Figure 4. Examples of low-tech coral out-planting methods used in Puerto Rico. From top left: (A) out-planting of *Acropora cervicornis* on top of dead coral heads; (B) and (C) creation of *A. cervicornis* thickets on formerly bombarded, flattened coral reefs in Culebra Island; (D) creation of semi-natural reef corridors of *A. cervicornis* using concrete bases and pvc plastic sticks to attach corals on open sandy bottoms; (E) *A. cervicornis* "flower pots" on open reef substrates; (F) reconstruction of reef's seascape with brain coral (*Pseudodiploria strigosa*) using concrete on natural open reef surface; and (G) 3-year-old out-planted *A. palmata* after wedging a fragment on a natural reef crack.

2.2. Limitation of previous efforts

The most significant limitation of previous efforts, not only in Puerto Rico, but also across the wider Caribbean region and across other locations around the world, has been the still small spatial scales impacted by ecological restoration efforts. Most projects have often been limited to the scale of 10–100 m^2. A combination of factors can limit the spatial scale of such endeavors, including: (a) limited economic, technological, and/or human resources; (b) limited source propagules; and (c) still limited success of larval-rearing methods. Many projects have often been experimental test beds for methods development, for methods demonstration, for mitigating specific environmental impacts (e.g., vessel grounding, storm/winter swell impacts, construction projects), or to address specific research questions. Many of these have shown promising success. However, their applicability to larger spatial scales still remains a question. This has led many detractors of ecological restoration to question, denigrate, or mock applied restoration ecologists, managers, practitioners, and NGO and community-based volunteers for "wasting money, time and efforts" in attempting to restore coral reefs through coral farming and other methods.

Coral farming and reef restoration have been successful at the scales so far implemented. But in order to become meaningful at ecological spatial scales, there is a need to improve spatial scales of future projects. This needs to include aspects such as: (a) increasing ecological and genetical connectivity to improve, for instance, fish assemblages spatial connectivity and functional redundancy; (b) rehabilitating benthic spatial heterogeneity to recover benthic microhabitats (e.g., shelter, foraging, and feeding grounds) for multiple species; and (c) rehabilitating overall coastal resilience, ecological services, and benefits (e.g., wave buffering role, nursery ground roles, landscape restoration to improve tourism and recreational value). This will require new fundraising strategies, as well as creative strategies for partnerships development, for engaging, educating, and training NGO and community-based volunteers, and for establishing a functional relationship between government agencies, academic institutions, industry, private business, and other community-based stakeholders. There might also be a need to combine low-tech with high-tech methods in an attempt to significantly improve the number of coral propagules for restoration. This may require novel international cooperative agreements for coral propagule sharing and exchange. This may even have the benefit of increasing local genetic diversity. Moreover, in many instances, there might still be a need to develop public policy and a vision in regards to coral reef conservation and ecological restoration to lead future efforts. Otherwise, successfully enhancing reef restoration spatial scales might remain a difficult task.

2.3. Benefits of reef rehabilitation

Low-tech coral farming and reef rehabilitation can have multiple local (**Table 2**) and regional (**Table 3**) ecological benefits, as well as multiple socio-economic and educational benefits (**Table 4**). Most of the benefits are derived from those previously described for marine protected areas [85]. The evaluation of reef rehabilitation benefits is often limited to immediate

Restoration criteria	Summary of benefits
Conservation	Propagation and reintroduction of largely depleted coral species which otherwise will have a very low probability of having successful sexual reproduction and colonization. Increased coral density to foster the recovery of coral reproduction potential at local scales and buffer the impact of reproductive isolation (Allee effects)
Reef accretion	Foster reef bio-construction by propagating and out-planting rapid-growing ecosystem engineer coral species. The reintroduction of rapid-growing coral species is aimed at helping local coral reef ecosystems to rapidly increase accretion rates, rehabilitate fish and invertebrate shallow-water nursery grounds, restore reef's wave buffering role, and adapt to projected rapid sea level rise
Habitat structural complexity	Rapid-growing coral species also help in the natural rehabilitation of benthic habitat structural complexity, which provides shelter to a myriad of reef demersal species
Biodiversity	Help replenish coral reef-associated biodiversity (e.g., fish, invertebrate species and functional groups) that use Acroporid biotopes as nursery, shelter and feeding grounds, and attract larger predators
Genetic resilience	Contribute to maintain and restore genetic diversity of targeted restored coral species by fostering reintroduction of multiple genetic clones, fostering genetic recombination on local scales, and promoting enhanced sexual recruitment on adjacent coral reefs by increasing ecological connectivity
Ecological functions	Rehabilitate coral functional redundancy as fish nursery grounds by improving benthic habitat complexity and restoring its function as fish and invertebrate habitat

Table 2. Local benefits from community-based coral farming and reef rehabilitation in face of climate change.

Restoration criteria	Summary of benefits
Uniqueness	Whether a restored area is "one of a kind" (e.g., habitats of endangered or rare species). Rehabilitated reef habitats with rare and/or threatened coral species can rapidly become a unique biological community with paramount significance across local, national, and regional scales because of its unique biological features, ecological functions, and critical genetic connectivity value
Naturalness	Degree to which the restored area helps in the recovery of reef's naturalness or lack of disturbance or degradation. Control or elimination of anthropogenic disturbance signals (e.g., overfishing, sedimentation, turbidity, pollution, anchoring, excessive SCUBA diving, rapidly declining coral cover). Areas with restored natural seascapes
Dependency	Degree to which a species or a group of species depend on a restored area. Degree to which an ecosystem depends on ecological processes occurring within the restored area. Enhanced ecological functions on local scales and recovered functional redundancy will benefit overall reef ecosystem functions
Representativeness	Degree to which a restored area represents a habitat type, ecological processes, biological community, geological features, or other natural characteristics, including the role as refuge for threatened or rare species
Integrity	Degree to which a restored area is a functional unit or an effective, self-sustaining ecological entity (e.g., a restored coral population undergoing annual sexual reproduction and functioning as recruiting and/or shelter, or feeding habitat for multiple species). Degree to which a restored area functions as a biological corridor between adjacent reefs, improving ecological connectivity
Productivity	Degree to which the productive processes within a restored area contribute benefits to adjacent reefs (e.g., fostering coral larval recruitment, fish spillover effects to other reef species). This can be achieved through enhancing recruitment, shelter, and/or feeding habitat. It can also benefit humans (e.g., ability of any given restored site to contribute to the sustainability of local fisheries either as a nursery ground or through the rehabilitation of important historical or traditional artisanal fishing grounds)
Connectivity	Degree to which a restored area is physically connected to other areas. Degree of connectivity between colonies of any given species at other areas via surface currents. The rehabilitation of critically located coral reefs will foster increased gamete and larval production of replenished coral species fostering potential higher recolonization of "downstream" reefs. Restored reefs will also foster similar effects for many fish and invertebrate species through spillover effects or mass spawning. The establishment of networks of restored reefs will improve the restoration success and connectivity effects on adjacent reefs
Regional significance	Degree to which the restored area represents a restored characteristic of the region or the degree to which the restored area fills a gap in a network of protected areas from the regional or sub-regional perspective. The larger the restored reefs network is, the larger the regional significance

Table 3. Regional benefits from community-based coral farming and reef rehabilitation in face of climate change.

components such as coral percent survival and growth rates. However, benefits to local components, such as conservation, reef accretion, habitat structural complexity, biodiversity, genetic resilience, and ecological functions, still remain poorly documented. Similarly, regional-scale factors, such as uniqueness, naturalness, dependency, representativeness, integrity, productivity, connectivity, and regional significance, are also limited. Furthermore, there are multiple socio-economic benefits, such as the role of reef rehabilitation as a climate change adaptation tool to increment carbon sequestration through calcium carbonate ($CaCO_3$) precipitation for

http://dx.doi.org/10.5772/intechopen.73151

Restoration criteria	Summary of benefits
Climate change adaptation	Rearing and propagation of high-temperature resistant, highly resilient, coral genetic clones with a higher ability to resist and recover from massive bleaching events will help improve overall reef ecosystem's resilience to future bleaching events
Reconstruction of physically destroyed reef structure	Foster the seascape-level reconstruction of bomb-cratered, physically demolished, and coral-depleted reefs, with the aim of fostering the rapid recovery of coral reef functions and ecological services
Ecological services	Increased coral densities will help to improve reef's greenhouse gases buffering role, its natural breakwater function, particularly during storm and hurricane swells, its natural pharmacy function (source of natural products of bio-medical significance), and will improve reef-based fisheries productivity. If reef rehabilitation is carried out within a no-take reserve, it will further foster larger fisheries productivity and a spillover effect favoring fisheries productivity across adjacent habitats open to fishing
Socio-economic benefits of improving costal resilience	Degree to which certain commercially important species depend on a restored area. Degree to which a restored area plays an important link to adjacent fisheries. Degree to which reef restoration will impact the local economy in the long term and improve existing or potential socio-economic value of an area for tourism and recreational activities. Degree to which reef restoration fosters the recovery of reef-based fisheries, improving catches on adjacent reefs, benefiting local artisanal fishers, and improving their livelihoods
Education and outreach	Reef rehabilitation provides a useful hands-on, transformative educational tool aimed at empowering local base-communities to manage their coral reefs and carry out coral farming and reef rehabilitation in face of projected climate change impacts
Community-based adaptation to climate change	Degree to which base communities educate, integrate into decision-making processes, become technically trained in coral farming and reef restoration methods, and become better adapted to manage their local coral reef resources under challenging scenarios of climate change

Table 4. Socio-economic and educational benefits from community-based coral farming and reef rehabilitation in face of climate change.

coral growth; to reconstruct a physically destroyed reef structure; to recover ecological services (e.g., essential fish habitat role), socio-economic benefits of improving coastal resilience, education and outreach, and community-based adaptation to climate change, which have been seldom addressed. These components should be thoroughly addressed in the future.

Knowledge regarding low-tech coral farming and reef rehabilitation has rapidly advanced in recent decades, but much more attention should be paid to expanding spatial scales of ecological rehabilitation, increasing the diversity of grown corals, exploring new methods, and improving the understanding of their long-term benefits as a tool to recover ecological functions in novel ecosystems and to restore coastal resilience in a time of major environmental and climate changes.

3. Lessons learned from coral farming in Puerto Rico

A suite of lesson-learning experiences in Puerto Rico has provided useful recommendations for managers to adaptively modify management actions, review, and amend existing marine protected areas management plans and to develop a set of minimum guidelines to drive future management-oriented decision-making processes, including reef restoration efforts. This will allow to maximize their ecosystem-level impacts, while at the same time address emerging threats and integrate challenging multi-disciplinary ecological paradigms into day-to-day management actions. The development of low-tech coral farming and reef rehabilitation methods has provided novel low-tech management tools to shape future efforts to recover fish communities, herbivore fish guilds, and long-spine sea urchin densities; reduce macroalgal cover; and recover coral densities and percent living cover. It has also provided guidance to reef managers and decision-makers regarding the ecosystem-level benefits of coral farming and reef restoration efforts. This can further allow the delineation of specific guidelines to implement future reef restoration efforts across the Caribbean region to promote cost-effective ecosystem-scale recovery. This will foster enhanced ecosystem resilience under increasing threats by local human-driven factors and climate change. Furthermore, it will provide the multi-disciplinary basis for addressing the emergent challenge of addressing novel coral reef ecosystem-based fisheries management, climate change-related impacts, community-based integration, and the development of conceptual models to address future multi-disciplinary social-ecological management challenges. Nevertheless, there are specific lessons learned directly associated to coral farming and out-planting activities that will provide specific guidance to managers and practitioners.

3.1. Siting of coral farms

Site selection for establishing either coral farms or out-planting locations is a critical step. In order to fully recover structural and functional characteristics of a degraded ecosystem, more research is needed for the selection of suitable transplant sites (e.g., optimum substrate characteristics; physical stability, exposure to wave action, optimum population λ). It is also paramount to address ecological factors that might stress out coral out-plants and affect their survival and growth (e.g., low percent cover or absence of red encrusting algae *Ramicrusta textilis*, sediment input and bedload, exposure to runoff and pollution). Another critical lesson learned in Puerto Rico was the need to avoid areas exposed to urban runoff, human trampling, and uncontrolled recreational impacts (e.g., snorkeling, SCUBA diving, kayaking, shore-based fishing, recreational navigation, and anchoring). Even "low-impact" nature-based recreational activities can have highly localized adverse impacts on shallow-water coral farms and reef restoration sites due to coral colony fragmentation and dislodgment.

3.2. Low-tech materials and design

A key component to low-tech approaches is maintaining a cost-effective operation, with multiple benefits and high success rates. Low-tech methods often involve the creative use of readily available, cheap materials to support *in situ* coral farms. Multiple coral farming unit designs have been successfully used in Puerto Rico involving the use of pvc plastic pipes, fishing lines, plastic-covered wire mesh, and concrete. There is not a specific universal method to meet all needs or that can be suitable for all locations. Factors such as wave action, surface current exposure, sediment dynamics, depth, visibility, and coral species to be used can be critical determinants of the methods to be implemented. However, there is evidence that horizontal line nurseries are highly successful in terms of coral colony percent survival rate, live tissue cover, and skeletal growth rate, when compared to colonies grown in wire mesh units [76, 80]. Coral colonies grown in line nurseries and other types of floating units often show faster growth rates, show lower living tissue lesions, and do better when out-planted to natural reef surfaces. Coral farming unit design is a function of specific local needs, available resources, projected number of coral propagules, projected coral reef restoration efforts, objectives of the restoration plan, size of source wild coral populations, and other logistical constraints. The latter may include: environmental conditions of the selected coral farming site, distance from source coral populations, availability of trained personnel, funding limitations, and other factors. But, in the long run, the local availability of materials can be the main factor influencing the final decision.

3.3. Timing of coral farming activities

It is critical that coral transplanting, unless necessary as an emergency restoration measure, avoids the warmest months. Survival rate shows a significant reduction during the late summer and early fall months due to a combination of impacts associated to high sea surface temperature, major runoff impacts, major risk of disease outbreaks, and the risk of bleaching. Most coral out-planting should be planned for winter and spring months to increase survival rates.

3.4. Collection of coral fragments

3.4.1. Avoidance of negative impacts on wild donor colonies

It is fundamental to reduce negative impacts of collections on wild donor coral colonies. Collection of coral fragments should be limited to 10–15% of the donor colony volume or surface area. No mortality or reduced growth should result in donor colonies due directly to fragmentation. In the case of branching corals, tissue regeneration at the breaking points should occur within 2–3 weeks, and branch growth should resume within a month. Impacts should be monitored at least for 3 months by direct comparison of a representative selection of donor colonies and adjacent control unaltered colonies and by looking at percent mortality, tissue regeneration rate, growth rate, and branchiness index (branch production). For larger foliose, plate, or massive colonies, donor colonies should be monitored for 6 months to a year as tissue regeneration, and skeletal regrowth is slower.

3.5. Transportation, handling, and out-planting

Transportation should always be conducted avoiding coral exposure to direct sunlight and warm temperatures. For short distances, colonies can be transported under subaerial exposure, but under humid conditions (e.g., under wet towels, under a saltwater sprinkler, etc.). But for longer distance travel, a water tank should be used provided with an air pump, water pump, and chiller to control temperature.

3.6. Local benefits

The major local relevance of coral farming activities in Puerto Rico has been the continuous expansion of staghorn coral (*A. cervicornis*) and elkhorn coral (*A. palmata*) low-tech restoration efforts in Culebra Island, Fajardo, Vega Baja, and multiple other locations. This has been achieved through an integrated effort of the different practitioners, NGOs, the Puerto Rico

Department of Natural and Environmental Resources (PRDNER) and NOAA-RC. Strengthening collaboration, communication, and sharing of lessons learning experiences among all engaged stakeholders has been a key for success, as well as for improving support and volunteer collaboration among groups. This has also allowed to significantly increase the number of harvested colonies available for future reef restoration efforts. In addition, during recent years, there has also been an increase in the number of new community-based volunteers technically trained in coral transplanting, coral harvesting, and farm maintenance to collaborate at all project sites. In the particular case of NGO SAM, this experience was also used to successfully train volunteers at the Dominican Republic, resulting in the development of a long-term coral farming program at Punta Cana.

Direct benefit	Added values
Enhance public presence and leading role of NGOs and the academia addressing coral reef conservation issues	Strengthen out the Caribbean-wide leading role of PR as a model for the development of effective strategies for the multi-disciplinary integration of different sectors in the mitigation and adaptation to climate change impacts. Presence on the community can be achieved by multiple outreach activities by project's personnel, including seminars, information dissemination through the web, and mass media articles
Increase number of trained and educated professionals and students	Provide direct support and the basic framework for providing hands-on education and technical training to graduate and undergraduate research students regarding marine biodiversity conservation, coral farming, and reef ecological rehabilitation in the context of climate change impacts. This may also provide theoretical and hands-on practical training on coral reef conservation, coral farming, reef rehabilitation and monitoring, and coral demographic data processing, analysis, and interpretation
Increase number of trained and educated stakeholders	Provide hands-on education and technical training to community-based volunteer personnel through community-based organizations regarding marine biodiversity conservation, coral farming, and reef ecological rehabilitation in the context of climate change impacts
Empowerment for collaborative management	Project's participants (academic, community-based) acquire the necessary knowledge, skills, and experience to support government efforts to manage coral reefs resources through a collaborative, participatory model
Advance the implementation of a no-take marine protected areas management plans	Advance the implementation of no-take MPA management plans in support of government efforts
Advance NOAA Habitat Focus Areas goals	In the particular case of Puerto Rico, reef rehabilitation can achieve the NOAA Habitat Focus Areas goal of sustaining resilient and thriving marine and coastal resources, communities, and economies by addressing a habitat-based issue/concern contributing to the loss or deterioration of coastal resiliency or marine habitats for target managed or protected coral species
Fill critical data gaps for resource managers and decision-makers	Advance knowledge and help fill critical qualitative and quantitative information gaps about ESA-listed coral species across the U.S. Caribbean to support the implementation of management strategies aimed at the recovery of their depleted populations
Coral reproduction (=net reef accretion)	Increase coral out-planting across reef-seascape scales to increase local reproductive populations of depleted species across different coral reef. In the long-term, this will foster increased reef accretion rates
Fish productivity	Increase reef accretion to enhance benthic spatial heterogeneity and the rapid rehabilitation of fish communities mostly by fostering increased fish recruitment and by enhancing herbivore guilds. These are important steps towards recovering connectivity and ecosystem resilience

Direct benefit	Added values
Connectivity (=buffer against further decline)	Increase number of rehabilitated reef patches to increase genetic connectivity across reef seascape scales. In the long term, this will increase connectivity with other coral reefs across ecological to regional scales
Coastal resilience	Increase coral density, wave buffering role, genetic connectivity, recover fish communities, and rehabilitate herbivory levels to help recover coastal resilience
Ecosystem resistance to future disturbances	Increase ecosystem resilience to foster an increased resistance to future disturbances (e.g., hurricanes). This is important for the sustainability of reef's ecological functions, goods, benefits, and services
Buffering of sea level rise (SLR) associated shoreline erosion	A rehabilitated coral reef should also recover its natural accretion rates to cope with increasing SLR and its concomitant shoreline erosion. Therefore, it should protect the shoreline from strong wave action and shoreline erosion
Socio-economic value of reef ecosystems	The recovery of reef's ecological functions, goods, benefits, and services should lead to increasing its net productivity and socio-economic value
Community revitalization	A healthy reef provides multiple benefits for local base communities and become instrumental in revitalizing local economies and societies, particularly, in small island scenarios
Food security & sovereignty	A rehabilitated reef will also recover its ability to produce fish protein. Therefore, increased fish biomass will contribute to increasing food security and sovereignty
Goods, benefits and services	Healthy recovered reefs will increase its multiple benefits to humans (e.g., production of food and natural compounds of bio-medical importance, natural breakwater, recreation, and tourism activities). This is fundamental for recovering the economy of small islands
Business opportunities	Successful coral reef rehabilitation has triggered a dramatic increase in low-impact tourism activities in Culebra Island with an informally estimated impact of at least $10 million USD annually. This project will contribute to recover other coral reef habitats, further representing new business opportunities and serving as a model for other locations in PR and the rest of the Caribbean
Recreational opportunities	Rehabilitated reefs and enhanced fish communities also become highly attractive for tourists, snorkelers, and SCUBA divers. This creates multiple new opportunities for the development of recreational activities
Sustainable tourism	Coral reef rehabilitation creates the basis for the development of small island sustainable tourism practices. In this sense, the academia and NGOs will have the unique opportunity to also become leaders in the development of environmentally and socio-economically sustainable activities for small islands
Carbon sequestration and offsetting	Exponentially increasing coral growth lead to an exponential increase in atmospheric carbon dioxide (CO_2) sequestration in the form of calcium carbonate ($CaCO_3$) precipitation during coral calcification. This creates the unique opportunity for developing a carbon offsetting business through low-tech coral farming and reef rehabilitation
Property values	Healthy thriving coral reefs adjacent to the shoreline help to increase adjacent properties values (e.g., landscape, shoreline erosion protection, source of recreation, and food protein)
Stakeholder livelihoods	Healthy reefs help to maintain sustainable livelihoods of local community residents by becoming a potentially sustainable source of food and revenue
People's security, happiness, and wellbeing	Increased livelihood, business, and recreation opportunities for local communities contribute for sustaining their quality of life, security, happiness, and wellbeing

Table 5. Summary of return of investment and added values of coral farming and reef rehabilitation projects.

But in summary, coral farming and reef rehabilitation have provided an important return of investment as well as multiple added values listed in **Table 5**. Basically, it has provided several important local benefits, including the basis for expanding the spatial scale of a sustainable, ecosystem-based model aimed at the recovery of coral reef's ecological functions and services. It has also fostered an improved integration and participation of community-based organizations, the academia, and government agencies to improve opportunities for community-based outreach, hands-on education, technical training, and empowerment. It has also contributed baseline information to support the development and implementation of a public policy in the Commonwealth of Puerto Rico for the conservation of marine biodiversity and the rehabilitation of coral reefs ecosystem resilience, functions, benefits, goods, and services.

In the long term, coral reef rehabilitation is a win-win for all local stakeholder sectors. For local managers, projects can enhance the public presence of leading community-based NGOs and the academia, can contribute to increase the number of trained and educated professionals and stakeholders, foster empowered collaborative management, advance the implementation of no-take MPAs, and habitat conservation goals, and can provide timely information for resource managers and decision-makers. Projects can also provide fundamental baseline information regarding factors such as coral reproduction and growth, fish productivity, connectivity, and coastal resilience. Reef rehabilitation strategies can also contribute to enhance local ecosystems resistance to disturbance, can contribute to buffer wave action, and in the long-term, shoreline erosion associated to sea level rise. Coral reef rehabilitation can also foster a myriad of socio-economic benefits such as increasing the socio-economic value of reef ecosystems; triggering community livelihood revitalization; recovering food security and sovereignty, goods, benefits, and services; fomenting the creation of business and recreational opportunities and the development of sustainable tourism practices; fomenting carbon sequestration and offsetting; and improving property values, multi-stakeholder livelihoods, and people's security, happiness, and wellbeing. Most of these impacts have never been addressed in the literature as they often fall outside the scope of most research and conservation grants, which fail to address multi-disciplinary and social-ecological components of coral reef restoration.

4. Lessons learned from maintenance and data collection

4.1. Maintenance

Regular maintenance of coral farms, and often of out-planted colonies, at least on their initial stages, is a critical process for the success of any project. Such activities can be easily coupled with regular monitoring of corals in farms and of out-planted colonies. Maintenance efforts should have the following objectives:

1. *Sustain health and survival of coral colonies*. This requires regular visits (e.g., depending on the coral farming method, location, trophic condition of the site, herbivory level, etc.), from monthly to at least 3-month interval. However, it is highly recommended to visit and inspect coral farms at least not later than 1 week after coral farming set up and to address any potential structural failure and any possible adverse impact of coral fragment mishandling

and/or transportation stress effect. If possible, corals should be revisited 2 weeks after transplanting. Then, they could be visited after a month, and then at 3-month interval, though this can vary depending on the method, distance from the shore, difficulty of access, etc. This will allow the frequent manual removal of algae, fouling, and opportunistic taxa (e.g., sponges, mat tunicates, hydroids, fire coral) that can potentially smother and/or overgrow coral fragments in farms. This will also allow to identify and remove injured or diseased colonies to prevent potential transmission to other healthy colonies.

2. *Repair potential mechanical damages on coral farming units.* Regular maintenance visits will allow to repair any potential mechanical failure of coral farming units as a result of strong wave action, storm impacts, or damages inflicted by human activities such as boating, anchoring, fishing gear, snorkelers, and recreationists.

3. *Allow qualitative and quantitative documentation of colony survival and growth.* The combination of regular maintenance and monitoring can allow regular qualitative (e.g., photography, video) and quantitative assessments of colony survival, growth, and health conditions.

4.2. Monitoring of coral farms

4.2.1. High survival of coral fragments in nurseries

An important goal would be to achieve a high percent survival (>80%) for coral fragments within the nursery, excluding stochastic events such as fragmentation by storm swells, disease outbreaks, massive bleaching, anchoring, fishing gear entanglement, or snorkeler/diver impacts. Percent colony survival rate should be quantified from the entire population on each farming unit, as well as assessing colony condition and source of mortality, if present (e.g., fragmentation, predation, disease, bleaching, etc.) in a representative sample of fragments. If different genetic clones are being grown, then information should be addressed for each specific clone with an appropriate replicate number of samples per clone.

4.2.2. High productivity and growth of coral fragments in nurseries

Coral fragment growth data, in combination with percent survival rate, are the most straightforward approach to address coral farming productivity success. Basic growth data can include high-resolution fragment's skeletal extension rate, total linear extension, branch abundance, branchiness index (number of harvestable branches above any given minimum size, say 10 cm), by addressing colony diameter or volume or by calculating weight of calcium carbonate produced ($CaCO_3$), either by direct measurements through buoyant weight methods or by geometric estimations. Growth data could be highly variable, depending on sampling size. Therefore, care must be taken to sample a representative number of fragments to minimize variance to achieve a precision >0.80. Depending on the scientific questions addressed and the specific needs of each project and coral species, sampling frequency could be variable, monthly, bi-monthly, seasonal or bi-annual. Sampling can also address colonies from different sources or genetic clones, different generations, and from different size categories. However, it must be noted that if sampling frequency is too low (say, 6-month interval), impacts associated to seasonal variability, pulse events (e.g., rainfall, runoff), predation, disease outbreaks, sea surface warming, or other ecological surprises can be overlooked and not appropriately addressed, therefore, missing critical timely information for managers to understand population dynamics.

High-resolution coral survival and growth data are also critical for parameterizing demographic models and addressing questions of demographic dynamics. However, to gather a basic understanding of coral survival and growth, a basic assessment of coral fragment sizes at the beginning of the project and at any given time later (e.g., 6 months, a year), standardized to initial size, will be enough to address productivity. Productivity will be calculated as annual growth/initial fragment size. Regardless of the approaches taken, it would be important to keep data from different sources/genotypes/generations separate.

5. Lessons learned from out-planting

5.1. Siting of out-planting

The selection of out-plants siting is paramount. The most important elements to consider are to conduct a prior evaluation of environmental history of the potential recipient sites. Is the area too close to urban centers? Are there any adjacent known sewage or storm water outfalls? Any adjacent river outlet? Is there heavy sediment resuspension from boating activities or wave action? Is it too shallow and the area is exposed to constant sea surface warm spells? Is the area highly frequented by snorkelers and trampling activities? Is it impacted by excessive fishing pressure? Is there any evidence of sediment bedload impacts (horizontal sediment displacement)? Is there any lack of juvenile coral colonies? Are there any wild surviving remnants of the targeted coral species? Are there too many standing dead colonies of the targeted coral species and no evidence of sexual or fragment recruitment? These are only a few important elements that must be taken into consideration when planning site selection for out-planting. But, the final selection of the out-planting site should be based on the following standard criteria:

1. Hard bottom substrate free of sediment bedload (=horizontal sediment transport) preferably exposed to moderate water circulation.
2. No observation of fire corals (*Millepora* spp.), sponges, or harmful algae in the vicinity of each out-planted point that could hamper coral colony survival and growth.
3. Select areas known to have previously supported the targeted coral species and that currently have adequate water quality (e.g., adequate transparency, low sedimentation input, no direct sewage effluents, far from river outlets).
4. If possible, select areas with high benthic topographic relief.
5. Avoid areas with high density of coral predators (e.g., corallivore gastropods, fireworms).
6. Avoid areas exposed to significant recurrent runoff effects.

7. Avoid coral out-planting in direct contact with other corals.

8. Prior to coral out-planting, scrap off the selected sites with a wire brush to remove any algal turf or minor sediment deposits, if any.

9. If out-planting massive corals, select reef outcrops for out-planting. If there is moderate-to-high wave energy or moderate-to-strong surface current conditions at the pre-selected recipient site, various masonry nails should be driven to the substrate as anchors. Then, a dense cement/sand/lime mixture should be placed over the cleaned area among each nail patch. The puck should be buried and secured in the cement, leaving the coral over the substrate.

5.2. Out-planting spatial design

Out-planting spatial design is fundamental for project success and for achieving specific coral reef conservation and rehabilitation goals. Out-planted coral density and spatial configuration can play a critical role in the formation of Acroporid coral thickets, although coral density may play an adverse role in coral survival and growth [60]. But demographic evidence from ongoing studies in Puerto Rico have not shown such trend. In contrast, locality and environmental conditions seem to be more critical factors than species [86] or density (Hernández-Delgado, unpublished data) in affecting out-planted coral survival rates. Increasing out-planted staghorn coral (*A. cervicornis*) (e.g., 1–4 colonies/m^2) can lead to faster recovery of fish assemblages than reefs with lower density (say, 1 colony per 4 m^2) or in control sites without out-plants (Hernández-Delgado, in review). If the objective of coral out-planting is to help rebuild over-exploited fish assemblages, then appropriate spatial designs are key to success. The faster a coral thicket can be formed the faster juvenile fish assemblages can re-establish. A key goal identified by NOAA was for the recovery of *A. cervicornis* populations when "*thickets are present throughout approximately 5 percent of consolidated reef habitat in 5 to 20 m water depth within the forereef zone. Thickets are defined as either a) colonies ≥0.5 m diameter in size at a density of 1 colony per m^2 or b) live staghorn coral benthic cover of approximately 25 percent*" [87]. Achieving such parameters will be dependent upon spatial design. Aspects such as natural reef spatial configuration, depth, wave exposure, presence of special features such as rocky outcrops, spur, and groove systems, aggregated reef patch density, etc. can influence spatial configuration. But as a general rule, depending on coral species, objective of the project and harvested coral abundance spatial configuration can be modified accordingly. In general, if the goal is to rehabilitate local fish assemblages, probably a mosaic of small-to-moderate restored thickets (e.g., 16, 25, 50, 100 m^2 per thicket, with densities of 1–2 colonies/m^2) can be constructed. But if the goal is to provide a long-term buffer against wave action and shoreline erosion, probably parallel lineal configurations of elkhorn coral (*A. palmata*) patches can be constructed along shallow depth contours, from the reef front to the back reefs (e.g., multiple patches 10 × 3 m, 20 × 3 m, 50 × 3 m). Or in such cases, restoration can follow natural existing outcrops contours and configurations. The take home message is to bear in mind what is the specific goal of the project and plan ahead the spatial arrange of corals in order to have an estimate of short- and long-term coral production goals.

5.3. Strategies for out-planting

Coral out-planting requires a thoroughly planned strategy to increase the probability of success. These include several categories. First, transportation needs to be appropriately planned. If harvested corals will be obtained from a local or nearby location (e.g., <1 h transportation), fragments can be transported even under subaerial exposure. However, care must be taken that corals remain humid, well oxygenated, and away from direct sunlight to prevent heat shock stress, possible bleaching, and mortality. But if corals will be transported from farther distances (e.g., 1–2 h or more), then corals must remain submerged, with battery-powered water pump, aerator, and chiller. This has proven a successful method for transporting harvested staghorn coral (*A. cervicornis*) over distances of 150 miles and over 10:30 h of transporting.

Timing of coral out-planting is also an important element to consider. Warmer months (early summer to late fall) must be avoided. At least, across the wider Caribbean region, the warmest period (June to November) coincides with the Atlantic hurricane season. Out-planting during higher sea surface temperature can create higher physiological stress to corals due to higher temperatures in combination with often higher dissolved nutrient concentrations associated to heavy rainfall and nutrient-loaded runoff impacts. These conditions also often foster increased macroalgal and cyanobacterial blooms in many locations, which could further harm recently out-planted colonies. In addition, corals undergo the final stages of gametogenesis during the summer months, therefore, adding extra stress to out-plants.

Finally, the selected strategy will largely have to do with project's goals and objectives. Therefore, components, such as coral size, number of colonies, spatial configuration, and genetic diversity, are important elements to take into consideration for future success.

5.4. Establish restoration benchmarks

Establishing quantitative benchmarks is a key element of any coral reef restoration project. This would imply defining a clear goal, often achievable within 1 year, but larger temporal scales must also be considered (say, 3, 5, 10 year goals). As a minimum, this would include benchmarks for the number of surviving out-planted fragments and for growth rates. Benchmarks would largely depend on the goal of the project. If a goal is limited to replenishing a depleted coral species at any given location or set of locations, then survival and growth rates would provide enough information to address success. For instance, if a 70% colony survival rate is established as a benchmark for out-planted corals, then a survival rate >80% would be considered a success and no actions or improvements would be necessary. If colony survival ranges say between 70 and 80%, caution should be taken and some adjustments should be made to ensure improved success in future efforts. But if colony survival falls below 70%, then action must be rapidly taken to improve methods, spatial design, or site selection.

However, if goals include enhancing ecological connectivity among adjacent reefs, then additional metrics would have to include addressing coral recruitment rates for the out-planted species and genetic connectivity. Also, if the goals include the ecological rehabilitation of reef's functions, then metrics regarding fish community structure, sea urchin populations, and/or herbivory rates would have to be included. The most important aspect of setting benchmarks is to keep in mind that experimental data from the restoration project must be compared to any given "control" or wild site. A wild site would be an ideal habitat with wild colonies of the same restored species, or a site with similar ecological/environmental conditions, but without the restoration intervention. An alternative would be to establish comparisons with different methods, among different locations, and/or to compare restoration performance metrics to those available in the literature from similar projects and from wild sites.

5.5. Data collection

5.5.1. High survival rates of out-planted corals

The survival rate of out-planted coral fragments would be expected to exceed 70% after 1 year in the absence of stochastic events (e.g., hurricanes, winter swells, extreme rainfall and runoff, massive coral bleaching, disease outbreaks, or other ecological surprises). It would be important to keep close track of mortality sources. Additional factors, such as predation, out-competition, disease, sediment bedload, turbidity, and changes in water quality, may play a critical role affecting colony survival and growth. This could vary from location to location, among seasons, and with stochastic disturbances. Percent survival rate should be addressed during each site visit. Also, if known, mortality sources should be documented. If too many colonies are out-planted, then a representative sample should be monitored. If possible, source/genotype/generation data should be kept separate.

5.5.2. High productivity and growth of out-planted corals

Similar to productivity and growth of nurseries, as a minimum, initial and final productivity data for out-planted colonies must be collected to set a benchmark range of coral colony parameters. Parameters addressed, as well as sampling size and frequency, would vary depending on the research questions and project's goals. However, sampling approaches would be similar to those outlined under Section 4.2.2. If possible, it would also be important to keep data from different sources/genotypes/generations separate.

6. Additional recommendations and needs for best management practices

There are other important elements to consider regarding the needs for best management practices (BMPs) for low-tech coral farming and reef rehabilitation. First, there is a need to improve the ability to demonstrate and communicate the socio-economic value and utility of coral reef rehabilitation in providing substantial ecosystem services (e.g., coastal protection, fisheries enhancement, resilience recovery, enhanced revenues from tourism, etc.) to local and state governing bodies, as well as the private sector. This will support decision-making processes on government institutions (across national, regional, local, municipal scales), as well as on cross-sectorial scales, and even on base-community level. This could be important to potentially unlock new funding avenues.

Secondly, there is a need to significantly increase the efficiency and scale of coral restoration to achieve the overall goal of establishing self-sustaining, sexually reproductive populations, and to enhance/restore genetic and ecological connectivity. Sexual reproduction is important to recover coral reef from crisis as it involves the evolutionary mechanisms, such as genetic recombination, that will enable adaptation to the future conditions that corals will face in the context of climate change. There is an implicit need to optimize current coral propagation techniques, including larval propagation, to improve out-planted coral colony survival rates and the efficiency of out-planting asexually derived coral fragments, a need to develop more efficient strategies to accelerate coral growth, and a need to determine the size and density of out-plants for a given coral species to rapidly reach the development of a functional thicket. Also, it would be central to improve genetic variability during out-planting, enabling long-term ecological adaptation to changing climate and environmental conditions and improving population resilience.

There is also a need to develop and implement standardized monitoring guidelines that cover various levels of information (e.g., coral skeletal dynamics, colony conditions, demographic dynamics, spatial extent of live coral cover, thicket development dynamics, genetics, ecosystem functioning). Long-term monitoring must also address different organizational scales, from individual coral colonies to ecosystem processes. Sharing data among practitioners across local, regional, and global scales is central to facilitate understanding of spatio-temporal variability of ecosystem status. Also, fostering inter- and trans-disciplinary dialog among practitioners, scientists, managers, base community volunteers, and other cross-sectorial stakeholders is fundamental, particularly to share lessons learned. There are also fundamental tools to better guide future development of coral reef rehabilitation projects, including: a) Cost-benefit analyses to improve investments in projects; b) Risk analyses to improve siting decision-making; c) the integration of cell automata models to address thicket development potential under different environmental scenarios; d) The coupling of numerical wave models with demographic and spatial heterogeneity models to project wave buffering impacts with reef rehabilitation; and e) the integration of genetic models aimed at the long-term increase of genetic diversity. This would allow prioritizing potential restoration candidate locations, funding allocation, research agendas, and developing targeted strategies for long-term coral persistence in a changing world.

7. Conclusions and recommendations

Low-tech coral farming and reef rehabilitation have become important community-based tools, particularly across multiple small island nations, to foster enhanced non-governmental participation in coral reef management. At least in the wider Caribbean region, these strategies have been successfully implemented to recover depleted coral populations, mostly of fast-growing, but increasingly rare, staghorn (*A. cervicornis*) and elkhorn coral (*A. palmata*). They have also been used with relative success, though with limited documentation, to recover depleted fish assemblages. Indirectly, coral reef rehabilitation has also resulted in enhanced benthic spatial heterogeneity, in providing multiple new microhabitat for fish and invertebrate species; has contributed to the recovery of coastal resilience, increasing the protection of shorelines against erosion; and has fostered an increased interest of the tourism sector as an enhanced attraction for visitors and recreationists. But in order to sustainably maintain such benefits, it would be important to pay attention to several metrics and/or components to ensure success.

First, it is important that there are no negative effects of collections from wild donor colonies. There should be no partial or total parental colony mortality, in comparison to adjacent control colonies, and collection of coral fragments should never exceed 15% of the parental colony volume. There should not be reduced growth observed in donor colonies either due directly to fragmentation, and parental colonies should show rapid tissue repair and skeletal regrowth (usually within less than a month). In addition, there should be high percent survival of nursery fragments within the first 1–2 years (>80% survival rate), with the exception of stochastic events such as hurricanes, winter swells, cold water events, disease outbreaks, severe bleaching, etc. If different genotypes are being tested, separate data from different genotypes must be assessed. It would also be important to address specific causes and time of mortality.

Another important component that must be thoroughly addressed is high growth and/or productivity of nursery-grown fragments. This should be calculated as total annual growth/initial fragment size, thus providing a standardized measure of productivity relative to the initial fragment size and helping to reduce some of the variability associated to variations in the initial size of fragments. This would be important for addressing differences among size categories and parameterizing demographic models. It could also allow testing for differences among sites, depth zones, treatments (e.g., along environmental stress gradients, MPAs vs. control non-MPA sites, etc.), and season. It would be difficult, however, to establish a baseline or a standardized range of parameter values due to high latitudinal, longitudinal, site, depth, environmental, genetic, fragment size, and seasonal variability. There can also be substantial variability associated to the use of different methods, even within the same sites, depth zones, and environmental regimes [79]. Therefore, each country, biogeographic zone, or individual project should establish its own monitoring strategies to establish their own baselines. It would also be important to provide frequent maintenance to coral farming units. Measures need to be taken to periodically address colony survival rates, address the physical structure of farming units, remove predators, remove algae or nuisance fouling taxa, and move the nursery to deeper waters in case of storms or hurricanes to improve productivity or projects. A final important metric that should be strongly enforced is to ensure a high survival (>70%) and high productivity of nursery-reared out-planted corals in the absence of stochastic events. This should be achieved through permanently tagging selected, representative out-planted fragments, and through a regular permanent monitoring program. This approach is fundamental for assessing demographic dynamics.

There are also important take home messages to foster increased success of low-tech coral farming. The first is the need to secure sustained, recurrent funding. This may allow to secure a continued input of harvested corals to reef rehabilitation projects, which may allow expanding the spatial scale of projects. Increased spatial scales is a second concern, with the particular aim of fostering enhanced ecological benefits, such as enhancing essential fish habitats, restoring juvenile fish nursery grounds, and recovering ecosystem and coastal resilience. In addition, there is a need to incorporate demographic modeling to coral reef rehabilitation projects. This may allow to improve the ability to address vital population dynamics and project population fate under variable environmental and climate change-related scenarios. Coral farming and reef rehabilitation have also shown to be a successful fishery management tool on local scales. Therefore, it can be used to integrate local communities and fisher villages to fishery management strategies. Under increased spatial scales, it should also become a tool to manage coral and fish connectivity, at least across ecological spatial scales. This would further foster the implementation of a participatory model to foster improved coastal resilience, MPA, and coastal resilience management.

Finally, it would be fundamental to foster the creation of functional partnerships among base communities, NGOs, the academia, government institutions, and the private sector. This would allow the development of stronger networks to improve volunteerism, outreach and education, and improve the possibility of securing continuous funding support. In a time of projected increases in climate change and sea level rise, low-tech coral farming and reef rehabilitation must be fortified and expanded across multiple localities. Only through the integration of multiple sectors of society, the goal of expanding the spatial scale and full community integration can be achieved. In a time of significant projected climate change impacts and sea level rise, improving the scale of coral farming and reef rehabilitation has become a critical tool for coral reef conservation. But multiple roadblocks must still be overcome. The future of coral reef productivity and its attractiveness for tourism can be sustained through proper participatory management for the enjoyment of future generations.

Acknowledgements

This publication was possible thanks to the support of the National Science Foundation (HRD #0734826) to the Center for Applied Tropical Ecology and Conservation and the support provided by the University of Puerto Rico Central Administration, both to E.A. Hernández-Delgado. Also, support was provided by NOOA Restoration Center and The Nature Conservancy to Sociedad Ambiente Marino through subaward MAR-SAM-110110. In addition, support was provided by the U.S. Fish and Wildlife Service Coastal Program, the University of Puerto Rico's Sea Grant College Program, the Ford Motor Company Foundation, Toyota Foundation, Roland Pesch, and by a myriad of community-based student volunteers through years of dedicated work. Our major appreciation to the passionate vision and pioneering support by the late fisher folks of Culebra Island, in particular, Don Ramón "Monchín" Feliciano and Don Anastasio "Taso" Soto.

http://dx.doi.org/10.5772/intechopen.73151

Author details

Edwin A. Hernández-Delgado[1,2,3]*, Alex E. Mercado-Molina[3,4] and Samuel E. Suleimán-Ramos[3]

*Address all correspondence to: edwin.hernandezdelgado@gmail.com

1 Center for Applied Tropical Ecology and Conservation, University of Puerto Rico, San Juan, Puerto Rico, USA

2 University of Puerto Rico, College of Natural Sciences, Interdisciplinary Program, San Juan, Puerto Rico, USA

3 Sociedad Ambiente Marino, San Juan, Puerto Rico, USA

4 Department of Marine Sciences, Florida International University, Miami, Florida, USA

References

[1] Thompson A, Schroeder T, Brando VE, Schaffelke B. Coral community responses to declining water quality: Whitsunday Islands, great barrier reef, Australia. Coral Reefs. 2014;**33**(4):923-938

[2] Cloern JE. Our evolving conceptual model of the coastal eutrophication problem. Marine Ecology Progress Series. 2001;**210**:223-253

[3] Rogers CS. Responses of coral reefs and reef organisms to sedimentation. Marine Ecology Progress Series. 1990;**62**:185-202

[4] De'ath G, Fabricius K. Water quality as a regional driver of coral biodiversity and macroalgae on the great barrier reef. Ecological Applications. 2010;**20**(3):840-850

[5] Uthicke S, Patel F, Ditchburn R. Elevated land runoff after European settlement perturbs persistent foraminiferal assemblages on the great barrier reef. Ecology. 2012;**93**(1):111-121

[6] Roberts CM. Effects of fishing on the ecosystem structure of coral reefs. Conservation Biology. 1995;**9**(5):988-995

[7] Jackson JB, Donovan MK, Cramer KL, Lam VV. Status and Trends of Caribbean Coral Reefs. IUCN, Gland, Switzerland: Global Coral Reef Monitoring Network; 2014. Available at http://cmsdata.iucn.org/downloads/caribbean_coral_reefs___status_report_1970_2012.pdf

[8] Fox HE, Caldwell RL. Recovery from blast fishing on coral reefs: A tale of two scales. Ecological Applications. 2006;**16**(5):1631-1635

[9] Precht WF, Aronson RB, Swanson DW. Improving scientific decision-making in the restoration of ship-grounding sites on coral reefs. Bulletin of Marine Science. 2001;**69**(2):1001-1012

[10] Hernández-Delgado EA, Montañez-Acuña A, Otaño-Cruz A, Suleimán-Ramos SE. Bomb-cratered coral reefs in Puerto Rico, the untold story about a novel habitat: From reef destruction to community-based ecological rehabilitation. Revista de Biología Tropical. 2014;**62**:350-367

[11] Rouphael AB, Inglis GJ. Increased spatial and temporal variability in coral damage caused by recreational scuba diving. Ecological Applications. 2002;**12**(2):427-440

[12] Dinsdale EA, Harriott VJ. Assessing anchor damage on coral reefs: A case study in selection of environmental indicators. Environmental Management. 2004;**33**(1):126-139

[13] Hoegh-Guldberg O, Mumby PJ, Hooten AJ, Steneck RS, Greenfield P, Gomez E, et al. Coral reefs under rapid climate change and ocean acidification. Science. 2007;**318**(5857): 1737-1742

[14] Hoegh-Guldberg O, Bruno JF. The impact of climate change on the world's marine ecosystems. Science. 2010;**328**(5985):1523-1528

[15] Baker AC, Glynn PW, Riegl B. Climate change and coral reef bleaching: An ecological assessment of long-term impacts, recovery trends and future outlook. Estuarine, Coastal and Shelf Science. 2008;**80**(4):435-471

[16] Eakin CM, Morgan JA, Heron SF, Smith TB, Liu G, Alvarez-Filip L, et al. Caribbean corals in crisis: Record thermal stress, bleaching, and mortality in 2005. PLoS One. 2010;**5**(11):e13969

[17] Brandt ME, McManus JW. Disease incidence is related to bleaching extent in reef-building corals. Ecology. 2009;**90**(10):2859-2867

[18] Randall CJ, Jordán-Garza AG, Muller EM, Van Woesik R. Relationships between the history of thermal stress and the relative risk of diseases of Caribbean corals. Ecology. 2014;**95**(7):1981-1994

[19] Aronson RB, Precht WF. White-band disease and the changing face of Caribbean coral reefs. Hydrobiologia. 2001;**460**:25-38

[20] Gardner TA, Côté IM, Gill JA, Grant A, Watkinson AR. Long-term region-wide declines in Caribbean corals. Science. 2003;**301**(5635):958-960

[21] Paddack MJ, Reynolds JD, Aguilar C, Appeldoorn RS, Beets J, Recent BEW. Region-wide declines in Caribbean reef fish abundance. Current Biology. 2009;**19**(7):590-595

[22] Anthony KR, Kline DI, Diaz-Pulido G, Dove S, Hoegh-Guldberg O. Ocean acidification causes bleaching and productivity loss in coral reef builders. Proceedings of the National Academy of Sciences. 2008;**105**(45):17442-17446

[23] Vega Thurber RL, Burkepile DE, Fuchs C, Shantz AA, McMinds R, Zaneveld JR. Chronic nutrient enrichment increases prevalence and severity of coral disease and bleaching. Global Change Biology. 2013;**20**(2):544-554

[24] Anthony K, Maynard JA, Diaz-Pulido G, Mumby PJ, Marshall PA, Cao L, Hoegh-Guldberg O. Ocean acidification and warming will lower coral reef resilience. Global Change Biology. 2011;**17**(5):1798-1808

[25] Perry CT, Harborne AR. Bioerosion on modern reefs: Impacts and responses under changing ecological and environmental conditions. In: Hubbard DK, Rogers CS, Lipps JH, Stanley GD Jr, editors. Coral Reefs at the Crossroads. Dordrecht, The Netherlands: Springer Science; 2016. pp. 69-101. DOI: 10.1007/978-94-017-7567-0_4

[26] Elsner JB. Evidence in support of the climate change–Atlantic hurricane hypothesis. Geophysical Research Letters. 2006;**33**:L16705

[27] Webster PJ, Holland GJ, Curry JA, Chang HR. Changes in tropical cyclone number, duration, and intensity in a warming environment. Science. 2005;**309**:1844-1846

[28] Emanuel K. Increasing destructiveness of tropical cyclones over the past 30 years. Nature. 2005;**436**:686-688

[29] Aronson RB, Precht WF. Physical and biological drivers of coral-reef dynamics. In: Hubbard DK, Rogers CS, Lipps JH, Stanley GD Jr, editors. Coral Reefs at the Crossroads. Dordrecht, The Netherlands: Springer; 2016. pp. 261-275. DOI: 10.1007/978-94-017-7567-0_11

[30] Bruno JF, Selig ER. Regional decline of coral cover in the indo-Pacific: Timing, extent, and subregional comparisons. PLoS One. 2007;**2**(8):e711

[31] De'ath G, Fabricius KE, Sweatman H, Puotinen M. The 27-year decline of coral cover on the great barrier reef and its causes. Proceedings of the National Academy of Sciences. 2012;**109**(44):17995-17999

[32] Sheppard CR, Harris A, Sheppard AL. Archipelago-wide coral recovery patterns since 1998 in the Chagos archipelago, central Indian Ocean. Marine Ecology Progress Series. 2008;**30**(362):109-117

[33] Golbuu Y, Victor S, Penland L, Idip D, Emaurois C, Okaji K, Yukihira H, Iwase A, Van Woesik R. Palau's coral reefs show differential habitat recovery following the 1998-bleaching event. Coral Reefs. 2007;**26**(2):319-332

[34] Adjeroud M, Michonneau F, Edmunds PJ, Chancerelle Y, De Loma TL, Penin L, et al. Recurrent disturbances, recovery trajectories, and resilience of coral assemblages on a south Central Pacific reef. Coral Reefs. 2009;**28**(3):775-780

[35] Gardner TA, Cote IM, Gill JA, Grant A, Watkinson AR. Hurricanes and Caribbean coral reefs: Impacts, recovery patterns, and role in long-term decline. Ecology. 2005;**86**(1):174-184

[36] Hughes TP. Catastrophes, phase shifts, and large-scale degradation of a Caribbean coral reef. Science. 1994;**265**(5168):1547-1551

[37] Rogers CS, Miller J. Permanent 'phase shifts' or reversible declines in coral cover? Lack of recovery of two coral reefs in St. John, US Virgin Islands. Marine Ecology Progress Series. 2006;**11**(306):103-114

[38] Hernández-Delgado EA, Rosado-Matías BJ. Long-lasting impacts of beach renourishment on nearshore urban coral reefs: A glimpse of future impacts of shoreline erosion, climate change and sea level rise. Annals of Marine Biology and Research. 2017;**4**(1):1021

[39] Hernández-Delgado EA, González-Ramos CM, Alejandro-Camis PJ. Large-scale coral recruitment patterns on Mona Island, Puerto Rico: Evidence of a transitional community trajectory after massive coral bleaching and mortality. Revista de Biología Tropical. 2014;**62**(3):283-298

[40] Bellwood DR, Hughes TP, Folke C, Nystrom M. Confronting the coral reef crisis. Nature. 2004;**429**(6994):827-833

[41] Hobbs RJ, Arico S, Aronson J, Baron JS, Bridgewater P, Cramer VA, et al. Novel ecosystems: Theoretical and management aspects of the new ecological world order. Global Ecology and Biogeography. 2006;**15**(1):1-7

[42] Hobbs RJ, Higgs E, Harris JA. Novel ecosystems: Implications for conservation and restoration. Trends in Ecology and Evolution. 2009;**24**(11):599-605

[43] Graham NA, Cinner JE, Norström AV, Nyström M. Coral reefs as novel ecosystems: Embracing new futures. Current Opinion in Environmental Sustainability. 2014;**7**:9-14

[44] Morse N, Pellissier P, Cianciola E, Brereton R, Sullivan M, Shonka N, et al. Novel ecosystems in the Anthropocene: A revision of the novel ecosystem concept for pragmatic applications. Ecology and Society. 2014;**19**(2):12

[45] Higgs E. Novel and designed ecosystems. Restoration Ecology. 2017;**25**(1):8-13

[46] Jaap WC. Coral reef restoration. Ecological Engineering. 2000;**15**:345-364

[47] Rinkevich B. Conservation of coral reefs through active restoration measures: Recent approaches and last decade progress. Environmental Science & Technology. 2005;**39**(12): 4333-4342

[48] Edwards A, Gomez E. Reef Restoration Concepts and Guidelines: Making Sensible Management Choices in the Face of Uncertainty. The Coral Reef Targeted Research & Capacity Building for Management Program: St. Lucia, Australia; 2007. 38 p

[49] Rinkevich B. Management of coral reefs: We have gone wrong when neglecting active reef restoration. Marine Pollution Bulletin. 2008;**56**:1821-1824

[50] Young CN, Schopmeyer SA, Lirman D. A review of reef restoration and coral propagation using the threatened genus *Acropora* in the Caribbean and western Atlantic. Bulletin of Marine Science. 2012;**88**(4):1075-1098

[51] Rinkevich B. Rebuilding coral reefs: Does active reef restoration lead to sustainable reefs? Current Opinion in Environmental Sustainability. 2014;**7**:28-36

[52] Lirman D, Schopmeyer S. Ecological solutions to reef degradation: Optimizing coral reef restoration in the Caribbean and western Atlantic. PeerJ. 2016;**4**:e2597

[53] Levy G, Shaisha L, Haimb A, Rinkevich B. Mid-water rope nursery—Testing design and performance of a novel reef restoration instrument. Ecological Engineering. 2010;**36**: 560-569

[54] Lirman D, Thyberg T, Herlan J, Hill C, Young-Lahiff C, Schopmeyer S, et al. Propagation of the threatened staghorn coral Acropora Cervicornis: Methods to minimize the impacts of fragment collection and maximize production. Coral Reefs. 2010;**29**:729-735

[55] Forrester GJ, O'Connell-Rodwell C, Baily P, Forrester LM, Giovannini S, Harmon L, et al. Evaluating methods for transplanting endangered elkhorn corals in the Virgin Islands. Restoration Ecology. 2011;**19**(3):299-306

[56] Herlan J, Lirman D. Development of a coral nursery program for the threatened coral *Acropora cervicornis* in Florida. Proceedings of the 11th International Coral Reef Symposium. Session no. 24. 2012. pp. 1244-1247

[57] Hollarsmith JA, Griffin SP, Moore TD. Success of outplanted *Acropora cervicornis* colonies in reef restoration. Proceedings of the 12th International Coral Reef Symposium. 20A Restoration of Coral Reefs. 2012. pp. 1-5

[58] Mbije NE, Spanier S, Rinkevich B. A first endeavour in restoring denuded, post-bleached reefs in Tanzania. Estuarine, Coastal, and Shelf Science. 2013;**128**:41-51

[59] Lirman D, Schopmeyer S, Galvan V, Drury C, Baker AC, Baums IB. Growth dynamics of the threatened Caribbean Staghorn coral *Acropora cervicornis*: Influence of host genotype, symbiont identity, colony size, and environmental setting. PLoS One. 2014;**9**(9): e107253

[60] Ladd MC, Shantz AA, Nedimyer K, Burkepile DE. Density dependence drive habitat production and survivorship in *Acropora cervicornis* used for restoration in a Caribbean coral reef. Frontiers in Marine Science. 2016;**3**(261):1-14

[61] Schopmeyer SA, Lirman D, Bartels E, Guilliam DS, Goergen EA, Griffin SP, et al. Regional restoration benchmarks for *Acropora cervicornis*. Coral Reefs. 2017;**36**:1-11. DOI: 10.1007/s00338-017-1596-3

[62] Taylor MD, Chick RC, Lorenzen K, Agnalt AL, Leber KM, Blankenship HL, et al. Fisheries enhancement and restoration in a changing world. Fisheries Research. 2017;**186**:407-412

[63] Bowden-Kerby A. Coral transplantation and restocking to accelerate the recovery of coral reef habitats and fisheries resources within no-take marine protected areas: Hands-on approaches to support community based coral reef management. International Tropical Marine Ecosystems Management Symposium. 2003;**2**:1-15

[64] Epstein N, Bak RPM, Rinkevich B. Strategies for gardening denuded coral reef areas: The applicability of using different types of coral material for reef restoration. Restoration Ecology. 2001;**9**(4):432-442

[65] Ammar MSA, Amin EM, Gundacker D, Mueller WEG. One rational strategy for restoration of coral reefs: Application of molecular biological tools to select sites for rehabilitation by asexual recruits. Marine Pollution Bulletin. 2000;**40**(7):618-627

[66] Akakura Y, Hanashiro S,Urabe S, Maehara H, Ono M, Mizoguchi T, et al. Various contrivances for restoration of coral assemblages in harbor development projects: A report. Proceeding of the 10th International Coral Reef Symposium. 2006. pp. 1651-1656

[67] Alfeche LR. Coral reef restoration through coral transplantation: The case of Duka Bay, Medina, Misamis Oriental. ITMEMS 2. 2003. 4 pp

[68] Bongiorni L, Giovanelli D, Rinkevich B, Pusceddu A, Chou LM, Danovaro R. First step in the restoration of a highly degraded coral reef (Singapore) by in situ coral intensive farming. Aquaculture. 2011;**322-323**:191-200

[69] Boch CA, ANC M. Testing the effectiveness of direct propagation techniques for coral restoration of *Acropora* spp. Ecological Engineering. 2012;**40**:11-17

[70] Borell EM, Romatzki SBC, Ferse SCA. Differential physiological responses of two congeneric scleractinian corals to mineral accretion and an electric field. Coral Reefs. 2010;**29**:191-200

[71] Baums IB. A restoration genetics guide for coral reef conservation. Molecular Ecology. 2008;**17**:2796-2811

[72] Baums IB, Johnson ME, Devlin-Durante MK, Miller MW. Host population genetic structure and zooxanthellae diversity of two reef-building coral species along the Florida reef tract and wider Caribbean. Coral Reefs. 2010;**29**:835-842

[73] Bowden-Kerby A. Low-tech coral reef restoration methods modelled after natural fragmentation process. Bulletin of Marine Science. 2001;**69**(2):915-931

[74] Bowden-Kerby A. Coral transplantation in sheltered habitats using unattached fragments and cultured colonies. Proceedings of the 8th International Coral Reef Symposium. Vol. 2. 1997. pp. 2063-2068

[75] Ortiz-Prosper AL, Bowden-Kerby A, Ruíz H, Tirado O, Cabán A, Sánchez G, et al. Planting small massive corals on small artificial reefs or dead coral heads. Bulletin of Marine Science. 2001;**69**(2):1047-1051

[76] Hernández-Delgado EA, Montañez A, Otaño A, Alejandro P, Fonseca JS, González C, et al. Expansion of the Puerto Rico Low-Tech Coral Aquaculture and Coral Reef Rehabilitation Project. NOAA-Restoration Center & The Nature Conservancy: San Juan, PR; 2013. 242 p

[77] Mercado Molina A, Hernández Delgado EA, Rivera Rivera JE, Rivera Rivera M, Suleimán Ramos SE, Olivo Maldonado I, et al. Protocolo para la propagación y la restauración de poblaciones del coral Cuerno de ciervo, *Acropora cervicornis*: Estrategias de bajo costo de la Sociedad Ambiente Marino. NOAA-Restoration Center & The Nature Conservancy: San Juan, PR; 2013. 102 p

[78] Hernández-Delgado EA, Suleimán-Ramos SE. E.S.A. coral species listing: A roadblock to community-based engagement in coral reef conservation and rehabilitation across the U.S. Caribbean? Reef Encounters. 2014;**29**(1):11-15

[79] Hernández-Delgado EA, Montañez-Acuña A, Otaño-Cruz A, Suleimán-Ramos SE. Bomb-cratered coral reefs in Puerto Rico, the untold story about a novel habitat: From reef destruction to community-based ecological rehabilitation. Revista de Biología Tropical. 2014;**62**(Suppl. 3):183-200

[80] Hernández-Delgado EA, Mercado-Molina AE, Alejandro-Camis PJ, Candelas-Sánchez F, Fonseca-Miranda JS, González-Ramos CM, et al. Community-based coral reef rehabilitation in a changing climate: Lessons learned from hurricanes, extreme rainfall, and changing land use impacts. Open Journal of Ecology. 2014;**4**:918-944

[81] Griffin S, Spathias H, Moore TD, Baums I, Griffin BA. Scaling up Acropora nurseries in the Caribbean and improving techniques. Proceedings of the 12th International Coral Reef Symposium. 20A. 2012. pp. 1-5

[82] Griffin SP, Nemeth MI, Moore TD, Gintert B. Restoration using *Acropora cervicornis* at the T/V MARGARA grounding site. Coral Reefs. 2015;**34**:885

[83] Bruckner AW, Bruckner RJ. Condition of restored *Acropora palmata* fragments off Mona Island, Puerto Rico, 2 years after the Fortuna reefer ship grounding. Coral Reefs. 2001;**20**:235-243

[84] Bruckner AW, Bruckner RJ, Hill R. Improving restoration approaches for *Acropora palmata*: Lessons from the Fortuna Reefer grounding in Puerto Rico. Proceedings of the 11th International Coral Reef Symposium. Session No. 24. 2010. pp. 1199-1203

[85] Salm RV, Clark JR, Siirila E. Marine and Coastal Protected Areas: A Guide for Planners and Managers. 3rd ed. International Union for Conservation of Nature and Natural Resources: Gland, Switzerland; 2000. 370 p

[86] Soto-Santiago FJ, Mercado-Molina A, Reyes-Maldonado K, Vélez Y, Ruiz-Díaz CP, Sabat A. Comparative demography of two common scleractinian corals: *Orbicella annularis* and *Porites astreoides*. PeerJ. 2017;**5**(e3906):1-18. DOI: 10.7717/peerj.3906

[87] National Marine Fisheries Service. Recovery Plan for elkhorn (*Acropora palmata*) and Staghorn (*A. cervicornis*) Corals. Acropora Recovery Team, National Marine Fisheries Service, NOAA: Silver Spring, MD, USA; 2015. 156 p

Chapter 11

Multi-disciplinary Lessons Learned from Low-Tech Coral Farming and Reef Rehabilitation: II. Coral Demography and Social-Ecological Benefits

Edwin A. Hernández-Delgado,
Alex E. Mercado-Molina,
Samuel E. Suleimán-Ramos and Mary Ann Lucking

Additional information is available at the end of the chapter

http://dx.doi.org/10.5772/intechopen.74283

Abstract

Low-tech coral farming and reef rehabilitation have become important tools to foster community-based participation in the management of coastal social-ecological systems. Lessons learned from coral demographic dynamics, ecosystem-level benefits, and sociological dynamics achieved in Culebra Island, Puerto Rico, are discussed. Important gaps regarding social-ecological interactions are also addressed. Coral reef rehabilitation efforts must be adaptive and focused on maximizing resilience as a long-term goal, with emphasis on managing non-linear dynamics, thresholds, environmental and climate uncertainty, and ecological surprises. In this context, coral demographic modelling becomes fundamental to address, not only ecological, but also sociological concerns. Only through sustained support and input of harvested corals restored populations, and by increasing the spatial scale of reef rehabilitation, restored populations can remain viable and grow under present and projected environmental and climate conditions. Understanding sociological dynamics, learning from others experiences, integrating visioning and scenario building, leadership building, multi-sectorial agents and actor groups, and strengthening cross-sectorial social networking are necessary adaptive approaches to cope with future environmental and climate changes, and are an integral part of reef rehabilitation. The combined benefits to social-ecological systems are multiple. With proper planning, design, funding, local support, and implementation, these can have long-lasting impacts in restoring coastal resilience.

Keywords: coral farming, coral reefs, ecological rehabilitation, lessons learned, Puerto Rico, Caribbean Sea, reef fish communities, social-ecological systems

1. Introduction

1.1. Coral reef decline and the erosion of social-ecological systems

Coral reefs have largely lost species diversity, ecological functions, ecosystem resilience, and socio-economic benefits across regional to global scales over the last four to decades. These have resulted from a combination of impacts from extreme natural hazards (e.g., hurricanes and tsunamis), from multiple localized human drivers [1–5], and from climate change [6, 7]. This has often resulted in a largely reduced ability to recover from acute, recurrent or chronic disturbances, compromising their capacity to sustain biodiversity, ecosystem services, local economies, and threatening the sustainability and resilience of social-ecological systems [8]. This is a particular concern for low-lying coastal communities and for small tropical islands, which often have very significant governance limitations, as well as limited socio-economic capital to cope with disasters. In this context, coral farming and reef rehabilitation efforts are becoming increasingly important strategies to incorporate in the coastal management toolbox, but which have never yet been implemented as strategies to address issues related to restoring the resilience of coastal social-ecological systems.

Ecological resilience can be defined as the buffering capacity or the "*ability of a system to absorb changes of state variables, driving variables, and parameters, and still persist*" [9]. In this context, resilience is a property of a system, and persistence or probability of extinction can be the result, depending on the system's trajectory and stability. Stability is the ability of a system to return to an equilibrium state after any disturbance [9]. The more rapidly a system returns, and with the least fluctuations, the more stable it is. However, under present rapidly declining of coastal social-ecological systems, mostly coral reefs, stability has also declined, so the ability of systems to absorb and recover from disturbance. Therefore, any long-term trend resulting in the net erosion of the stability of a social-ecological system can threaten its long-term resilience, and may result in a combined loss in the ecological (e.g., biodiversity, functional redundancy, ecosystem services) and social persistence (e.g., local economy, livelihoods). The geographical isolation of small islands, in combination with historical socioeconomic and political constraints, increasing threats by sea level rise (SLR) and climate change, can often magnify vulnerability to such impacts [8], and may result in a net erosion of social resilience.

Social resilience was defined as "*the ability of groups or communities to cope with external stresses and disturbances as a result of social, political and environmental change*" [10]. Social resilience is a fundamental characteristic of ecosystems which has still remained poorly addressed, but which is critical to maintain ecosystems functions and services in the face of disturbance, including extreme weather events (e.g., hurricanes), natural hazards (e.g., tsunamis), chronic human-driven environmental degradation, SLR, and climate change. There is a clear link between social and ecological resilience, particularly for coastal communities, and small island nations that are largely or fully dependent on ecological and environmental resources for their economy and local livelihoods.

Exposure and sensitivity to hazards of coastal social-ecological systems are largely dependent on the ecological status and vulnerability to disturbance of coral reefs. Reducing exposure and sensitivity requires maintenance and enhancement of reef ecosystems functions through sustainable management and use [11]. In this context, coral farming and reef rehabilitation have become fundamental management tools to engage base communities in sustainable social-ecological systems management. It would also be important to maintain the local memory of resource use, and foster the development of learning processes for responding to environmental feedback and social cohesion [11]. Therefore, base-community engagement in coral farming and reef restoration must also be coupled with hands-on education and training, with the aim of achieving long-term local empowerment, stewardship and support. The other critical element of vulnerability of coastal social-ecological systems is adaptive capacity. Sustaining coastal social-ecological systems requires the recovery of biodiversity in ecological systems, and expanding the diversity of the local economic livelihood portfolio. Both alternatives can be readily achieved through low-tech, community-based coral farming and reef rehabilitation. However, an important challenge is the need to empower local to national governance structures and social capital, bridging gaps among local communications, academia, private organizations, and government for integrative responses, and building horizontal, cross-sectorial networks in society for social learning.

Nevertheless, bridging the gap between decision-makers, natural resource managers, empirical academic research in regards to coral farming and reef rehabilitation, and the socio-economic component of these efforts have remained poorly explored, and still remains as a top challenge to overcome across local, national and regional scales. Understanding the critical value of integrating question-driven research in reef rehabilitation efforts is paramount to advance knowledge and to communicate that technical knowledge to base communities and local stakeholders. One such component is to integrate coral demographic dynamics and modeling into active reef rehabilitation efforts. But also, the integration of lessons learned from sociological dynamics in regards to coral farming and reef rehabilitation is a highly necessary added value that should contribute to improve future projects.

Therefore, lessons learned regarding the need to understand the mechanisms of improving the management of both, the ecological and the social components of coastal tropical systems is essential to improve management success, and to foster an improved education, stewardship and participation of base communities in coastal management. This makes necessary to examine the role of assisted recovery of depleted coral diversity, restoring coral functional groups, and the rehabilitation of coral reefs at the reefscape, functional level, as a new strategy to buffer and restore present declining trends.

1.2. Goals and objectives

The goal of this chapter is to briefly update the state of knowledge regarding applied coral demographic dynamics to low-tech coral farming and reef rehabilitation efforts, mostly using case studies of restored populations of endangered Staghorn coral, *Acropora cervicornis*. Important elements associated with coral demographic and oceanographic modeling have also been addressed as decision-making tools regarding its application to large-scale restoration efforts. Also, sociological lessons learned, which are often overlooked, have been discussed, including: volunteer work, team assemblage, building local support and stewardship, and socio-economic benefits. Most of the discussed examples are derived from lessons learned through the *Community-Based Coral Aquaculture and Reef Rehabilitation Program* established in 2003 in Culebra Island, Puerto Rico, by non-governmental organization (NGO) Sociedad Ambiente Marino.

2. Lessons learned from coral demographic dynamics

2.1. Coral demographic modeling in reef rehabilitation

Whether any given population increases, decreases, remain stable or face extinction depends upon the rates at which an individual grows, die, and reproduce. Most conservation and management-oriented efforts are intended to increase population growth rate of the targeted species until reaching a growing or stable state and identifying which vital stage(s) is (are) essential for the conservation and management of an endangered or threatened species. Demographic-based population models are convenient and efficient tools that not only allow to perform population viability analyses, but also allow a detailed examination of the relationship between demographic traits/rates and population growth rate (λ). One of the strengths of demographic-based population models is that they take into consideration the influence of the developmental stage (age, size, and stage) on individual's vital rates and link it to the population level [12]. At the same time, demographic models can be subjected to prospective (e.g., sensitivity and elasticity) and retrospectives (Life Table Response Experiment) analyses to examine the relative importance of each of the vital rates on λ and to investigate the effects of physical and biological disturbances on the population dynamics of a target species [13, 14]. Prospective analyses (e.g., sensitivity and elasticity) looks at how λ would respond when a particular life cycle transition is perturbed [12]. Life Table Response Experiment analysis, on the other hand, provides information on how much variation in a particular life cycle transition contribute to the observed differences in λ between treatments (e.g., restoration vs. no restoration). Another advantage is that models can be manipulated to assess how any given population would respond to changes in any of the vital rates (e.g., reproductive failure, mass mortality) or any given restoration initiative; thereby providing the basis for the design of future restoration and conservation projects under variable environmental conditions, climate change-related scenarios, etc.

In the last couple of decades, demographic-based population models (e.g., population matrix models and integrated population models) have become an essential part of conservation studies [15]. However, few coral biologists have applied demographic-based population models to answer specific conservation questions (but see [16–19]). Vardi et al. [17] and Mercado-Molina et al. [18] used size-based population matrix models have been used to describe the demography and population dynamics of threatened coral species *Acropora palmata* [19] and *A. cervicornis*, respectively [18]. They found that the demographic transition that contributes the most to local λ is the survival of large colonies. Thereby, providing evidence that restoration and conservations efforts of these corals species should be focused on enhancing the probability of large colonies to survive. Mercado-Molina et al. [20, 21] found that both growth and branching rates of *A. cervicornis* increase with size. Therefore, positive contribution of large colonies to λ, at least for *A. cervicornis*, can be partly explained by (1) a rapid growth that can allow the colony to reach a refuge size in which mortality associated with diseases, predation, and bleaching can be considerably reduced; and (2) an increase in the number of branches with the potential to be detached from the parental colony and become established as an independent colony, contributing to faster formation of thickets.

Demographic transitions and population growth rates of *A. palmata* were relatively stable over a 6-year study period, except for a hurricane year which naturally caused a significant decline in population growth rate [17]. In contrast, spatiotemporal differences both in the transitions rates and λ of *A. cervicornis* were observed [18]. These contrasting results indicate that even when *A. cervicornis* and *A. palmata* share similar life cycles, their demography varies considerably. Therefore, conservation and restoration activities should be designed at the species-specific level whenever possible, with separate specific goals and objectives. The spatiotemporal differences in demographic transitions displayed by *A. cervicornis* [18] also suggest that restoration efforts should be partitioned among several locations rather than allocating all the resources into one site. This action will enhance the persistence of the species if localized extirpation occurs due to spatial variability.

2.2. Modeling as decision-making tools

The current limitation in human, technical, and economic resources, together with multiple frequent logistical constraints, have made coral conservation a difficult task, especially in small islands, and when long-term studies are not feasible or available. The most basic demographic parameter that can be obtained using demographic modeling is the intrinsic population growth rates, lambda (λ). When $\lambda = 1$, the population is stable, if $\lambda < 1$ the population is decreasing in size, and when $\lambda > 1$ then the population is growing positively. Thus, by directly estimating λ practitioners can determine whether a given population needs management attention. Using population models to project population size over time is one of the most attractive alternatives to understand how any given wild or restored population would behave under different restoration and conservations scenarios (see [17, 18, 20]). Studies by Mercado et al. [18, 20] coincided in that without human intervention (e.g., coral out-planting) local *A. cervicornis* population growth in the wild is not granted. However, contrary to the *A. palmata* populations [17], which are expected to remain stable (no significant growth, no significant decline) over time, *A. cervicornis* populations can go extinct in less than two decades [18, 20]. Such contrasting population trajectories may be the result of *A. cervicornis* being more susceptible to low and moderate environmental changes [18, 22]. Therefore, *A. cervicornis* is more at risk of localized extinction than it is congeneric. Indeed, *A. cervicornis* colonies are much more ephemeral than *A. palmata* (e.g., they suffer a high rate of complete mortality and complete colony dislocation) [22]. This implies that continuous low-tech coral farming and out-planting efforts are fundamental to sustain restored populations in the wild.

The initial size of the coral transplant needs to be taken into consideration to assure the success of any restoration project [17, 18]. Both studies concluded that transplanting large colonies will result in higher population growth rates than transplanting small colonies, as many standard coral farming operations do, at least across the Caribbean. However, transplanting large colonies pose some challenges; specifically, regarding the time necessary for a nursery-reared coral fragment to reach the effective transplantation size [18]. However, numerical simulations have demonstrated that increasing the number of small-sized transplants of *A. cervicornis* enhances the probability of population stability [18]. For an initial population size of 150 *A. cervicornis* colonies, transplanting 50–75 fragments ≤100 cm in Total Linear Length (TTL) annually would result in a positive population growth rate.

Population models can also be used to test the effectiveness of different management regimes (e.g., intensity, environmental and climate change scenarios) such as alternating times of out-planting and nonout-planting. So far, only Vardi et al. [17] have published results by taking such approach. They projected *A. palmata* population size under a scenario that alternate 2 years with no out-planting and 5 years in which 1000–3000 colonies are transplanted, followed by an additional 13 years with no out-planting. Under such management design, populations will grow positively over time. Therefore, it can be argued that such management plan is appropriate to assure the persistence of the impacted population. **Figure 1** shows a simulation based on the model developed by Mercado-Molina et al. [18] for two wild populations of *A. cervicornis* in northeastern Puerto Rico, in which the effect of two out-planting scenarios on local population abundance was simulated. The scenarios considered the out-planting of 1000 colonies for two and five consecutive years, respectively. These scenarios were based on the fact that most restoration projects are funded for less than 5 years and for the number of fragments that in our experience can be produced in 1 year in our nursery units. The results indicated that under such out-planting regimes populations would

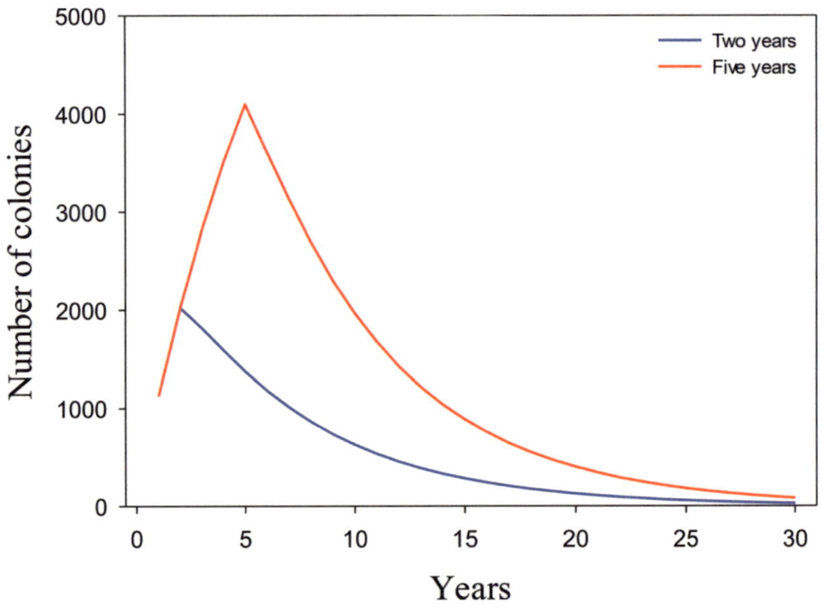

Figure 1. Numerical simulation based on the model developed by Mercado-Molina et al. [18] for two wild populations of Staghorn coral (*Acropora cervicornis*) in northeastern Puerto Rico, in which the effect of two out-planting scenarios on local population abundance were assessed.

not be able to persist over time. This outcome, together with simulations run by Mercado-Molina et al. [18], led us to conclude that restoring populations of *A. cervicornis* by out-planting coral fragments is a feasible strategy, but one that requires sustained human intervention.

2.3. Increasing the spatial scale of reef rehabilitation

One of the major limitations of coral reefs restoration is that all projects so far are small in spatial scale, often varying from tens to hundreds of m^2, with a limited number of projects ranging between 100 s m^2 to less than 1 km^2. Increasing the spatial scale of reef rehabilitation is essential, at least for *A. cervicornis,* because its demography varies considerably in time and space. Increasing the spatial scale of population rehabilitation will increase the probability of species persistence. Coral out-plant spatial array is also critical for the formation of thickets. Before the 1980s, *A. cervicornis* used to dominate the seascape of shallow-water reefs by monopolizing vast areas of the substrate [23]. It is necessary to take conservation and/or restoration initiatives directed at re-establishing the large thickets this coral used to form. The *Acropora* Recovery Plan [24] established the development of *A. cervicornis* thickets as a major goal of restoration projects. However, there is still scarce information regarding the demographics and dynamics of thicket formation that could be used as a basis for the design of management strategies. But model thicket formation can use novel approaches such as individual-based dynamical automaton models (IBDA) [25, 26], and use the predictions of the model to determine the number and spatial arrangement of out-plants that will maximize the likelihood of thicket formation, and improve reef restoration strategies and spatial designs.

Increasing the spatial scale is also important for increasing the recovery of fish assemblages and rehabilitating reef processes. However, there is still a significant information gap regarding the role of coral reef restoration on enhancing essential fish habitats and fish assemblages. Fish assemblages are sensitive to the spatial heterogeneity of the benthos [27] and habitat condition [28]. Any disturbance resulting in mass coral mortalities [29], benthic community regime shift [30], and in loss of benthic spatial heterogeneity [31] should adversely affect coral reef fish assemblages. Therefore, management strategies aimed at rehabilitating depleted fish assemblages should include coral out-planting at increasing spatial scales and/or focused on developing habitat mosaics as a mechanism to restore benthic spatial heterogeneity.

Also, increasing coral reef rehabilitation spatial scale is a fundamental step necessary to achieve progress in restoring and managing coastal resilience (e.g., wave buffering, reducing shoreline erosion rates). But fundamental questions associated with coral reef restoration projects at sites, where wave energy reduction is an important design criterion. What is the degree of wave attenuation that can be expected from out-planting Elkhorn coral (*Acropora palmata*) at the selected sites? What factors (how big should colonies be at out-plant, how far apart, thicket size, shape, and spatial arrangement, water depth, local wave climate, etc.) should be considered when designing a coral reef restoration project in order to maximize wave energy dissipation? What are the expected costs for downscaling numerical wave models for different locations? What data should be collected to successfully simulate the performance of the proposed coral restoration activities? What other data on coral health should be collected to better inform the modeling efforts? These are areas which are being currently investigated at the University of Puerto Rico, which should provide new light in regards to the potential application of coral reef restoration as a novel coastal resilience management strategy.

2.4. Next steps in coral reef restoration

The next logical steps in coral reef restoration have mostly to do with improving *ex situ* propagation of coral larvae, enhancing the effectiveness of micro-fragmentation techniques to foster a higher number of small colonies and faster initial growth rates of massive coral species, improving the ability to discriminate and propagate different genetic clones, and improving the spatial scale of coral out-planting to achieve faster and functional coral patch or thicket formation. Furthermore, there is a critical need to figure out: (a) how to distribute available funding more fair and evenly, and how to achieve economic auto-sustainability; (b) how to shift the standard institutional short-term, isolated vision of projects to a long-term goal-driven program (c) how to develop some standardized farming, maintenance, and out-planting practices; (d) how to implement standardized integrative metrics of success (e.g., from colony to ecosystem level); (e) how to foster, achieve and support community-based and NGOs participation in these projects; and (f) how to foster the creation of functional partnerships among government institutions, the academia, NGOs, base communities, and other private sectors. But there are still a few limiting components associated with low-tech coral reef restoration efforts that still must be quickly addressed.

2.4.1. Lack of knowledge of Staghorn coral (Acropora cervicornis) branching dynamics

Despite colony branching dynamics is the basis for *Acropora cervicornis* restoration projects, little is known about branch production of the species. Branching rates in this species growing in nurseries and in colonies out-planted to the reef, respectively, increase with colony size [19, 20, 32]. Thus, growing large-sized colonies in nurseries, as well as those colonies out-planted to the reef, may result in greater number of branches available for a restoration project, increasing the potential coral propagule abundance available for future restoration efforts. At the same time, out-planting large colonies would result in colonies with higher levels of branching complexity in relatively shorter time than transplanting small single-branched fragments, which favor faster thicket formation. It is known that more complex coral colonies promote reef biodiversity [33, 34]. Still missing, however, is information about the intrinsic (i.e., genetics) and extrinsic (i.e., temperature; light) factors that stimulate/limit branch production. Such information is essential, for example, to select the sites for the deployment of nursery units, select the most appropriate sites for restoration, and estimate the number of branches that can be produced for restoration purposes and future natural asexual propagation in the wild.

2.4.2. Increasing the spatial scale of reef rehabilitation

Increasing the spatial scale of population rehabilitation will increase the probability of species persistence for most corals. Nevertheless, the process of selecting the sites to be restored is not based on empirical data about the demographic performance of targeted corals, but rather on the assumption that the historical or current presence of any given species (e.g., *A. cervicornis*) reflects the appropriate conditions for the development of the species. Also, site selection might often be based on the perceptions that water transparency, deeper environments, or high distance from potential pollution sources represent the most suitable habitat conditions for out-planting. Site selection can be critical for coral restoration success as poor site condition can be detrimental [35]. Even low to moderate differences in local biotic and abiotic conditions can have profound effects on λ [12]. Also, preliminary results by Hernández-Delgado (unpub. data) suggest that the abundance and widespread dispersion of invasive red encrusting algae *Ramicrusta textilis* (Rhodophyta, Peyssonneliaceae) is a critical factor affecting the survival and growth of *A. cervicornis*, even under remote conditions and high water transparency. Accordingly, a better criterion of restoration success should be the local population growth rate of *A. cervicornis*, rather than presence/absence of the species.

2.4.3. Most restoration projects are not firmly grounded on quantitative demographic analyses

Because population growth rate is inevitably linked to individuals' survival, growth, and reproduction, effective conservation initiatives require knowledge on how variation in vital rates relate to variations in population growth. Population studies focused on restored populations of *A. cervicornis* have not been firmly grounded on quantitative demographic analyses [19]. Several population studies have estimated rates of colony growth and survival [36–38]. None, however, identified how spatiotemporal variations in outplants survival, growth, and rates of recruitment (e.g., number of outplants) affect λ of restored populations. The lack of studies that directly evaluate the population response to demographic variability limits our capacity to develop effective restorations initiatives. Very few studies have attempted to address essential questions such as: How long restored populations would last without human intervention? How many fragments would be necessary to keep populations viable? How often out-planting activities need to be carried out to assure the persistence of the restored populations? Which is the effective colony size of transplantation? The answers to these questions are fundamental for the development and success of restoration activities. And demographic modeling can lead the way to answer them.

2.4.4. Short-term funding: a roadblock to long-term success

Funding is a major factor limiting the development and success of restoration projects. Most of the projects are funded for 1–3 years [39]. This short period of economic support certainly limits the amount of spatiotemporal demographic data that can be used to parameterize population models. Indeed, the low spatiotemporal resolution is one of the main criticisms raised by many researchers against the use of population modeling for conservation purposes. More data is always better; however, "limited" data must not discourage the use of demographic and population modeling as a tool for the development of restoration initiatives. Collecting data for an undetermined amount of time waiting to obtain "robust" demographic data to parameterize any given model might just be too late for a threatened species whose populations are declining very rapidly.

One year of demographic data is the minimum amount necessary to perform a basic population model based on estimates of population growth rates. The demography of many marine clonal/modular organisms has been successfully described using ≤2 year of demographic data [13, 16, 40–44]. The relatively "short time frame" of these studies has not impeded making a significant contribution to our understanding of coral demography. In fact, most of the studies focusing on the demography of corals are short-term (≤4 year). This is not surprising, given the limited resources available to monitor populations of conservation concern. On the other hand, studies considered "long-term" (5> year) have been focused, with the exception of Vardi et al. [17], on massive species such as *Porites astreoides, Pseudodiploria strigosa,* and *Orbicella* (*Monstastraea*) *annularis,* which contrary to *Acropora cervicornis,* are characterized by low growth rates and therefore require higher temporal resolution to detect changes in demographic transitions [45, 46].

If the intention is to conduct demographic analyses that take into consideration environmental variability, both in space and time, then at least 2 years of demographic data are necessary. It is well established in the demographic literature that two temporal points (2 years) are sufficient to perform the stochastic analyses (e.g., population viability analysis, stochastic population growth). Morris et al. [47], in their book "A Practical Handbook for Population Viability Analysis (PVA)," stated that *demographic data on a subset of life stages for only 1–2 years*" is sufficient to make a population viability analysis "*profitable.*" Fieberg and Ellner [48] recognized that "*[Stochastic matrix] models are typically parameterized using two or more sets of estimated transitions rates between age/size/stage classes.*" Likewise, using two annual transitions to perform demographic analyses (e.g., PVA, stochastic population growth) is more suitable [12]. It is important to note, however, that demographic and population models are not crystal balls that predict the future of a population under a certain set of conditions. Nature cannot be replicated, and as such the results of any given model need to be considered as possible population outcomes which should be combined with the best information available to take educated conservation decisions for this species.

2.4.5. Coral reef rehabilitation to restore ecological connectivity

Depending on the configuration of coral out-planted patches, its spatial distribution and the temporal extension of coral reef rehabilitation efforts it may become a critical tool to manage ecological connectivity among adjacent reef systems. The whole concept has to do with fostering enhanced depleted coral stocks, therefore, increasing local populations' reproductive potential and output. This will allow increased gamete release, reduced gamete waste, reduced Allee effect, and enhanced probabilities of sexual reproduction and recruitment. In theory, this would allow to enhance genetic recombination, improve population fitness, and allow for increased connectivity with downstream reef systems. For this to be successful, understanding local to regional oceanographic dynamics is fundamental. Thus, numerical wave model development, as well understanding often complex surface circulation patterns, is very important as a planning tool to shape future long-term coral reef restoration initiatives. Indirectly, this can also become a very important indirect component of reef fish conservation and restoration management as restored coral reefs can restore benthic spatial heterogeneity and rehabilitate essential fish habitat functions across ecologically connected scales fundamental for reef fish dispersal.

3. Lessons learned from fish community dynamics

3.1. Impact of community-based reef rehabilitation on fish communities

Coral reef rehabilitation results in increased benthic spatial heterogeneity, which enhances micro-habitats for fish shelter on local scales. Post-larval and juvenile grunts (*Haemulon* spp.) have shown up to 10-fold increase or more in abundance in areas where *Acropora cervicornis* has been out-planted (Hernández-Delgado and Suleimán-Ramos [49]). But also, multiple other taxa show significant increases in fish abundance and biomass. Ongoing studies by Hernández-Delgado have shown that juvenile guilds of multiple families, such as parrotfishes (Scaridae), wrasses (Labridae), damselfishes (Pomacentridae), blue tangs and doctorfishes (Acanthuridae), and predators, such as snappers (Lutjanidae) and groupers (Serranidae) can increase in abundance and biomass, in comparison to adjacent control sites without out-plants, or in comparison to restored sites before out-planting. There is also an increase in fish abundance and biomass with increasing thicket age, comparing 1-, 2-, and 4-year-old patches. Further, areas located within the Canal Luis Peña no-take Natural Reserve showed higher fish density and biomass, in comparison to control sites outside the reserve exposed to fishing. Therefore, preliminary evidence already points out at the emerging role of low-tech community-based coral reef rehabilitation as a highly useful tool to restore and rebuild coral reef-based fisheries.

3.2. Impacts on herbivory

Ongoing studies by Hernández-Delgado have also shown increased abundance and biomass of fish and invertebrate herbivore guilds. As mentioned above, parrotfishes (Scaridae) and acanthurids are among the most abundant fish taxa across reef rehabilitation sites, in comparison to areas with no coral out-planting. Further, *Acropora cervicornis* out-planting has resulted in increased abundances of the Long-spine urchin (*Diadema antillarum*). This has resulted in increased herbivory upon macroalgae and algal turf, and in increased percent cover of crustose coralline algae (CCA). Over temporal scales of 5–10 years, this has resulted in higher coral sexual recruitment rates across restored areas.

4. Sociological lessons learned

4.1. Building local support and stewardship of social-ecological systems

Building local support and stewardship of social-ecological systems is a critical process for achieving success in any community-based marine protected area (MPA) participatory management or co-management effort. Community-based coral farming and reef rehabilitation also requires such support and stewardship. Multiple environmental problems frequently raise concern on residents of coastal communities, and a few highly concerned people assume the community leader role hoping to find solutions. However, at least in Puerto Rico, most base-community members lack the technical and scientific resources to meet the minimum and urgent needs of their community. Therefore, a basic step for successfully achieving solutions is to organize, establish a goal and delineate a functional plan to achieve objectives. But this may often require seeking technical and scientific support from the academia and NGOs. Integrating multiple stakeholders in coral farming and reef rehabilitation efforts is a key for overcoming such obstacles.

Community-based leaders can often provide a fundamental historical background that can provide valuable information to understand and resolve problems. Traditional ecological knowledge has been significant for success in Culebra Island and at Vega Baja, Puerto Rico. Particularly, old fisher folks can provide very detailed information regarding the ecological history of local coral reefs that can help rebuild local environmental history and identify coral reef rehabilitation strategies. In addition, the interaction among base communities, NGOs, the academia, the private sector, and the government can allow and strengthen the development of trust. This is a critical element for achieving successful transparent collaboration in social-ecological systems. Building up such local partnerships will foster building stronger functional networks, with the support and respect from agencies and private institutions. It can also strengthen outreach and educational efforts through a combination of hands-on training activities, workshops, and other methods to generate commitments among the stakeholders who traditionally adopt roles as volunteers as they feel confident and dominate different skills.

Another key element to build local stewardship and support are exchanges and cross-sharing of experiences with sister organizations and base communities to share knowledge, and lessons learned in support of each other's work. Networking, among different sectors, can further allow strengthening communication and sharing of experiences.

4.2. Building a volunteer network

Building up a strong and consistent volunteer network is another key to success. This can be achieved through proper organization, direction, well-established goals, and a functional, realistic work plan. There is also a need to integrate educational and hands-on training to develop and strengthen theoretical and technical skills, build stewardship and compromise, assign roles and tasks, etc. Even the difference in personalities and needs can provide a wide range of opportunities for participation. Individuals have different needs, from basic nutrient supplementation, to self-realization. Different needs function as motivation in performing tasks beyond satisfying personal needs. The collective need of volunteers represents the necessity of their environments or communities.

A transparent dialog between volunteers and collaborators can help build up cooperative working links serving different needs for the same adversity. Further, building up large teams of volunteers can help to have always people available for labor-intensive field work, preventing burning out the same group of people. It is therefore important to know about your volunteers, their interests, needs, their chemistry as a group, their personalities, and their strengths and weaknesses.

4.3. Team assemblage

A fundamental step in achieving team success is the selection process of proper members of a coral farming or reef rehabilitation team. Team technical leadership is important to provide direction during planning and field work. Personality issues, individual responses and performance to different specific tasks and roles, and differences in strength and weaknesses are also important elements to consider. Understanding the profile of volunteers, their needs, and the different characters and temperaments can allow making a good distribution of the workforce, avoid conflicts that impede the growth of the organization, as well as the fulfillment of goals and objectives.

4.4. How to overcome lack of funding?

Lack of long-term commitment by funding sources can be a major obstacle for advancing project's goals and achieving success. Lack of commitment by government agencies and funding institutions, indifference by private businesses and tourism industry, and the lack of a long-term vision of projects goals can lead to rapid failure. Therefore, the need to engage local community, build stewardship, volunteerism, integration of university students through research and first laboral experience programs, etc., becomes instrumental to buffer limited funding, and to strengthen management of coastal social-ecological systems. Nevertheless, in a time of significant socio-economic constraints, there is a need to explore alternative funding avenues from multiple auto-sustainable economic strategies. These might include alternatives such as: (1) "Adopt a coral" program—aimed at the general public and the private sectors, including options such as: adopting an individual coral, a determined number of colonies, a coral thicket, a reef patch or an entire reef; (2) Develop a "Reef sponsoring program" for private corporations—aimed at developing a sponsoring program that may also have different levels of support; (3) Develop crowd funding strategies through the web—aimed at using the world wide web to develop a cyber-campaign for raising awareness about coral farming, reef conservation and restoration, and for fundraising for any given project, with usually a goal-driven funding limit for a specific project; (4) Establish a system of green taxes—aimed at auto-sustaining natural resource management, including activities such as MPA management, mooring buoy maintenance, patrolling, outreach and education, guided tours, coral farming, and reef rehabilitation, among others. Green taxes can be derived from multiple tourism-based activities such as airplane landing fees, cruiseship taxes, private yacht taxes, SCUBA diving and snorkeling charter boat operations, kayaking, vehicle rental, hotels, etc.; and (5) Establish different sources of funding from different government revenue collection systems—this may include through specific taxes to luxury yachts, vehicles and properties, from liquor and cigarette expenses, from industrial revenues, etc.

Under current local, regional, and global socio-economic decline, it is paramount to develop and implement creative strategies for seeking financial support. But to achieve this, strengthening local organizations, building up strong partnerships with different sectors, and fostering community-based participation are fundamental steps.

4.5. How to overcome other roadblocks?

Even successful community-based and academic projects can face multiple roadblocks in their day to day work. Aspects such as permitting bureaucratic processes, access to restoration sites, beach access issues, privatization and roadblocks, conflicts with other uses (e.g., tourism, charter boats, kayaking, fishing, and navigation), lack of prioritization of coral reef rehabilitation by local/national government institutions, lack of local community stewardship and support, indifference by private businesses, etc., can all be deleterious for project's success. If any or at least some of these factors are present in any project there will be a need to improve outreach and educational campaigns to strengthen project's pertinence to local stakeholders and institutions, and to strengthen social-ecological systems resilience. Also, it would be important to build up communication channels with private entities and show the benefits that successful coral farming and reef rehabilitation can bring to their businesses. Achieving such collaborative support would be important to strengthen economic support.

4.6. How to overcome uncertainties and change?

Management of uncertainties and change under projected environmental and climate changes constitute a major challenge. For instance, increasing frequency and/or strength of hurricanes fuelled up by increasing sea surface temperature (SST), if combined with weak governance, can result in major crisis. In situations where uncertainties and change are key features of the social-ecological landscape, critical factors for sustainability and rapid recovery are resilience, the capacity to cope with crisis and adapt, and the conservation of sources of innovation and renewal [50]. Such is the case of the impact of extreme weather events and ecological surprises impacting coral farming and reef rehabilitation. However, interventions in social-ecological systems with the aim of altering resilience immediately confront issues of governance. Who decides what should be made resilient to what is a critical question for any reef rehabilitation program. For whom is resilience to be managed, and for what purpose are also two key elements that must be decided during the planning stages of any project, always bearing in mind the long-term goal of managing uncertainties and change.

4.7. Socio-economic benefits of coral farming and reef rehabilitation can be offset by lack of governance

A major lesson learned from the Culebra Island coral farming and reef rehabilitation experience has been that the rapid increase in socio-economic benefits from increased nature-based tourism does not always contribute to support social-ecological systems under a weak governance structure. Increasing tourism and business opportunities (e.g., kayaking, shore-based SCUBA/snorkeling, charter vessels, beach swimming, hotel lodging, vehicle rental, bus services, etc.) have resulted in a significant boom in gross revenues for local and external private businesses. This has resulted in increasing alternative job opportunities. But a weak governance structure still allows the leak of revenues from the local community, favoring external businesses, and the total lack of economic support of the local MPA, and local coral farming and reef rehabilitation efforts. Therefore, strengthening governance is a critical step to support the ecological and socio-economic recovery of social-ecological systems resilience, stability and persistence, and a mechanism to foster increased local participation and sharing of benefits.

A second important benefit in Culebra Island has been increasing fish densities on rehabilitated reefs, therefore contributing to enhance fishing on adjacent areas, through fish spillover effects. Also, reef rehabilitation has resulted in increased recovery of shoreline protection from wave action and erosion. Therefore, the combined benefits are multiple and, with proper planning, design, funding, governance, local support, and implementation, this can have long-lasting impacts in restoring coastal social-ecological resilience, and overall ecosystem services and productivity.

4.8. The challenge of engaging the youth: lessons learned from marginalized small island communities

Coral farming and reef rehabilitation in Culebra Island have also contributed to educate local children and modify local resident's behavior favoring coral reef conservation. Local NGO Coralations has developed for nearly two decades a highly successful educational engaging program called "*Exploradores Marinos*" or *Marine Explorers*. This has allowed approaching local kids with an understanding of their community relationship with the coastal resources (e.g., recreation and sustenance), and introduce planned, inquiry-based discoveries that sprout from that identity origin, as opposed to introducing a totally different perspective (e.g., "welcome to your ocean laboratory"). Second, it is important to understand that planning is compromised for families living on financial brink and that time must be budgeted to compensate for disorganization, lack of preparation, competing programs, transport, last-minute emergencies, health, and poor-diet related illnesses. Such conditions become critically magnified due to the small size of Culebra Island (<70 km^2), its location 27 km off northeastern Puerto Rico, and its small population size (<2000 residents). Also, programs need to be no cost for economically compromised participants, however, engagement must require compensation for programs to be valued. Required community service is one option, but always rewarded and never treated as punishment.

All developing humans seek attention. They quickly learn that attention is rewarded for both positive and negative behaviors. Many at-risk youth are conditioned to negative behavioral awards from a very young age but ocean therapy allows them to be removed from their familiar territory for rapid and constructive positive reward programming. However, the positive reward scenarios need to be well thought out, safe and many times staged in advance (e.g., collaborative removal of derelict fishing gear from the reef, recovering of lose coral fragments at risk to support coral farmers, etc.). Medical disclosures from juvenile community volunteers are sometimes dishonest because parents are concerned their child would be stigmatized or prevented to participate in the project. This is dangerous for seizure-related illnesses and inquiries have to be conducted discretely with parents in a climate of trust. This shows that parents consider coral farming and reef conservation-oriented education as unique, novel, enriching experiences for their kids, that they would do anything to ensure their participation in the project. But such risks need to be addressed in a case by case scenario to prevent kids with potentially threatening conditions to engage into risky in-water activities.

An important lesson of working with kids has been to focus activities on accomplishing missions, and refocus anxious students on a defined mission. It is also important to keep groups small and develop excursions that force interdependent collaborations. This increases cost of outreach and educational programs but reaps the rewards 10-fold in many benefits, including greater probability of interesting animal encounters and less opportunity for accidents. This can be done by matching the student to skill level contribution in team activity, and while the skills of some exceed that of others, emphasis must be kept on the importance of all individual contributions to the success of the overall defined mission. Young adult behaviors are conditioned by peer to peer interactions. Everyone is not equal, but it is important to try and find where the kid excels and expose that to other kids.

When dealing with outreach and educational activities for kids, if it is not fun, there will be no long-term engagement. Even a specifically defined work mission has to be fun in order to maintain youngsters engaged. The most be beneficial asset to engaging younger students to the program is mixing with fully engaged slightly older students. Lessons introduced cannot feel like lessons and one of the best methods has been found to be ensuring information is redundant interrogative approach. One of the most successful approaches was when a science teacher at school was offering almost identical information onto what the kids were seeing, feeling, and experience in water during their out-of-school participation in coral reef conservation educational activities.

But a roadblock to success is the rapidly increasing use of electronic devices by young students. Heads must be out of their apps. Electronics have to be confiscated with the understanding that kids, like many adults, are addicted to these devices, or there is competing attention. Also, kids behavior and attention during educational activities are affected by diet. Sometimes kids can exhibit uncontrolled or even violent reactions if allowed to eat food items with excessive content of sugar, high fructose corn syrup, and artificial colorants which they are often conditioned to eat. Nevertheless, hands-on in-water educational experiences are always behavior-transforming experiences that have shown to have multiple positive benefits for local kids. The most important elements have been getting to know their backyard marine resources, improving their respect for adults, their behavior with other kids, and their appreciation for their own resources and the benefits they derive from them.

In the long term, building up a local meaningful voice—ownership—pride of their resources, their natural reserve, and their island has always been a key mission focus of the community service projects in the hope it may help prevent the slippery slope tragedy of the commons rationalizations which could lead to continuing resource overexploitation. These approaches can probably be the only hope for a natural reserve with well-documented enforcement problems associated to poor governance, patrolling and compliance, and lack of public education and outreach. Also, in the long term, participation in local snorkeling activities moved to Friday mornings in Culebra Island has shown that participant kids perform better in school throughout the rest of the day. Therefore, such behavior-modifying activities can contribute to improve performance in schools and overall attitudes. This is an aspect that deserves to be studied, as youth represent the future of base-community participation in the management of social-ecological systems.

The "*Educadores Marinos*" program has produced a few successful stories with kids pursuing careers in Environmental Sciences and other professional disciplines in academic institutions. For many others, the end point for *explorers* is intervention/interpretation jobs and the collaboration/participation in pilot research projects being conducted in Culebra. However, expectations are something that need to be addressed. When working with at-risk youth hope that at the end of the years, one will at least have forged a relationship/dialog with the soon-to-be young adult. Respect has to be earned with their kids and their parents to ensure engagement and support.

5. Conclusions and recommendations

Low-tech coral farming and reef rehabilitation have become important tools to foster community-based participation in the management of coastal social-ecological systems. But, there are still important gaps that need to be addressed in order to integrate the technical and scientific components of coral farming and reef rehabilitation with the sociological components. Preliminary evidence of Acroporid corals demographic dynamics has already shown important lessons learned. First, conservation and restoration activities should be designed at the species-specific level whenever possible, with separate specific goals and objectives for individual species. Each coral species, and even different genetic clones within any given species, can respond differently to environmental variation. Also, species-specific variability in acclimation responses to changes in environmental conditions suggests that there can be different vulnerabilities and, as such, restoration projects should always consider such variation among species. This could trigger multiple nonlinear responses to environmental and climate variation. Therefore, coral reef rehabilitation efforts must be adaptive and focused on maximizing resilience as a long-term goal. It should also look forward to develop strategies and techniques to propagate multiple coral species with different life traits to buffer against future nonlinear impacts. The resilience approach emphasizes on managing nonlinear dynamics, thresholds, environmental and climate uncertainty, and ecological surprises [51]. It also pays attention to how periods of slowly evolving, gradual change interplay with periods of rapid, stochastic change, and how such dynamics interact across different temporal and spatial scales. In this context, demographic modeling becomes fundamental to address such concerns.

Second, spatiotemporal differences in demographic transitions displayed by corals such as *Acropora cervicornis* suggest that restoration efforts should be partitioned among several locations rather than allocating all the resources into one site. Further, it also suggests that a combination of *in situ* (e.g., underwater) and *ex situ* (e.g., land-based coral aquaculture farms) strategies should be implemented to cope with potential impacts of extreme weather events and ecological surprises. These actions will enhance the persistence of the species if localized extirpation occurs due to any significant disturbance (e.g., recurrent runoff events, hurricanes). It should also foster the propagation of multiple coral species in support to coral biodiversity restoration and seascape enhancement efforts. Another fundamental lesson learned is that addressing differences in population dynamics among coral colony size categories is important for parameterizing demographic models. This may allow addressing contrasting species-specific, size-specific, genetic clone-specific, or condition-specific population trajectories. Such contrasting population trajectories may be the result of different life traits and different susceptibilities to low and moderate environmental changes. Identifying spatiotemporal variation patterns in such elements may imply that continuity in low-tech coral farming and out-planting efforts is fundamental to sustain restored populations in the wild. In addition, transplanting large colonies will result in higher population survival and growth rates than transplanting small colonies, as many standard coral farming operations do, at least across the Caribbean. Also, transplanting large colonies can achieve faster ecological benefits (e.g., thicket formation, enhanced-essential fish habitat role). However, transplanting large colonies pose some challenges; specifically, regarding the time necessary for a nursery-reared coral fragment to reach the effective transplantation size. This means that efforts need to be taken to improve coral farming techniques to accelerate coral growth, ensure high survival rates, and use methods that trigger faster colony growth rates.

It is also central to establish auto-sustainable funding mechanisms to support coral farming and reef rehabilitation projects. Demographic evidence has already proved that only through sustained input of harvested corals restored populations can remain viable and grow under present and projected environmental and climate conditions. Therefore, supporting the continuous operation of such projects becomes increasingly important, but at the same time increasingly challenging, in particular for developing countries, economically and politically constrained colonies, and small island nations. Depending only on funding through government agencies has shown to be a poor mechanism for support. Government support could be even nonexistent in many countries. Government institutions have often highly changing agenda, which typically respond to highly fluctuating political steering and philosophies in regards to natural resource conservation and climate change. Therefore, funding programs can frequently change goals and objectives, which could risk local support of projects, regardless of their historical trajectory and success. Coral farming and reef rehabilitation need to be incorporated into natural resource conservation and restoration public policies, and also into climate change and SLR adaptation strategies and policies. A potential sustainable strategy for economic support could be implementing a green tax at local or national levels (e.g., tourism activities, cruise-ship visits, airplane landings, hotel room rental, vehicle sales, property construction tax, and industrial revenues). This may include measures such as those implemented in the state of Hawai'i, USA, where coastal development projects are required to economically support State-led coral farming and reef restoration operations as part of mandatory environmental mitigation regulatory requirements.

On the other hand, increasing the spatial scale of reef rehabilitation is essential for reef rehabilitation to become an important strategy to restore coastal social-ecological systems resilience. Population demographic dynamics in *Acropora palmata* and *A. cervicornis* vary considerably in time and space. Increasing the spatial scale of population rehabilitation will increase the probability of species persistence, and will enhance its ecosystem functions (e.g., fish nursery ground, buffering wave action). Coral out-plant spatial array is also critical for the formation of thickets. In this sense, coupling demographic modeling with oceanographic numerical models is a highly promising tool to support planning, designing, and implementing future coral reef rehabilitation efforts.

But, the sociological and economic components of coral reef rehabilitation have still remained out of the formula. Understanding sociological dynamics should become an absolute priority to improve the success of future coral farming and reef rehabilitation efforts. Projects developed in Culebra Island, Puerto Rico, since year 2003 have contributed to educate local children and modify local residents' behavior favoring coral reef conservation. Particularly, addressing behavior-modifying activities and learning how to overcome roadblocks to success are fundamental to develop sustainable strategies to educate, train, and empower local residents to participate in social-ecological systems management. It is critical to foster the creation of strong, functional, cross-sectorial partnerships, which respect the integration of base communities and small non-governmental organizations (NGOs) in the planning and implementation of projects. The stronger the environmental governance collaboration, the improved the success in addressing problems in social-ecological systems [52].

But also, understanding of social processes like social learning and social memory, mental models, and knowledge–system integration are a critical trans-disciplinary integration to improve projects success and social-ecological systems resilience [51]. Further, integrating visioning and scenario building, leadership building, multi-sectorial agents and actor groups, and strengthening cross-sectorial social networking are necessary adaptive approaches to cope with future environmental and climate changes. Another particular challenge of social-ecological systems is how to deal with institutional and organizational inertia and change, with adaptive capacity, transformability, and systems of adaptive governance that allow for management of essential system services [51]. Further, strengthening adaptive governance capabilities is essential to overcome stochastic events and crisis (e.g., natural disasters; ecological surprises). Strong governance connects individuals, organizations, agencies, and institutions at multiple organizational levels [53]. Further, building vision, leadership, and trust are also important features of resilient social-ecological systems [54]. Strengthening the organization of base communities can empower key persons to provide leadership, trust, vision, meaning, and they help transform management organizations toward a learning environment, and can foster the participation of at-risk youth, and the integration of adaptive, participatory co-management efforts. A resilient social-ecological system may make use of crisis as an opportunity to transform into a more desired state.

In addition, the following 10 components have been shown to be fundamental to address sustainable and resilient social-ecological systems [55]: (1) *Size of resource systems*—in our case, the spatial scale of reef rehabilitation becomes a major element of concern to achieve sustainability and meaningful impacts on resilience; (2) *Productivity of system*—increasing the spatial scale of reef rehabilitation also fosters an increase in ecological and social benefits; (3) *Predictability of system dynamics*—the incorporation of restored coral demographic models, coupled with oceanographic numerical models, should be the next step to improve our ability to predict system dynamics; (4) *Resource unit mobility*—corals are not mobile entities, but reef-associated biota can be, therefore, improving governance regarding management of mobile links such as reef fisheries is important to improve management success; (5) *Collective choice rules*—fostering increased local participation in planning and decision-making processes will increase local stewardship, support, and compliance with management, and may reduce cost and difficulties of enforcement; (6) *Number of users*—the impact of group size and determining limits of acceptable change (from the perspective of recreational and tourism uses) must be incorporated into management; (7) *Leadership/entrepreneurship*—developing leadership and entrepreneurship skills in members of local base communities is paramount to improve stewardship, support and trust, and would likely result in the protection of local livelihoods and business opportunities; (8) *Norms/social capital*—need to build up shared moral and ethical standards, and common trust in resource users to facilitate decision-making and monitoring processes; (9) *Knowledge of social-ecological systems/mental models*—knowledge of social-ecological systems is central to share common knowledge among different user sectors, to understand carrying capacity and limits of acceptable change of the resource, its attributes of resilience, and to prevent failure to organize and destroy the system; and (10) *Importance of resource*—understanding the value of the resource to local environmental, ecological, and socio-economic sustainability, to the support of sustainable livelihoods, and for sustaining food security and sovereignty. The take-home message is that reef managers and reef rehabilitation practitioners need to engage social scientists to support their efforts as a strategy to foster improved local support, stewardship, compliance, and success.

Coral reef rehabilitation in Culebra Island, Puerto Rico, has resulted in a rapid increase in benefits for local communities. Increasing tourism and business opportunities have resulted in a significant boom in gross revenues for private businesses, and in improved, and diversified livelihoods. This has resulted in increasing alternative job opportunities. But, leakage of revenues needs to be reverted to enhance sustainability, local benefits, stewardship, and support. Coral reef rehabilitation has also resulted in increasing fish densities on rehabilitated reefs, therefore attracting further nature-based tourism, and in contributing to enhance fishing on adjacent areas, through fish spillover effects. In addition, it has resulted in increased recovery of shoreline protection from wave action and erosion. Therefore, the combined benefits to social-ecological systems are multiple, and with proper planning, design, funding, local support, and implementation this can have long-lasting impacts in restoring resilience and overall services and productivity of coastal social-ecological systems.

Acknowledgements

This publication was possible thanks to the support of the National Science Foundation (HRD #0734826) to the Center for Applied Tropical Ecology and Conservation, as well as by the support provided by the University of Puerto Rico Central Administration, both to E.A. Hernández-Delgado. Also, support was provided by NOOA Restoration Center and The Nature Conservancy to Sociedad Ambiente Marino through sub-award MAR-SAM-110110. In addition, support was provided by the U.S. Fish and Wildlife Service Coastal Program, the University of Puerto Rico's Sea Grant College Program, the Ford Motor Company Foundation, Toyota Foundation, Roland Pesch, and by a myriad of community-based and student volunteers through years of dedicated work. Our major appreciation to the passionate vision and pioneering support by the late fisher folks of Culebra Island, in particular, Don Ramón "Monchín" Feliciano, and Don Anastasio "Taso" Soto.

Author details

Edwin A. Hernández-Delgado[1,2,3]*, Alex E. Mercado-Molina[3,4], Samuel E. Suleimán-Ramos[3] and Mary Ann Lucking[5]

*Address all correspondence to: edwin.hernandezdelgado@gmail.com

1 Center for Applied Tropical Ecology and Conservation, University of Puerto Rico, San Juan, Puerto Rico, USA

2 University of Puerto Rico, College of Natural Sciences, Interdisciplinary Program, San Juan, Puerto Rico, USA

3 Sociedad Ambiente Marino, San Juan, Puerto Rico, USA

4 Department of Marine Sciences, Florida International University, Miami, Florida, USA

5 Coralations, Culebra, Puerto Rico, USA

References

[1] Cloern JE. Our evolving conceptual model of the coastal eutrophication problem. Marine Ecology Progress Series. 2001;**210**:223-253

[2] Rogers CS. Responses of coral reefs and reef organisms to sedimentation. Marine Ecology Progress Series. 1990;**62**:185-202

[3] De'ath G, Fabricius K. Water quality as a regional driver of coral biodiversity and macroalgae on the Great Barrier Reef. Ecological Applications. 2010;**20**(3):840-850

[4] Roberts CM. Effects of fishing on the ecosystem structure of coral reefs. Conservation Biology. 1995;**9**(5):988-995

[5] Jackson JB, Donovan MK, Cramer KL, Lam VV. Status and trends of Caribbean coral reefs. Global Coral Reef Monitoring Network, IUCN, Gland, Switzerland. 2014; Available at http://cmsdata.iucn.org/downloads/caribbean_coral_reefs___status_report_1970_2012.pdf

[6] Hoegh-Guldberg O, Mumby PJ, Hooten AJ, Steneck RS, Greenfield P, Gomez E, et al. Coral reefs under rapid climate change and ocean acidification. Science. 2007;**318**(5857): 1737-1742

[7] Hoegh-Guldberg O, Bruno JF. The impact of climate change on the world's marine ecosystems. Science. 2010;**328**(5985):1523-1528

[8] Hernández-Delgado EA. The emerging threats of climate change on tropical coastal ecosystem services, public health, local economies and livelihood sustainability of small islands: Cumulative impacts and synergies. Marine Pollution Bulletin. 2015;**101**(1):5-28

[9] Holling CS. Resilience and stability of ecological systems. Annual Review of Ecology and Systematics. 1973;**4**(1):1-23

[10] Adger WN. Social and ecological resilience: are they related? Progress in Human Geography. 2000;**24**(3):347-364

[11] Adger WN, Hughes TP, Folke C, Carpenter SR, Rockström J. Social-ecological resilience to coastal disasters. Science. 2005;**309**(5737):1036-1039

[12] Caswell H. Matrix Population Models. 2nd ed. Sunderland MA: John Wiley and Sons, Ltd., Sinauer Associates

[13] Hughes TP. Population dynamics based on individual size rather than age: A general model with a reef coral example. The American Naturalist. 1984;**123**(6):778-795

[14] Fong P, Glynn PW. A dynamic size-structured population model: does disturbance control size structure of a population of the massive coral *Gardineroseris planulata* in the Eastern Pacific? Marine Biology. 1998;**130**:663-674

[15] Salguero-Gómez R, De Kroon H. Matrix projection models meet variation in the real world. Journal of Ecology. 2010;**98**(2):250-254

[16] Linares C, Doak DF, Coma R, Díaz D, Zabala M. Life history and viability of a long lived marine invertebrate: the octocoral *Paramuricea clavata*. Ecology. 2007;**88**(4):918-928

[17] Vardi T, Williams DE, Sandin SA. Population dynamics of threatened elkhorn coral in the northern Florida Keys, USA. Endangered Species Research. 2012;**19**:157-169

[18] Mercado-Molina AE, Ruiz-Diaz CP, Pérez ME, Rodríguez-Barreras R, Sabat AM. Demography of the threatened coral *Acropora cervicornis*: implications for its management and conservation. Coral Reefs. 2015a;**34**:1113-1124

[19] Soto-Santiago FJ, Mercado-Molina A, Reyes-Maldonado K, Vélez Y, Ruiz-Díaz CP, Sabat A. Comparative demography of two common scleractinian corals: *Orbicella annularis* and *Porites astreoides*. PeerJ. 2017;**5**(e3906):1-18. DOI: 10.7717/peerj.3906

[20] Mercado-Molina AE, Ruiz-Diaz CP, Sabat AM. Demographics and dynamics of two restored populations of the threatened reef-building coral *Acropora cervicornis*. Journal for Nature Conservation. 2015b;**24**:17-23

[21] Mercado-Molina AE, Ruiz-Diaz CP, Sabat AM. Branching dynamics of transplanted colonies of the threatened coral *Acropora cervicornis*: Morphogenesis, complexity, and modeling. Journal of Experimental Marine Biology and Ecology. 2016;**482**:134-141

[22] Williams DE, Miller MW, Kramer KL. Demographic Monitoring Protocols for Threatened Caribbean *Acropora* spp. Corals. Vol. 543. NOAA Technical Memorandum NMFS-SEFSC. 93 pp

[23] Goreau TF. The ecology of Jamaican coral reefs I. Species composition and zonation. Ecology. 1959;**40**(1):67-90

[24] National Marine Fisheries Service. Recovery Plan for Elkhorn (*Acropora palmata*) and Staghorn (*A. cervicornis*) Corals. Prepared by the. Silver Spring, MD, USA: Acropora Recovery Team, National Marine Fisheries Service, NOAA; 2015. 156 p

[25] Judson OP. The rise of the individual-based model in ecology. Trends in Ecology & Evolution. 1994;**9**(1):9-14

[26] Sleeman JC, Boggs GS, Radford BC, Kendrick GA. Using agent-based models to aid reef restoration: Enhancing coral cover and topographic complexity through the spatial arrangement of coral transplants. Restoration Ecology. 2005;**13**:685-694

[27] Graham NJ, Nash KL. The importance of structural complexity in coral reef ecosystems. Coral Reefs. 2013;**32**(2):315-326

[28] Newman MJ, Paredes GA, Sala E, Jackson JB. Structure of Caribbean coral reef communities across a large gradient of fish biomass. Ecology Letters. 2006;**9**(11):1216-1227

[29] Jones GP, McCormick MI, Srinivasan M, Coral EJV. decline threatens fish biodiversity in marine reserves. Proceedings of the National Academy of Sciences of the United States of America. 2004;**101**(21):8251-8253

[30] Ainsworth CH, Mumby P. Coral–algal phase shifts alter fish communities and reduce fisheries production. Global Change Biology. 2015;**21**(1):165-172

[31] Alvarez-Filip L, Côté IM, Gill JA, Watkinson AR, Dulvy NK. Region-wide temporal and spatial variation in Caribbean reef architecture: is coral cover the whole story? Global Change Biology. 2011;**17**(7):2470-2477

[32] Hernández-Delgado EA, Mercado-Molina AE, Alejandro-Camis PJ, Candelas-Sánchez F, Fonseca-Miranda JS, González-Ramos CM, et al. Community-based coral reef rehabilitation in a changing climate: Lessons learned from hurricanes, extreme rainfall, and changing land use impacts. Open Journal of Ecology. 2014;**4**:918-944

[33] Vytopil E, Willis B. Epifaunal community structure in *Acropora* spp. (Scleractinia) on the Great Barrier Reef: implications of coral morphology and habitat complexity. Coral Reefs. 2001;**20**:281-288

[34] Untersteggaber L, Mitteroecker P, Herler J. Coral architecture affects the habitat choice and form of associated gobiid fishes. Marine Biology. 2014;**161**:521-530

[35] Ferse SC. Poor performance of corals transplanted onto substrates of short durability. Restoration Ecology. 2010;**18**(4):399-407

[36] Bowden-Kerby A. Low-tech coral reef restoration methods modelled after natural fragmentation process. Bulletin of Marine Science. 2001;**69**(2):915-931

[37] Garrison V, Ward G. Storm-generated coral fragments – A viable source of transplants for reef rehabilitation. Biological Conservation. 2008;**141**:3089-3100

[38] Hollarsmith J, Griffin S, Moore TD. Success of outplanted *Acropora cervicornis* colonies in reef restoration. In Proceedings of the 12th International Coral Reef Symposium. 2012; **20A**(3):1-5

[39] Young CN, Schopmeyer SA, Lirman D. A review of reef restoration and coral propagation using the threatened genus *Acropora* in the Caribbean and Western Atlantic. Bulletin of Marine Science. 2012;**88**(4):1075-1098

[40] Babcock RC. Comparative demography of three species of scleractinian corals using age- and size-dependent classifications. Ecological Monographs. 1991;**61**(3):225-244

[41] Gotelli NJ. Demographic models for *Leptogorgia virgulata,* a shallow-water gorgonian. Ecology. 1991;**72**(2):457-467

[42] McFadden CS. A comparative demographic analysis of clonal reproduction in a temperate soft coral. Ecology. 1991;**72**(5):1849-1866

[43] Edmunds PJ. The effect of sub-lethal increases in temperature on the growth and population trajectories of three scleractinian corals on the southern Great Barrier Reef. Oecologia. 2005;**146**:350-364

[44] Smith LD, Devlin M, Haynes D, Gilmour JP. A demographic approach to monitoring the health of coral reefs. Marine Pollution Bulletin. 2005;**51**(1):399-407

[45] Edmunds PJ, Elahi R. The demographics of a 15-year decline in cover of the Caribbean reef coral *Montastraea annularis*. Ecological Monographs. 2007;**77**(1):3-18

[46] Hernández-Pacheco R, Hernández-Delgado EA, Sabat AM. Demographics of bleaching in the Caribbean reef-building coral *Montastraea annularis*. Ecosphere. 2011;**2**(1):art9. 1-13

[47] Morris W, Doak D, Groom M, Kareiva P, Fieberg J, Gerber L, et al. A Practical Handbook for Population Viability Analysis. The Nature Conservancy; 2009. 75 pp

[48] Fieberg J, Ellner SP. Stochastic matrix models for conservation and management: a comparative review of methods. Ecology Letters. 2001;**4**(3):244-266

[49] Hernández-Delgado EA. Community-based low-tech coral reef rehabilitation impacts on Culebra Island HFA fish assemblages: A BACI approach. (Manuscript in preparation)

[50] Lebel L, Anderies JM, Campbell B, Folke C, Hatfield-Dodds S, Hughes TP, Wilson J. Governance and the capacity to manage resilience in regional social-ecological systems. Ecology and Society. 2006;**11**(1):19

[51] Folke C. Resilience: The emergence of a perspective for social–ecological systems analyses. Global Environmental Change. 2006;**16**(3):253-267

[52] Bodin Ö. Collaborative environmental governance: Achieving collective action in social-ecological systems. Science. 2017;**357**(6352):eaan1114:1-8

[53] Folke C, Hahn T, Olsson P, Norberg J. Adaptive governance of social-ecological systems. Annual Review of Environment and Resources. 2005;**30**:441-473

[54] Olsson P, Folke C, Berkes F. Adaptive comanagement for building resilience in social–ecological systems. Environmental Management. 2004;**34**(1):75-90

[55] Ostrom E. A general framework for analyzing sustainability of social-ecological systems. Science. 2009;**325**(5939):419-422

Section 4

Industrial Products from Corals

Chapter 12

Structure-Activity Relationship (SAR) Studies to Maximize the Activity of Compounds Isolated from Octocorals

Carmenza Duque, Leonardo Castellanos and Edisson Tello

Additional information is available at the end of the chapter

http://dx.doi.org/10.5772/intechopen.74686

Abstract

This chapter presents some significant study cases on octocoral organisms (*Eunicea succinea, Eunicea mammosa, Eunicea knighti, Pseudoplexaura flagellosa, Eunicea laciniata, Antillogorgia elisabethae, Muricea austera, Paragorgia* sp., *Lobophyton* sp., *Sarcophyton glaucum* and *Sinularia lochmodes*) that have been identified as a source of promising bioactive compounds and whose results have further been used for studies on structure-activity relationship (SAR) as a strategy to increase the value of the activity initially detected. The scientific literature data discussed here were obtained with the SciFinder tool during the period 2000–2016 and from the additional results here presented for the biofilm inhibition activity of compounds and synthetic analogs for the cases related with *Eunicea knighti* and *Pseudoplexaura flagellosa* (until now unpublished data of the authors of this chapter).

Keywords: bioprospecting, marine natural products, octocorals, structure-activity relationship, synthetic and natural analogs, bioactivity

1. Introduction

Animals, plants and microorganisms of marine origin are suitable sources for the discovery and development of numerous medicines and industrial products. Among the 36 animal phyla described to date, 34 are represented in the marine environment, with about half being exclusively marine [1, 2]. Such statistics demonstrate the immense potential of marine biodiversity from which only a small percentage has been studied as a source of compounds useful to humans. The next few lines will briefly describe a very short history of marine natural products drug discovery, until today one of the major fields of application of marine natural products (extensive reviews on the subject can be found in Refs. [3–8]).

Studies began in the early 1950s with the discovery of spongouridine and spongothymidine isolated from the marine sponge *Tehya crypta* [9–11], natural compounds later transformed by synthesis into vidarabine and cytarabine, respectively, and approved as drugs by the USA Food and Drug Administration (FDA) [3]: vidarabine as an anticancer agent in 1969, and cytarabine as an antiviral drug in 1976. Although the latter has recently been withdrawn from the market, it is still being used in ophthalmic treatments. However, it was only in the mid-1970s that the isolation of prostaglandins from octocoral species such as *Plexaura homomalla* [12], the structural determination of several of them and their application in human medicine encouraged many scientists to conduct research in this field. In such a way, by the end of the last century, thousands of compounds had been isolated as major metabolites of conspicuous marine organisms, easy to collect and easy to study.

Since then, thanks to tremendous efforts by scientists, to the improvement and easy access to better techniques of collection of organisms (snorkeling, scuba, submersibles, remote operated vehicles (ROVs)), to the discovery of modern techniques of chemical analysis and of biological activity, to the emergence of the "omic" approaches (genomics, proteomics, metabolomics, transcriptomics), to the recent genome mining approaches (exploitation of genome public data) for the discovery of new natural products, and to the use of molecular biology in the field of bioengineering, today approximately 25,000 bioactive marine compounds with novel structure are known, many of them with potential industrial use [13]. Focusing only on the pharmaceutical industry, 8 marine compounds approved by the FDA and/or by the European Medicines Agency (EMEA) are on the market as therapeutic agents, 11 in different clinical phases, 1458 in the preclinical phase (the data on the compounds in the clinical and preclinical phases were taken as reported in recent references [3, 13, 14] and from A.M.S. Mayer's website in the USA (http://marinepharmacology.midwestern.edu, accessed: 2017-01-30). In addition, there are many other compounds that remain in the laboratories of academic research groups or research centers waiting for the opportunity and adequate funding to enter the commonly so-called marine pharmaceutical pipeline, which allows them to start the path toward their conversion to new drugs.

Furthermore, it is worth mentioning that in parallel with the aforementioned studies, toward the first decade of the present century, numerous researchers turned their attention to look for new sources of bioactives, that is, marine microorganisms (cyanobacteria, marine fungi and other classes of Eubacteria) hoping to find new compounds and new activities, and because it began to be known that many compounds previously isolated from macroorganisms were actually produced by their associated microbes, as described in [4].

The aforementioned shows that marine organisms really are a fascinating source of molecules with unique structures and exploitable biologic activity. The following are some of the compounds of marine origin established in the market as therapeutic agents or used as industrial products [3, 13, 14]: compounds used in cancer treatment such as cytarabine (CytosarU™, Depocyt™) (mentioned above), trabectedina (Yondelis™), complex tetrahydroisoquinoline alkaloid obtained from tunicate *Ecteinascidia turbinata*; Eribulin mesylate (Halaven™) isolated from sponge *Halichondria okadai*; Brentuximav vedotin (Adcetris™) isolated from sea hare

Dollabela auricularia and Plitidepsin (Aplidin™) isolated from ascidian *Aplidium albicans* which last year received an orphan drug designation by EMA and by FDA; compounds used as antivirals: vidarabin (Vira A™) (mentioned above), and the recent developed iota-carrageenane (carragelose™) obtained from red algae (ready to enter the market); compounds used to treat pain: ziconotide (Prial™) very powerful product, isolated from marine snail *Conus magnus* ziconotide (Prial™); compounds used in hypertriglyceridemia treatment: omega-3-acid ethyl esters (Lovaza®) and some other products such as the skin cream called Resilience™ whose active base is a crude extract of pseudopterosines and *seco*-pseudopterosines; and although these ingredients ran their race as antiinflammatories, they were unsuccessful in reaching the final stage; on the other hand, they retained their status as cosmeceutical products of great demand for the care of sun-induced skin irritations.

Among marine organisms source of bioactives, octocorals—a sub-class of Anthozoa—are a diverse group of colonial animals with 8 tentacle polyps and 8 internal mesenteries comprising about 3000 species (1.5% of all marine animals) of soft corals, gorgonians (sea fans, sea whips), sea pens and blue corals [15, 16] found throughout the world's oceans. They have proven to be a prolific source of natural products having new structures, many of them without terrestrial-counterpart with relevant biological activities, which have been arising enormous interest both in the academic world and in the industry in the last 50 years. The first publication on octocorals came out in 1958 [17]. Since then, many studies on metabolites from octocorals have been published in the chemical literature and biological activity and high-quality reviews have appeared on the subject. Among such important contributions, it is worth mentioning the article written by Coll in 1992 [18], the annual reviews initiated by Faulkner in 1984 until 2002 [6] and continued by Blunt and his New Zealand group since 2003 [7], as well as the reviews by Rodríguez in 1995 [19], Berrue and Kerr in 2009 [20], Berrue et al. in 2011 [21], Almeida et al. in 2014 [22], Hu et al. in 2015 [8] and by Lei et al. in 2016 [23]. According to all these published data, the chemical constituents of octocorals are mostly steroids, acetogenins, sesquiterpenes and numerous diterpenes (with at least 40 skeletal classes) and diterpenes glycosides (compounds unique to gorgonians), exhibiting biological activities such as ichthyotoxic, antimicrobial, anticancer or/and cytotoxic, antiviral, antiinflammatory, antiproliferative, feeding stimulation, feeding deterrent, antipredatory, antifouling, antileishmanial, antiplasmodial and antiHIV-1, among others.

All the studies mentioned above clearly show that the natural products from Cnidaria (mostly corals) and from Porifera (mostly sponges) accounting for 56.89% of the total reported marine bioactives [19] have become a very attractive source of study for scientists, with the added value of being exploited industrially, particularly in pharmacology. However, for a compound discovered in a laboratory to be transformed in an industrial product, it is first necessary to maximize its biological activity and face the big problem of the sustainable supply (as mentioned in the literature [13]: whatever the use given to the compound, *several grams to hundreds of grams are required for preclinical development, multikilogram quantities are needed for clinical phases and tons for industrial uses—figures that contrast with the minimum quantities isolated commonly in the research laboratories*).

These current problems are critical in the industrial development of natural products and have lead to the development of new alternate ways such as preparation of synthetic or hemisynthetic analogs, among other applications, enhancing the activity and designing pharmacophores of lower complexity that can then be synthesized by faster and easier routes.

For this reason, this chapter aims to show some studies of the scientific literature in the last 15 years, where octocorals emerge as an excellent source of bioactive compounds and how the increase in their activity has been achieved through the use of the structure-activity relationship (SAR) strategy. This method has become a powerful tool for the discovery of new bioactive compounds and to promote the activity by converting bioactive compounds through synthesis into chemical analogs. Furthermore, we will discuss how the preparation of analogs could also be a way of helping the key current problem of material supply in a sustainable manner. Finally, we will present some recent unpublished experimental data from our laboratory where the isolation of terpenoids and some of their natural homologs from octocorals, and their conversion by chemical synthesis into compounds with higher biological activity have been a good strategy to achieve the aforementioned purposes.

2. SAR study cases in octocorals (2000–2016)

The studies highlighted in this item show the results of the literature survey using the SciFinder tool between 2000 and 2016, of some relevant studies reported, describing natural analogs and semi-synthetic derivatives prepared as a strategy to promote biological activity of compounds isolated from octocorals. The analyzed cases are chronologically organized throughout the chapter and each SAR study appears in the chemical literature grouped under subheadings with the name of the corresponding species. **Figure 1** presents some of octocoral species discussed here.

2.1. *Eunicea succinea* and *Eunicea mammosa*

Octocorals of the *Eunicea* genus are one of the most interesting gorgonians because they are a source of abundant and diverse cembranoids and dolabellane diterpenoids, including some glycosylated [20]. On the Scopus database, there are more than 56 search results of chemical studies on *Eunicea* species. Biological activities described for cembranoids isolated from *Eunicea* extracts include antiplasmodial, quorum sensing inhibition, antiviral and cytotoxic activity against several cancer cell lines [20]. It has been described that cembranolides containing cyclic ethers possess potent antileukemic activities [24], that is, euniolide **1**, 12,13-bisepieupalmerin **2** and eupalmerin acetate **3** cembranoids, isolated in large quantities from *E. succinea* and *E. mammosa*, collected in shallow waters of Mona Island (Puerto Rico). These natural compounds showed strong cytotoxic activity against several cell lines, being the oxygenated C-13 compounds (**2** and **3**) more active than the euniolide **1** [25].

In 2000, Puerto Rican scientists [25] synthesized a series of unusual analogs of natural cembranolides **1–3** containing cyclic ether ring systems. They conducted some saponification reactions using KOH from euniolide **1** to obtain eight derivatives, including the well-known

http://dx.doi.org/10.5772/intechopen.74686

Figure 1. Some octocorals reviewed in this chapter in relation to the SAR studies of their active isolated metabolites.

crassin acetate; from compound **2** to obtain five derivatives and from compound **3** to obtain four compounds. Natural compound **3** was also treated with CH_2N_2, Ac_2O, photolysis, ozonolysis and H_2O_2 to obtain 18 more derivatives. In this way, the authors obtained a large variety of structural diverse diterpenoids with multiple oxygen bridges, nitrogen atoms and hydroxyl groups. Finally, the 3 natural compounds together with 11 derivatives were tested in the NCI-60 human tumor cell line screen. Natural compounds **1** and **2** had strong cytotoxic activity (IC_{50} values of 0.1–43 μg/ml) while eupalmerin acetate **3** showed to be less active (IC_{50} values from 0.3 to 16 μg/mL). Unfortunately, the synthetic compounds screened were less cytotoxic than the natural diterpenoids prototypes, only α-methylene-δ-lactones **4** and **5** obtained from **2** and **3,** respectively, showed a characteristic pattern of differential cytotoxicity and were approximately equipotent than the natural products from which they were obtained.

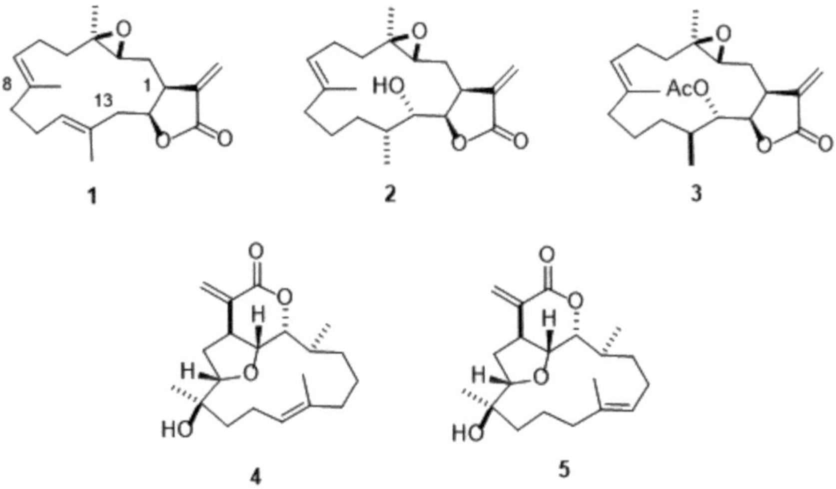

Cembranoids 12-*epi*eunicin **6**, 4-epijeunicin **7** and 13-epieupalmerin **8** were isolated from gorgonian octocoral *E. mammosa* collected in Bahamas, and their structure and their anticancer activity were determined [26]. Compounds **6** and **7** showed moderate cytotoxic activity with A549 (human lung carcinoma), H116 (human colon carcinoma), PSN1 (human pancreatic adenocarcinoma) and T98G (human caucasian glioblastoma); in contrast the activity of compound **8** was higher (IC_{50} ranging from 0.5 to 5 μg/ml). In order to evaluate structure-activity relationships, analogs **9–13** were prepared by chemical transformations of the natural compounds and their activity were evaluated in the same mentioned assays. The chemical modifications introduced to the natural compounds potentiated the activity (excepting in compound **10**), being the most active and selective compound **13** against A-549, H116 and PSNI. It is noteworthy that synthetic analogs **11** and **13** exhibited greater potential than their parent natural products. According to the mentioned results, it could be said that the introduction of cyclic ether linkages across the cembrane skeleton enhances the activity, as well as the introduction of an extra epoxide, that is, in compound **13**, significantly increases the activity against H116 and PSN1 (IC_{50} = 5 μg/ml in **8** to IC_{50} = 0.5 μg/ml in **13**). These results suggest, as mentioned by the authors of this study [26], that the analogs of this series appear to be attractive targets for the development of antitumor agents.

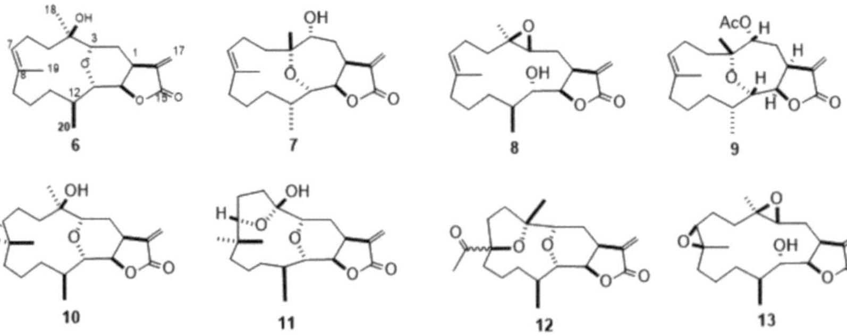

http://dx.doi.org/10.5772/intechopen.74686

2.2. *Eunicea laciniata*

As there is not much published information about the antiviral activity of dolabellanes isolated from soft corals, in 2014 Colombian and Brazilian researchers studied the dolabellanes diterpenoids 13-keto-1(R),11(S)-dolabella-3(E),7(E),12(18)-triene (**14**) and β–araneosene (**15**) for their antiviral properties. These diterpenoids were isolated in multigram scale from the Caribbean octocoral *E. laciniata* collected at Santa Marta bay. The antiviral data showed that they exhibited low antiHIV-1 activity and low toxicity. Supported by the fact that oxygenated dolabellanes, isolated from brown algae, showed good antiviral activity [27, 28], the authors of this article obtained derivatives **16–18** by epoxidation, by epoxide opening and by allylic oxidation, respectively. These oxygenated compounds showed significant improvement in the antiHIV-1 potency (100-fold) [29]. Their high antiviral activity, along with their low cytotoxicity, makes them promising antiviral compounds, and is a good example of the usefulness of this strategy to improve the biological activity of marine natural products. Currently, the researchers are obtaining more oxygenated dolabellane derivatives in a continuous work to improve the antiviral activity of natural dolabellanes.

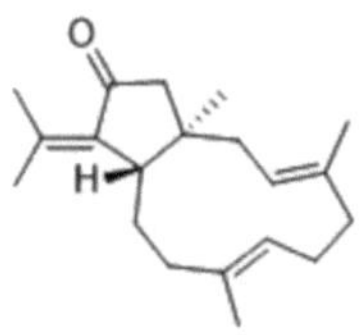
14

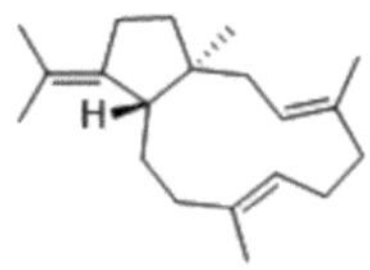
15

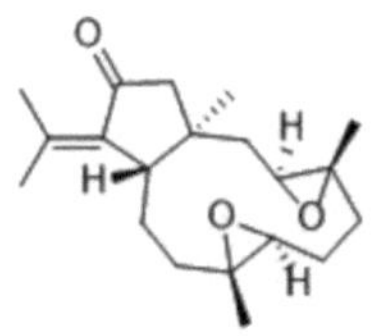
16

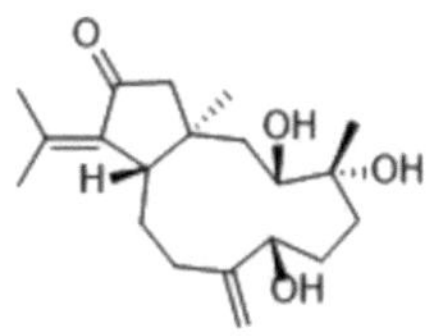
17

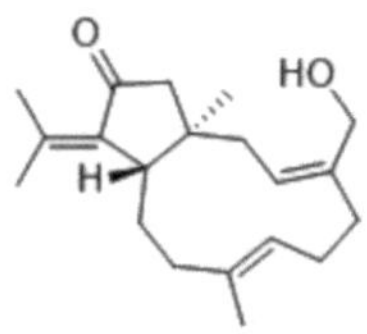
18

2.3. *Eunicea knighti* and *Pseudoplexaura flagellosa*

In this section, we show our recent published results on quorum sensing inhibition (QSI) and our until now unpublished biofilm inhibition data, related as antipathogenic activity of natural compounds isolated from *E. knighti* and *P. flagellosa* and some of their synthetic analogs prepared in our laboratory. However, first we would like to provide an introduction on bacterial biofilms, quorum sensing inhibition and their relationship with the recently used term antipathogenic activity.

Quorum sensing (QS) is defined as a phenomenon related to the gene expression of bacteria in function of the density of their population, allowing the synchronization of phenotypes through bacterial communication. Recently, quorum sensing has been recognized as one of the main factors that regulates phenotypes such as bioluminescence, transfer of tumor-inducing plasmids (Ti plasmids), antibiotic production, swarming motility, biofilm maturation (assembled bacterial communities that coordinate themselves for the expression of different phenotypes that change over time and with the environment) and the production of virulence factors [30]. Many bacteria do not express virulence factors until the population density is high enough to overwhelm host defense and establish infection. Compounds with QS inhibitory activity are capable of preventing bacterial communication and suppress some virulence factors. These compounds have been termed as antipathogenic drugs [30]. Furthermore, some QS inhibitor makes biofilms susceptible to antimicrobial treatments and can reduce mortality and virulence in experimental models of infection. Thus, compounds with QSI and biofilm inhibition activity can be considered as leads to antipathogenic drugs [30].

In the last 10 years, many researchers have focused their studies on marine metabolites, mainly from octocorals [31], that exhibit antipathogenic activity, which involve, as mentioned, QSI and biofilm inhibition activity. As previously described by Tello and colleagues in 2009, 2011 and 2012 [32–34], octocorals *E. knighti* and *P. flagellosa* collected in Santa Marta Bay (Colombian Caribbean Sea) were extracted with organic solvents followed by fractionation on vacuum column chromatography and reverse-phase HPLC to afford 16 pure compounds, and their stereostructures were elucidated by means of spectroscopic features. Their activity as QS inhibitors was evaluated against *Chromobacterium violaceum* (ATCC 31532)—a recognized biosensor using a standard disk-diffusion assay, following the parameters described by Tello et al. 2012 [34] and 2013 [35]. Whatman filter paper disks (5.2 mm diameter) were initially sterilized and then loaded with 2.5, 5.0, 7.5, 15.0 and 30.0 μg/disk of each compound. The disks were placed on agar dishes plated with 100 μL of *C. violaceum* culture grown in trypticase soy broth (106 cfu/mL, 0.5 Mac Farland) and finally the agar plates were incubated for 48 h at 26°C. This QSI assay is based on inhibition of QS pigment production (violet color) without interfering with bacterial growth. Kojic acid was used as a positive control, as it is a known inhibitor of quorum sensing systems [34, 35]. The biofilm inhibition assay was performed on polystyrene multi-well plates (96 wells), the pre-inoculates of the bacterial strains *Vibrio harveyi*, *Pseudomonas aeruginosa* and *Staphylococcus aureus* were grown in the Luria-Bertani (LB) culture medium with an optical density (OD) of 0.2–0.3 A at 600 nm. All pure compounds were evaluated at five different concentrations (0.5, 1.0, 2.5, 10.0 and 100.0 ppm). Finally, each well was filled with LB culture medium up to 200 μL. Control of growth inhibition was monitored by measuring the absorbance of each well at 621 nm before and after incubation as was described by Tello and colleagues in 2013 [36]. Biofilm inhibitory activity is reported in **Table 1**, as IC_{50} (ppm).

Based on the results of QS inhibition (QSI) and biofilm inhibition, on their high amounts in the gorgonians and on the diverse reactive functional groups present in their structures (e.g. epoxide groups in the C-7 and C-8, hydroxy groups in the C-2 and C-18, reactive double bonds between the C-3/C-4 and C-11/C-12, and keto or hydroxy reactive groups in C-3, C-6 and C-11) six of the natural compounds were selected as lead compounds to improve their QSI activity and to establish their biofilm inhibition activity via preparation of synthetic analogs using regioselective, straightforward and reproducible reactions such as epoxide ring opening, oxidations, treatment with iodine, photochemicals, methylation and acetylation, and synthesis of cyclic hemiketals [35]. In total, we had in hand 50 cembranoids (natural and synthetic) which were assayed for their QSI and biofilm inhibition activities. The results displayed in **Table 1**

Assay	Biofilm Inhibition IC_{50} (ppm)[a]			Quorum sensing Inhibition[b] (µg/disk)[c]
Compounds	*Pseudomonas aeruginosa*	*Staphylococcus aureus*	*Vibrio harveyi*	*Chromobacterium violaceum*
19	5.0	0.01	80.2	—
20	12.8	0.3	>100.0	—
21	52.9	15.7	7.8	—
22	3.7	1.8	5.8	2.5
23	23.8	11.7	17.3	5.0
24	56.1	2.2	>100.0	2.5
25	4.0	10.1	17.1	—
26	4.5	10.0	69.7	—
27	11.5	>100.0	11.0	7.5
28	17.2	>100.0	>100.0	—
29	9.2	20.9	0.3	—
30	6.4	1.0	>100.0	7.5
31	12.2	5.7	9.5	30.0
32	6.8	1.4	53.8	—
33	10.1	2.0	0.3	—
34	50.0	1.3	9.8	7.5
35	8.3	3.2	1.0	15.2
36	4.1	0.01	>100.0	—
37	9.3	0.8	>100.0	7.5
38	14.7	0.5	>100.0	—
39	52.3	3.2	21.4	7.5
40	37.4	0.13	34.5	—
41	57.3	0.04	1.2	7.5
42	21.6	1.3	16.8	7.5
43	49.8	35.5	>100.0	—
44	>100.0	>100.0	>100.0	—
45	>100.0	>100.0	>100.0	—
46	52.5	0.03	53.1	7.5
47	56.0	3.1	>100.0	7.5
48	>100.0	1.2	>100.0	—
49	>100.0	>100.0	>100.0	30.0
50	>100.0	0.8	>100.0	—
51	>100.0	8.5	>100.0	—
52	>100.0	0.6	>100.0	30.0

Assay	Biofilm Inhibition IC_{50} (ppm)[a]			Quorum sensing Inhibition[b] (μg/disk)[c]
Compounds	*Pseudomonas aeruginosa*	*Staphylococcus aureus*	*Vibrio harveyi*	*Chromobacterium violaceum*
53	47.2	0.01	6.2	7.5
54	46.2	0.02	62.5	30.0
55	51.4	0.07	14.5	30.0
56	>100.0	20.0	10.4	—
57	>100.0	81.5	>100.0	—
58	>100.0	>100.0	>100.0	—
59	33.6	3.1	38.3	—
60	28.5	0.3	2.6	—
61	43.5	4.2	52.5	—
62	>100.0	2.1	>100.0	—
63	>100.0	>100.0	>100.0	15.0
64	>100.0	>100.0	>100.0	7.5
65	>100.0	>100.0	>100.0	30.0
66	>100.0	>100.0	84.4	—
67	>100.0	>100.0	>100.0	7.5
68	>100.0	>100.0	>100.0	—
Kojic acid	17.2	24.7	>100.0	90.0
Solvent	NI	NI	NI	NI

[a]Concentration that inhibits 50% of the biofilm. ND, not determined; NI, no inhibition of biofilm.
[b]The activity was measured taking into account the inhibition of the violet pigment.
[c]Minimum amount required in μg per disc of compound to inhibit the production of violet pigment. There is no inhibition zone even at 30 μg per disk.

Table 1. IC_{50} of natural and synthetic analogs (50 cembranoids) in the bacterial biofilm and quorum sensing inhibition assays.

estimate the correlation between QS and biofilm inhibition, demonstrating the potential antipathogenic effect of the 50 cembranoids evaluated, as discussed below.

The results demonstrated that half of the synthetic tested cembranoid analogs showed QSI activity without toxicity against the biosensor bacteria, results worth being highlighted, mainly because 16 active synthetic analogs were obtained from 5 non-active natural compounds (in QSI bioassay). The synthetic compounds with the best QSI activity were **22** (2.5 μg/disk), **23** (5.0 μg/disk) and **24** (2.5 μg/disk), presenting similar structure features, scilicet: C-7*R* hydroxy methine group, a double bond with *E* configuration between C-8 and C-9, and an acetyl group at C-18. The above allows us to infer that the presence of a *E*-hydroxy allylic moiety is highly relevant for the activity [33]. It is worth noting that the three synthetic cembranoids (**22**, **23**, and **24**) were obtained from three natural inactive natural cembranoids (**19, 20** and **21**, respectively). Finally, these compounds were not able to inhibit the *C. violaceum* growth, thus, it suggests that inhibition on violacein pigment production is triggered by the disruption of QS systems. Therefore, there is no selective pressure for the development of resistance in bacteria.

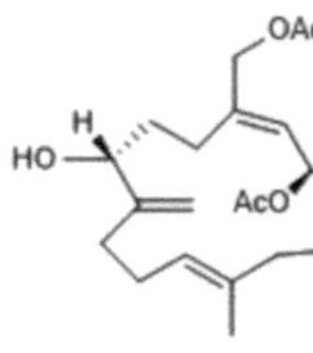
46

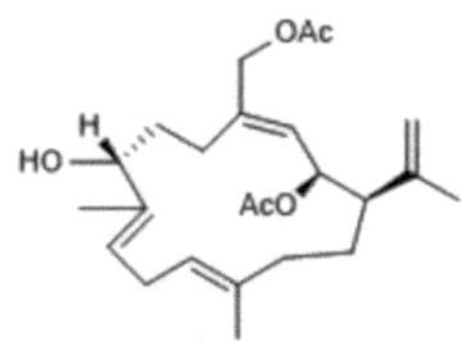
47

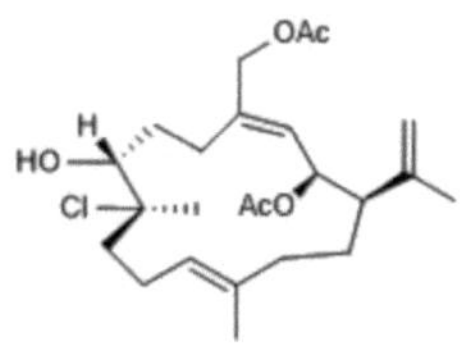
48

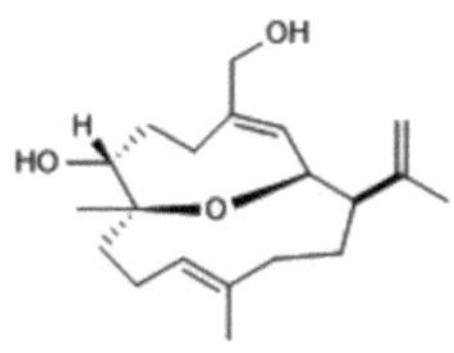
49

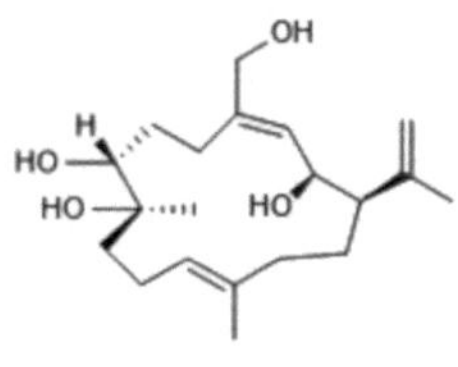
50

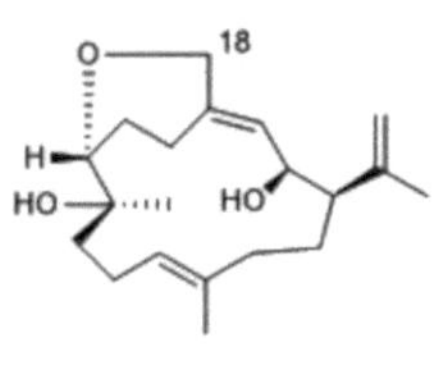
51

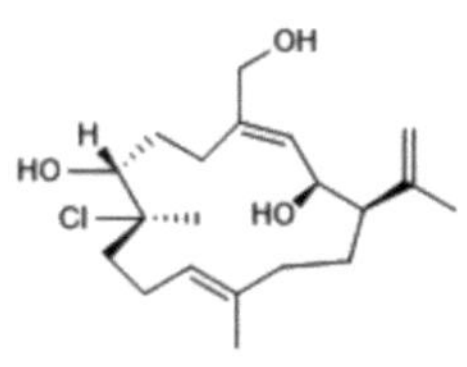
52

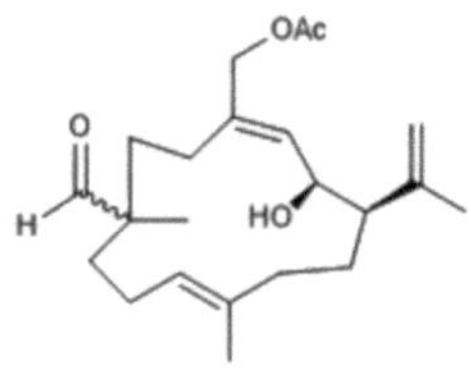
53

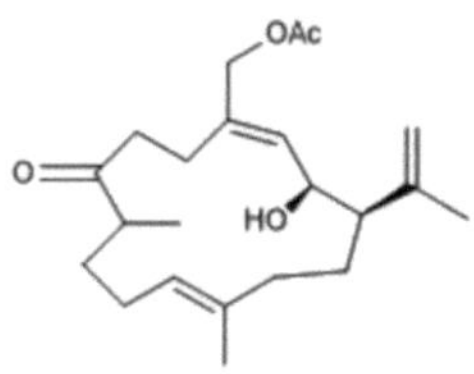
54

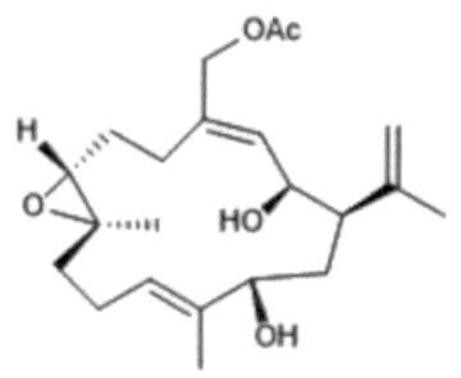
55

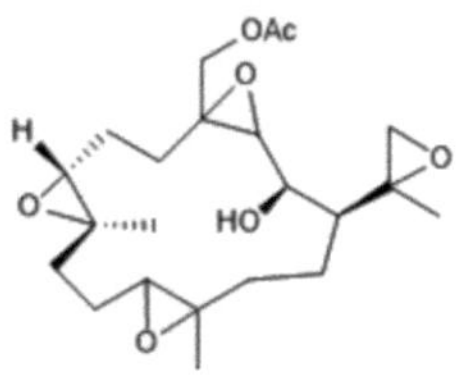
56

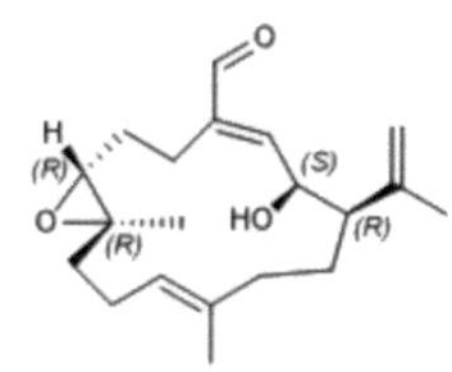
57

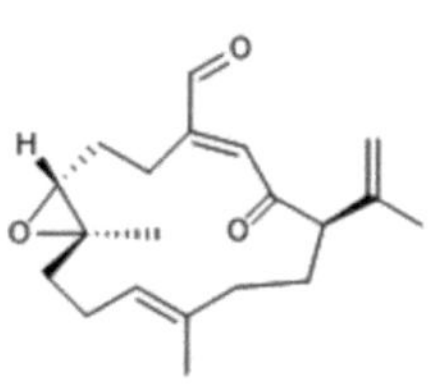
58

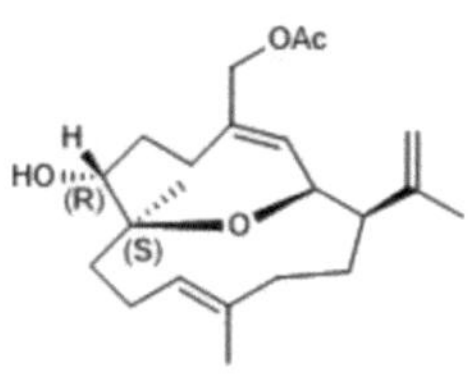
59

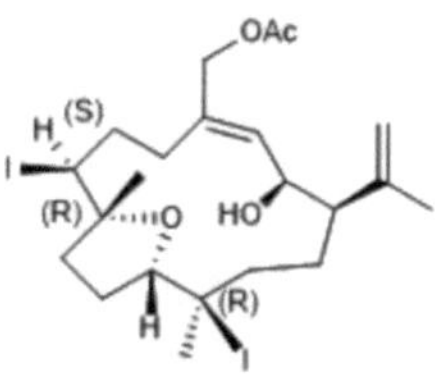
60

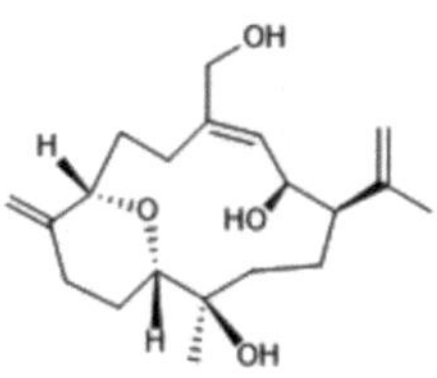
61

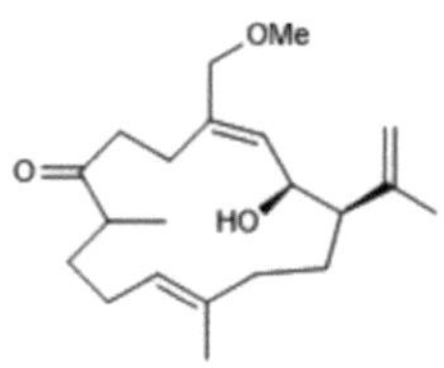
62

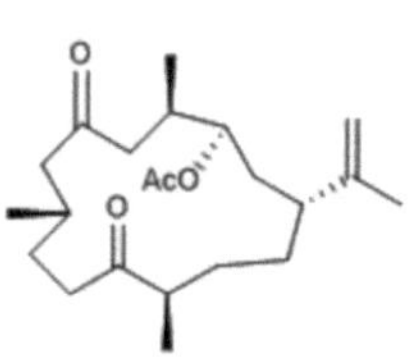
63

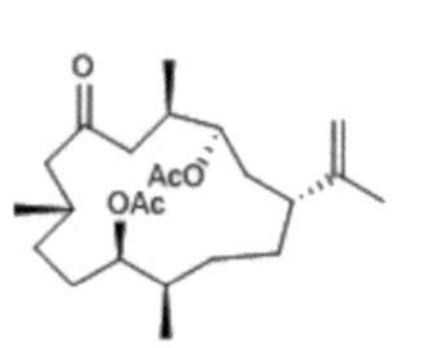
64

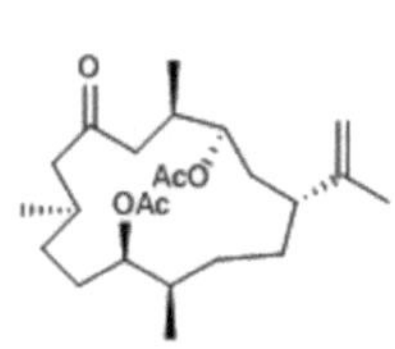
65

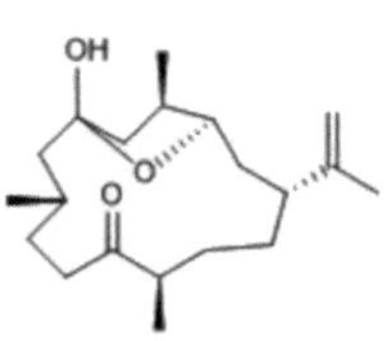
66

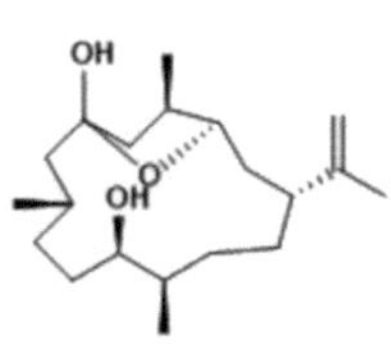
67

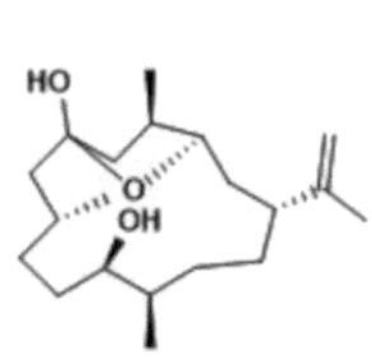
68

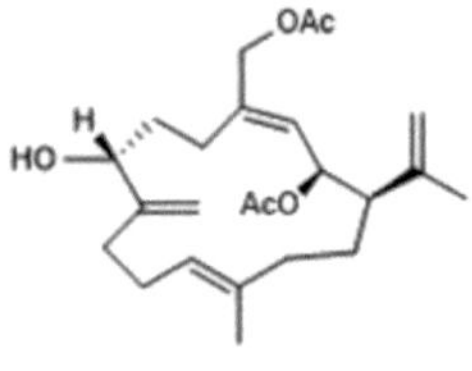

46

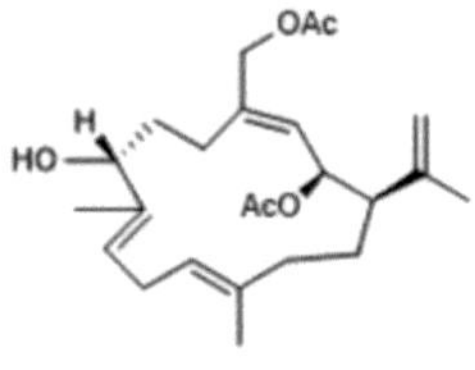

47

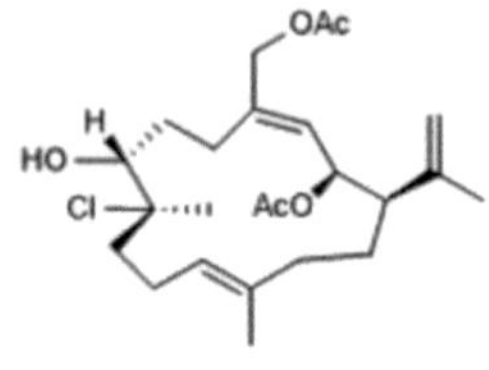

48

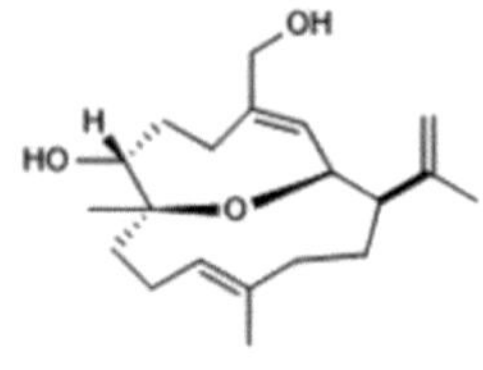

49

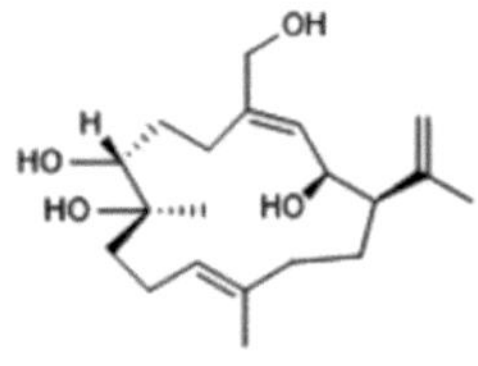

50

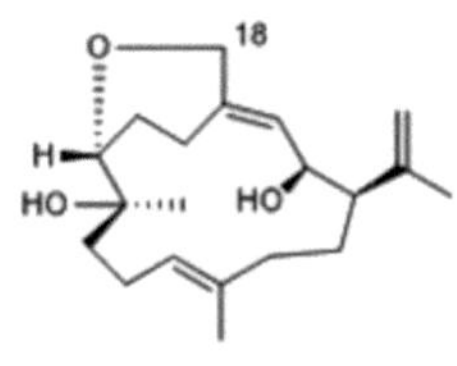

51

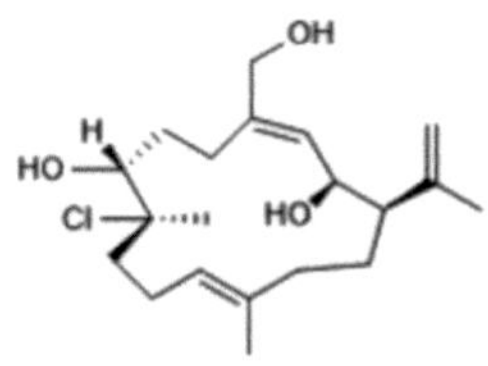

52

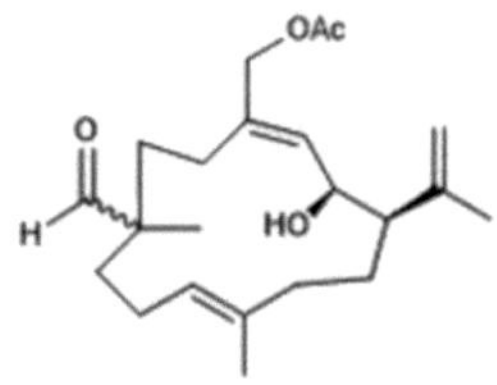

53

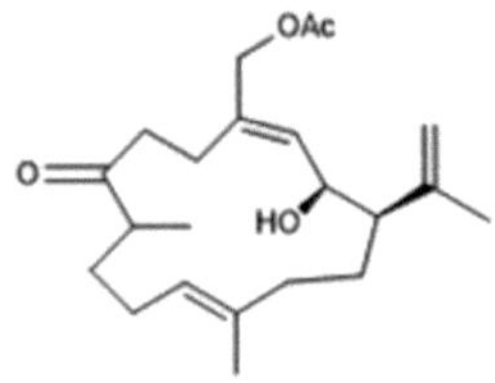

54

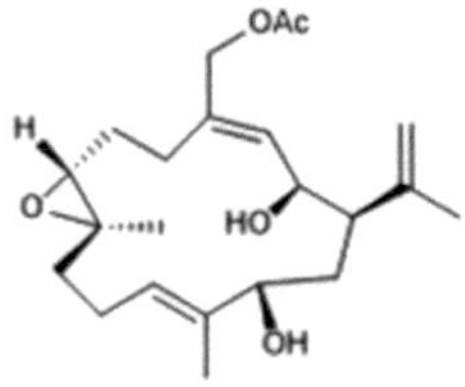

55

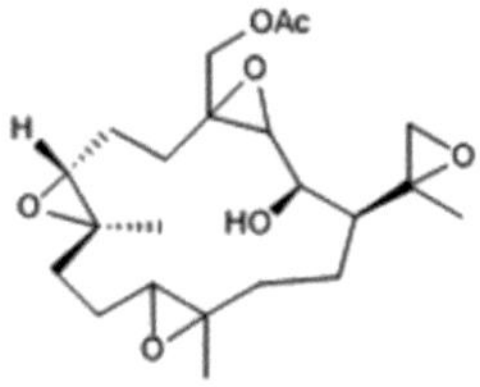

56

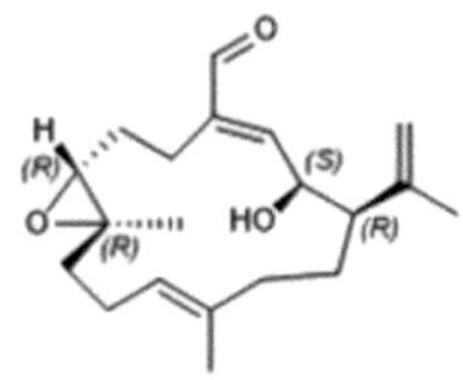

57

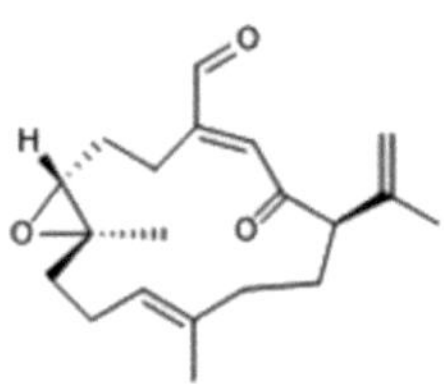

58

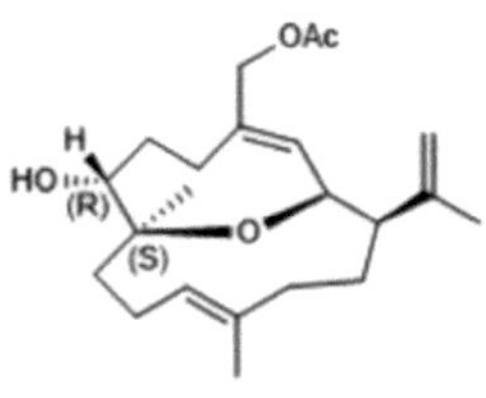

59

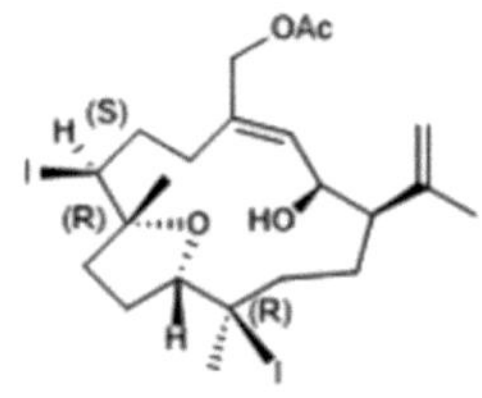

60

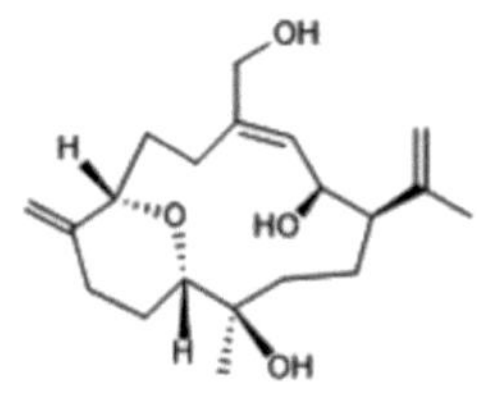

61

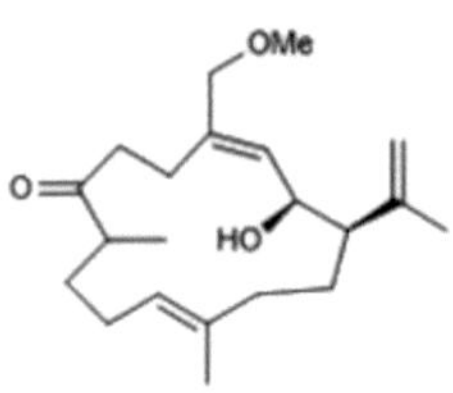

62

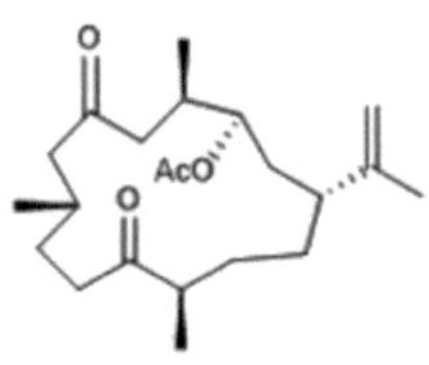

63

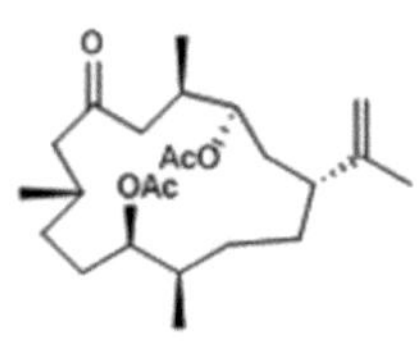

64

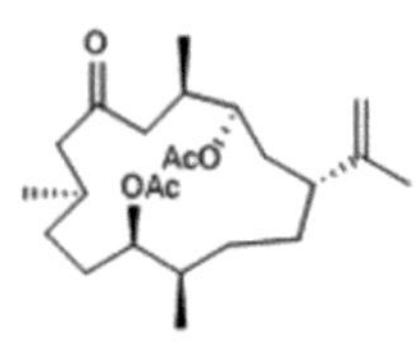

65

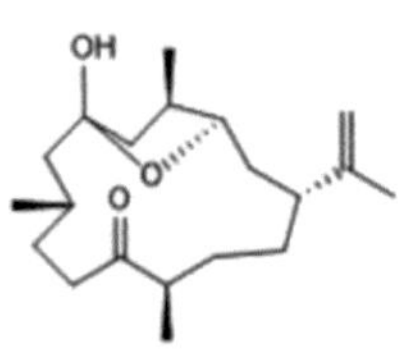

66

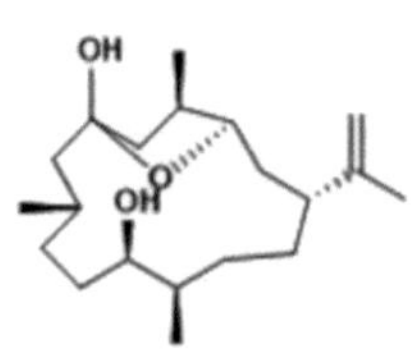

67

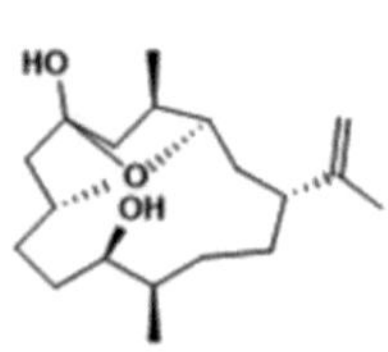

68

The biofilm inhibition results showed that half of the synthetic analogs inhibited the formation of biofilm in the three bacterial strains used at concentrations lower than 100.0 ppm. It was found that several compounds that did not exhibit QSI did not show inhibition of biofilm formation either in the three biosensor strains evaluated, for example, compounds **44, 45, 58** and **68**; however it was not a generalized trend. Otherwise, all compounds having QSI showed inhibition of bacterial biofilm in at least one of the three strains used, except for compounds **49, 63–65** and **66**. It was also evident that some of the synthetic analogs showed better biofilm inhibition activity than their natural precursors used as leads (**19–21, 25, 27** and **28**), in several cases achieving or increasing the activity. It was observed that most of the synthetic cembranoid type analogs exhibited an excellent activity to inhibit the formation of biofilm against Gram-positive *S. aureus* bacteria, although good results were also obtained against Gram-negative bacteria *P. aeruginosa* and *V. harveyi*.

In particular, 9 synthetic analogs inhibited *S. aureus* bacteria at concentrations less than 1.0 ppm, 10 inhibited the same strain at concentrations between 1.0 and 10.0 ppm and 5 between 10.0 and 100.0 ppm (**Table 1**). The compounds that showed the best potency against this biosensor strain were **40, 41, 46, 50, 52–54** and **60**, out of which compounds **41** (0.04 ppm), **46** (0003 ppm), **53** (0.01 ppm), **54** (0.02 ppm) and **55** (0.07 ppm) were the most potent ones since they were up to 2000 times more active than the recognized kojic acid QS inhibitor, which presented an IC_{50} of 24.7 ppm. Against *P. aeruginosa* bacterium, only compound **22** (3.7 ppm) had a lower IC_{50} value than kojic acid (17.2 ppm), but it is important to mention that 16 synthetic cembranoids analogs had IC_{50} values between 10.0 and 100.0 ppm. Finally, four synthetic analogs (**22, 41, 53** and **60**) were shown to be active at concentrations lower than 10.0 ppm against *V. harveyi* bacteria, being compounds **41** and **60** the most active with IC_{50} values of 1.2 and 2.6 ppm, respectively. In addition, 11 compounds had IC_{50} values between 10.0 and 100.0 ppm, being still more active than the positive control used, which had an IC_{50} value greater than 100.0 ppm against this biosensor strain. It is noteworthy that biofilm inhibition was achieved in all cases without interruption of bacterial growth in the strains used, even at the highest concentration evaluated (100.0 ppm). The fact that compounds **22, 24, 39, 41, 42, 46, 47, 51–54** showed biofilm and QSI activity suggests that these compounds interfere in bacterial communication, preventing the development or maturation of bacterial biofilm and successive development of bacterial communities, and therefore may serve as potential antibiotics.

It is worth highlighting that the comparison of the results of the bacterial biofilm inhibition with the results found in the literature showed that the synthetic cembranoids analogs present an excellent activity and low toxicity compared to other natural products reported, for example, oroidin had an IC_{50} of 0.26 ppm against *P. aeruginosa* and an IC_{50} of 35.0 ppm against *Acinetobacter baumannii*, dictiol C inhibited the formation of biofilm in the bacterial strain *Pseudoalteromonas* sp. D41 at an IC_{50} of 9.2 ppm and finally agelasine D oxime showed an MIC of 0.027 ppm against *Staphylococcus epidermis*. These IC_{50} values were higher than those reported by us here for some of the synthetic analogs against the *S. aureus* biosensor bacteria, for example, compound **41** (0.04 ppm), compound **46** (0.03 ppm), compound **53** (0.01 ppm), compound **54** (0.02 ppm) and compound **55** (0.07 ppm) (**Table 1**).

Some considerations about the structure-activity relationship of the compounds evaluated in this bioassay must be taken into account, for example, the presence of an electronegative group on C-7 (in most of the compounds "oxygen") is highly relevant for the activity, since the most active compounds presented this functionality. Also, the formation of a double bond with *E*-configuration between carbons C-8 and C-9 is one of the most important rearrangements to increase or induce biofilm inhibitory activity, since its absence or formation of the *Z*-isomer decreases the activity, the presence of an electronegative group on C-2 and C-18 (oxygen) enhances this activity, as can be observed in all active compounds in this bioassay, except for compound **23**. Finally, the formation of an exomethylene between carbons C-8 and C-19 also has a positive effect on the activity, as can be seen in compounds **39** and **46**.

In summary, six natural compounds were selected as lead compounds (**19–21**, **25**, **27** and **28**) in an attempt to induce or enhance their antipathogenic activity by selectively applied chemical transformations at different active sites of the cembrane nucleus. Thus, 33 analogs of cembranoids (**22–24**, **39–68**) were obtained, being half of them remarkably active in the QSI bioassay against the *C. violaceum* biosensor strain and in the bacterial biofilm inhibition against *P. aeruginosa, S. aureus* and *V. harveyi* strains, all without interfering the bacterial growth. Finally, a select group of structurally related cembranoids (**23**, **23**, **39–42**, **46**, **53–55**) were obtained as QS and bacterial biofilm inhibitors, making them excellent candidates to be used as antipathogenic drugs by the pharmaceutical industry, since the relationship between biofilm inhibitors and QS inhibitors is often associated with potent antipathogenic agents.

2.4. *Antillogorgia elisabethae* (Syn. *Pseudopterogorgia elisabethae* [37])

Pseudopterosins, *seco*-pseudopterosins and amphilectosins constitute an important class of diterpene glycosides found in the gorgonian *Antillogorgia elisabethae* [37] (known before as *Pseudopterogorgia elisabethae*), except for *seco*-pseudopterosins A-D isolated from *Pseudopterogorgia Kallo,* collected from different regions of the Caribbean (Bahamas, Bermudas, Florida, Providencia and San Andrés). The pseudopterosins were discovered first by Fenical et al. in the late 1980s [38, 39] and since then numerous members of the scientific community have been involved in this attractive area finding many other compounds of this kind, most of them with potent biological activity [40, 41], that is, antiinflammatory and analgesic, wound healing, antibacterial, anticancer, antiviral, antimalaria and antituberculosis, in *in vitro* and *in vivo* assays with a new mechanism of action. Very recently, a protection of synaptic function and potential as a neuromodulatory agent for PsA has also been reported [42]. In addition, those compounds have demonstrated efficacy in Phase II clinical trials as an antiinflammatory and wound healing agents and are the first commercial licensed natural product for use as an additive in Estée Lauder skin care and cosmetic products. To date, 31 pseudopterosins, 11 *seco*-pseudopterosins and 2 amphilectosins are known from nature. A recent review about this topic can be consulted in [21].

We have had a special interest in pseudopterosins G, and P-U, 3-*O*-acetyl-PsU, *seco*-Ps J and *seco*-Ps K isolated by Duque and collaborators [43–45] from specimens of *P. elisabethae* collected at Providencia Island. After their isolation and chemical structure determination, we soon discovered their high chemical diversity (natural analogs) and their potent therapeutic activity [46–48] (antiinflammatory, cytotoxic and antimicrobial activity).

http://dx.doi.org/10.5772/intechopen.74686

Their antiinflammatory activity was evaluated by us using *in vitro* experiments as myeloperoxidase (MPO) assay, nitric oxide release (cell-based assay) and scavenger activity on this radical [46]. Our results reported in [46] showed that compounds PsG, PsK, PsP, PsQ, PsS, PsT, PsU and *seco*-PsK are promising molecules with an interesting and potent antiinflammatory activity. In our experiments, they displayed more potent action than indomethacin, a clinical drug used currently to treat inflammation and with different mechanism of action. Furthermore, the results for the different MPO inhibition values obtained provided us with preliminary insights toward their structure-activity relationship, that is, the activity depends on the kind of sugar moiety, on whether sugar moiety is in a free form or acetylated, on the acetylation position within the sugar moiety and on the glycosylation position. In addition, regarding the results of NO release in J-774 cell-based assay, we found a greater activity for the pseudopterosins than for the seco-pseudopterosins, clearly showing that the non-glycosylation improves the inhibition of NO release. And finally, by comparing the different NO inhibition values for individual compounds, the inhibitory activity apparently depends on the glycosylation position, on the stereochemistry of the aglycone and on the type of the skeleton. For example, the amphilectane skeleton (PsP) has more inhibitory activity than the serrulatane skeleton (*seco*-PsK). However, more experiments are needed in order to support structure-activity relationships among these kinds of compounds.

The results of the cytotoxicity of the natural homologous compounds PsG, PsP, PsQ, PsS, PsT, PsU, 3-O-acetyl-PsU, *seco*-PsJ and *seco*-PsK (**69–77**), evaluated using human cancer cell lines (HeLa, PC-3, HCT116 and MCF-7) showed moderate and non-selectivity activity between the lines used. After examining the mentioned cytotoxic activity results, it could be seen that some SARs were evident. According to the results that we have published in [47], the position of glycosylation on the terpene skeleton appears to affect the inhibitory activity profile, for example, PsG (glycosylated in C-9 with fucopyranose) is more active than PsP (glycosylated in C-10 with fucopyranose). Further, the type of sugar moiety also influences the activity, for instance, PsP, which is glycosylated with fucopyranose, is more active than PsT, which is glycosylated with arabinopyranose. Likewise, PsQ (C-4′ mono-acetylated fucose as sugar moiety) is more active than PsU (C-4′ mono-acetylated arabinose as sugar moiety), and *seco*-PsK (non-acetylated fucose as sugar moiety) is more active than *seco*-PsJ (mono-acetylated arabinose as sugar moiety).

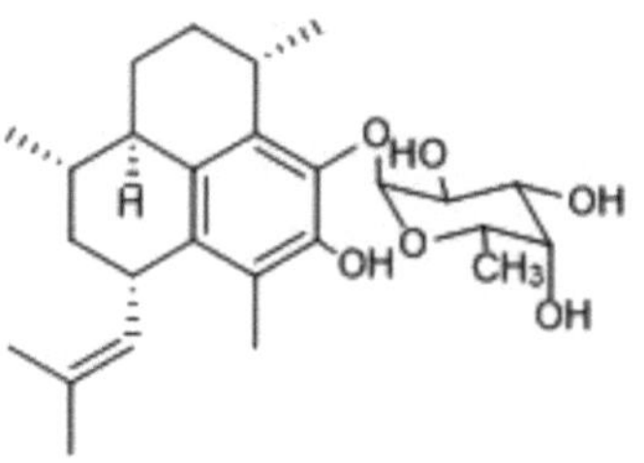

69

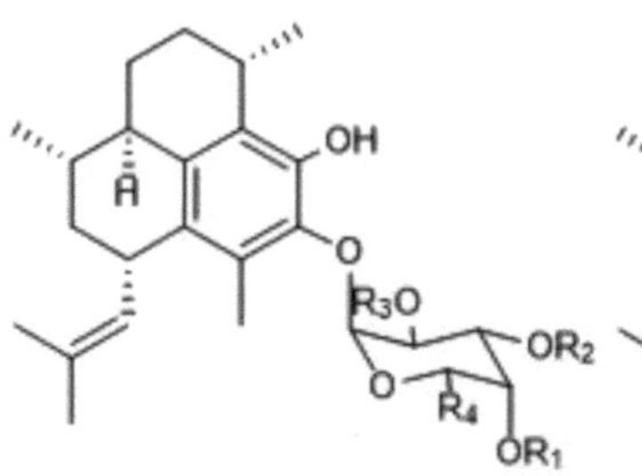

70 R_1, R_2, R_3= H, R_4=CH_3
71 R_2, R_3= H, R_1= Ac R_4=CH_3
72 R_1, R_2= H, R_3= Ac R_4=CH_3
73 R_1, R_2, R_3, R_4= H,
74 R_2, R_3, R_4= H, R_1= Ac
75 R_1, R_2, R_4= Ac, R_3= H

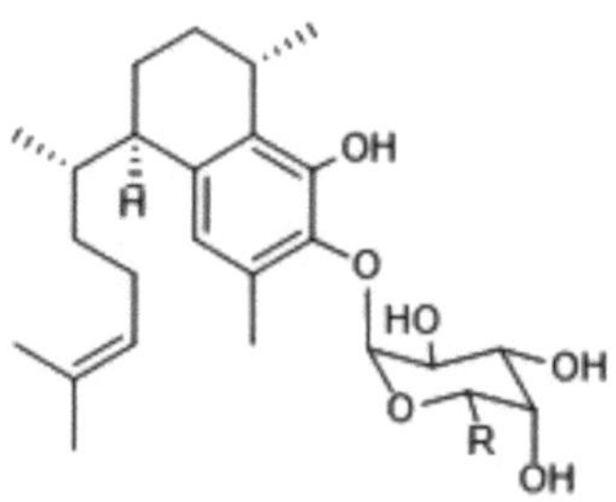

76 R = H
77 R = CH_3

Regarding the antimicrobial activity for the natural homologous **69–77**, we found good and selective activity against Gram-positive bacteria, *Staphylococcus aureus* and *Enterobacter faecalis*, being the most active PsG, PsU, PsQ, PsS and *seco*-PsK. Additionally, they did not show activity against the Gram-negative bacteria or the yeast used in our assay and, more importantly, their antimicrobial potency was comparable to the reference drug vancomycin. In examining our just mentioned data, published in [47], the following SARs could be noted: fucopyranose glycosilation at C-9 instead at C-10 seems to increase the activity (PsG vs. PsP); arabinopyranose instead of fucopyranose glycosilation favors the activity (PsT vs. PsP); likewise, mono-acetylated arabinose as sugar moiety increases the activity (PsU vs. PsQ and PsS (mono-acetylated fucose as sugar moiety)). In contrast, this behavior initially observed in pseudopterosins is not consistent when the results are applied to the *seco*-pseudopterosins *seco*-Psk (glycosilated with fucopyranose), which is more active than seco-PsJ (glycosilated with arabinopyranose).

Additionally, it is important to mention that in our experiments published in Ref. [48], we assayed pseudopterosins and *seco*-pseudopterosins (natural analogs **69–77**) as antifouling agents against marine bacteria isolated from heavily fouled marine surfaces. These compounds at a 30 μg dose showed moderately to highly active against all Gram-positive microfouling bacteria assayed, and non-active against the Gram-negative bacteria used. Tetracycline and kanamycin reference antibiotics used in the assay showed similar values of activity with doses of 30 μg as well. Furthermore, we performed assays of natural compounds **69–77**, kanamycin and tetracycline on bacterial growth and on biofilm disruption (% of inhibition) of *Pseudomonas putida* IsoF used as a positive control for biofilm formation and of six marine bacterial strains associated with fouled surfaces. The natural analogs tested showed no activity (did not inhibit bacterial growth and did not promote biofilm formation) against Gram-negative bacteria *Pseudomonas putida* Iso F, *Alteromona Macleodii* and *Ochrobactrum pseudopgringonense* strains 1 and 2. In contrast, they inhibited both growth and biofilm formation of Gram-positive bacteria (*Oceanobacillus iheyensis, Bacillus* sp. and *Kocuria* sp.).

Finally, it is worth mentioning the many studies carried out using the chemical synthesis (total synthesis and semi-synthesis) in order to increase the activity and to solve at least partly the problem of the sustainable supply of pseudopterosins. Those studies were conducted by Broka in 1988, Corey in 1989, 1990, 1998 and 2000, McCombie in 1990 and 1991, Buszek in 1995, Schmalz in 1997, Kociensk in 2001, and Harroweven in 2004 (complete information on this topic can be found cited and widely commented in Ref. [21]). Unfortunately, the mentioned syntheses have not yet been used, perhaps due to the complexity or/and non-economically ways of the synthetic routes applied. However, those efforts have provided information on improvement of their biological activity, pharmacophore and mechanism of action. Moreover, it is worth noting that semi-synthetic alkoxy or phenoxy substitution such as ether and acetate derivatives of pseudopterosins are under patent protection [21].

At this point, we want to mention the recent studies reported in [40] where simplified synthetic analogs of pseudopterosins **78–87** were prepared by Fenical and colleagues using a new and efficient synthesis taking into account the following general structural modifications: degree of substitution of the hexahydrophenalone core, different configurations at C-4 and at C-7, and several sugar moieties and place of the glycosidation. Nine of the 10 compounds evaluated as racemic mixtures were active in the mouse-ear assay (the most active one was twice more active than PsA) and no statistical differences were identified among compounds. Additionally, the synthetic route involving only six steps leads to derivatives without substitutents at C-1 and C-3 (reducing the number of stereoisomers) and allows for the preparation of multigram amounts of them.

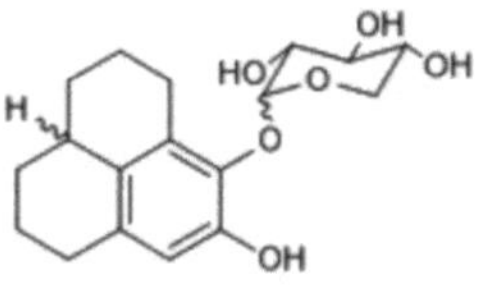

78

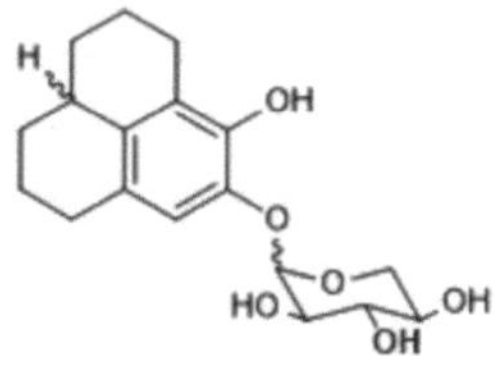

79

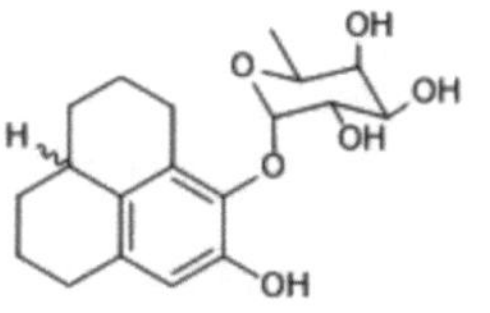

80

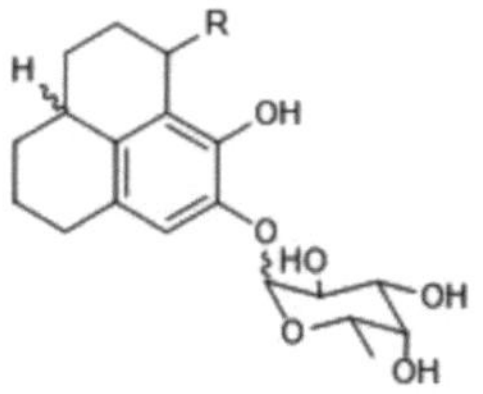

81 R=H
82 R= O
83 R= dimethyl
84 R= alpha methyl
85 R= beta methyl

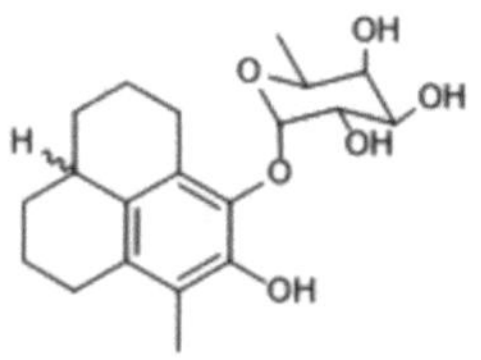

86

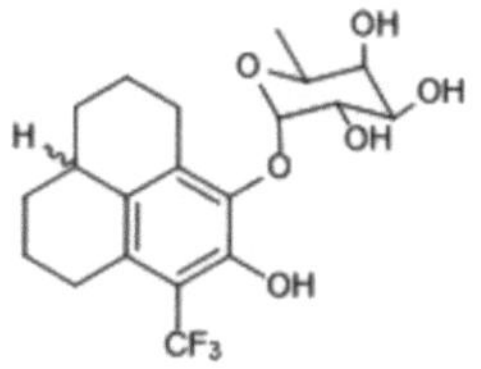

87

2.5. *Muricea austera*

Specimens of *Muricea austera* were collected in the Pacific coast of Panama during an expedition of the Smithsonian Tropical Research Institute [49]. The MeOH extract of *M. austera* showed *in vitro* activity against chloroquine-resistant *Plasmodium falciparum*. Bioassay-guided fractionation using vacuum liquid chromatography followed by flash chromatography and normal-phase HPLC purification yielded six compounds: three tyramine derivatives (**88–90**) and three steroidal pregnane glycosides (**91–93**). The structures of the compounds were determined based on their spectroscopic data. Several synthetic analogs were obtained under basic hydrolysis and perbenzoylation reactions. All natural compounds and synthetic analogs were evaluated against a drug-resistant *Plasmodium falciparum* and intracellular form of *Trypanosoma cruzi*.

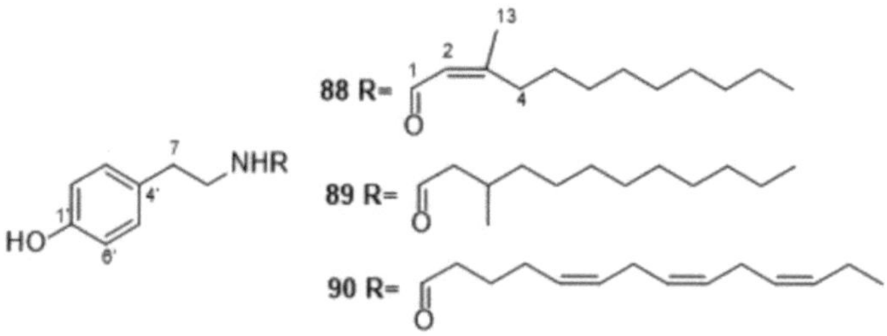

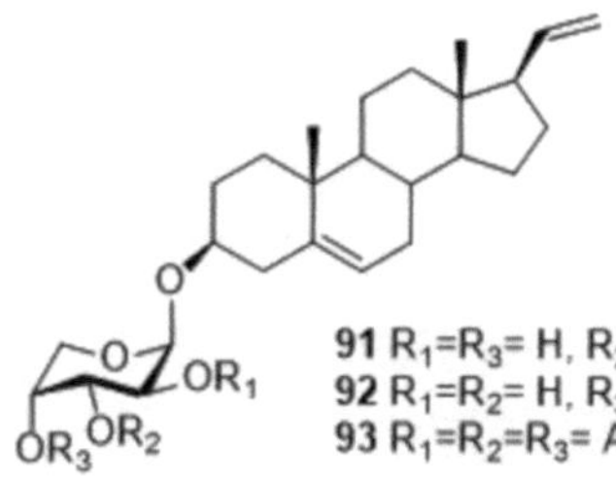

91 R_1=R_3= H, R_2= Ac
92 R_1=R_2= H, R_3= Ac
93 R_1=R_2=R_3= Ac

Natural compounds **88–92** showed moderate activity, being compounds **88** (IC_{50} 36 μM), **89** (IC_{50} 45 μM) and **90** (IC_{50} 45 μM) the most active ones. The antiplasmodial activity of glycosides **91** and **92** (IC_{50} 67 and 80 μM) was increased in their peracetylated natural analog **99** (IC_{50} 28 μM) [49]. Arabinopyranosides synthetic analogs **94** and **95** were also evaluated, showing that perbenzoylated derivatives **94** (IC_{50} 35 μM) and **95** (IC_{50} 21 μM) were more active against *P. falciparum* than natural compounds **91** and **92**. The antiplasmodial activity of analogs with stereochemistry as D-arabinopyranose **96** and **97** and D-galactosides **98** and **99** were also evaluated. Interestingly, compounds **96–98** displayed antiplasmodial activity, being compound **98** (IC_{50} 29 μM) the most active one, while perbenzoylated methyl β-D-galactoside **99** was inactive.

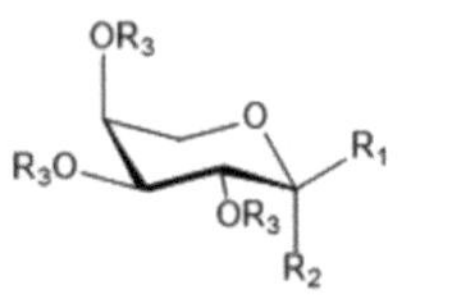

94 R_1= H, R_2= OMe, R_3= Bz
95 R_1=OMe, R_2= H, R_3= Bz

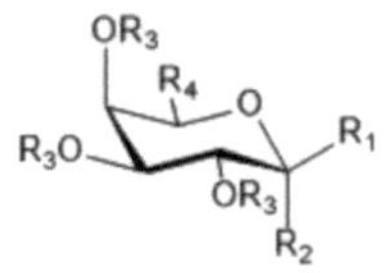

96 R_1= H, R_2= OMe, R_3= Bz, R_4= CH_3
97 R_1= OMe, R_2= H, R_3= Bz, R_4= CH_3
98 R_1= H, R_2= OMe, R_3= Bz, R_4= CH_2OBz
99 R_1= OMe, R_2= H, R_3= Bz, R_4=CH_2OBz

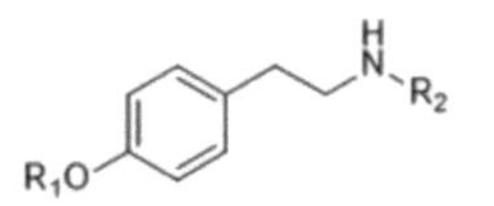

100 R_1= H, R_2= $CO(CH_2)_8CH_3$
101 R_1= H, R_2= $CO(CH_2)_{10}CH_3$
102 R_1= H, R_2= $CO(CH_2)_{12}CH_3$
103 R_1= H, R_2= $CO(CH_2)_{10}CHOH(CH_2)_5CH_3$

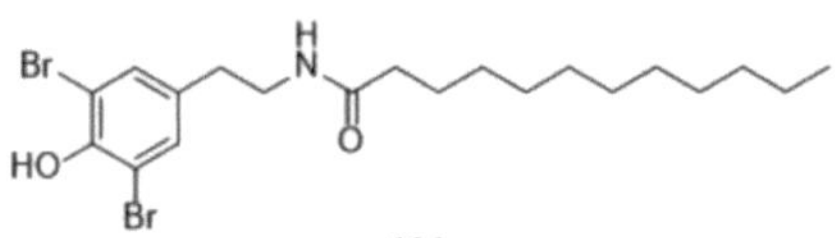

104

Given the antiplasmodial activity displayed by natural tyramine derivatives **88** (IC_{50} 36 μM), **89** (IC_{50} 45 μM) and **90** (IC_{50} 38 μM), thirteen synthetic analogs were evaluated. The results indicated that the derivatives with a fatty acid moiety **100** (IC_{50} 72 μM), **101** (IC_{50} 47 μM) and **102** (IC_{50} 34 μM) showed similar activity to those of their natural analogs **88–90**, suggesting that the increasing of the number of carbons of the fatty acid chain produces an increase in the activity, while the presence of polar groups decreases the activity as in compound **103** (IC_{50} 62 μM). Finally, the presence of a bromine group on the tyramine aromatic ring as in **104** (IC_{50} 17 μM) substantially enhances the antiplasmodial activity [49].

2.6. *Paragorgia* sp.

The octocoral genus *Paragorgia* has been barely studied; however, some diterpenoids and steroids were reported in 1984 [50]. In 2008, Spanish researchers collected *Paragorgia* sp. by bottom trawling near the Madagascar Island at a depth of 790 m. The sample was extracted with isopropanol, and a bioguided isolation procedure allowed to isolate three novel cytotoxic steroids derivatives named parathiosteroids A-C **105–107**. The structures incorporate an A-ring with different degrees of unsaturation, and a side chain containing both a thioester and an acetamide groups. These structural novelties do not have precedents in marine natural products chemistry [51]. Natural compounds **105–107** displayed cytotoxic properties against colon (HT-29), lung (A-549) and breast (MDA-MB-231) tumor cell lines with GI_{50} values in the micromolar range. Interestingly, parathiosteroid B **106** showed a selective cytotoxicity against HT-29 with a GI_{50} of 6.5 μM. Related compounds were detected in an aerobic degradation study of bile acid cholate by a *Pseudomonas* sp. [52].

In addition the authors obtained by simple and fast synthesis in the laboratory, the three natural products **105–107**, starting from commercially available 20-(hydroxymethyl)-pregnan-1,4-dien-3-one **108** and (+)-estrone **109**. The synthesis includes oxidation of C-20 hydroxymethylene to carboxylic acid followed by thioesterification of the carboxylic acid with N-acetylcysteamine. The unsaturation pattern of A-ring at **106** was obtained by Birch reduction of **108** followed by bromination at C-2 and subsequent dehydrohalogenation.

Furthermore, to obtain different synthetic analogs, the authors used $XCH_2CH_2NHCOCH_3$ (X = O or NH) instead of sulfur derivate and prepared analogs with different oxidation patterns at the A-ring. In this way, more than 20 steroids were prepared. These analogs were assayed for their cytotoxic activity against HT-29, A-549 and MDA-MB-231 cell lines. Analog **110** showed to be more active than natural products (GI_{50} MDA-MB-231: 39 μM; A-549: 79 μM; HT-29: 72 μM); this compound has no double bonds at A-ring and neither thioester at side chain. Structure-activity relationship studies [51] showed that the presence of $XCH_2CH_2NHCOCH_3$ moiety (X = S, O and N) in the side chain is essential for the antiproliferative activity, and that a low oxidation degree on an A-ring results in a higher bioactivity.

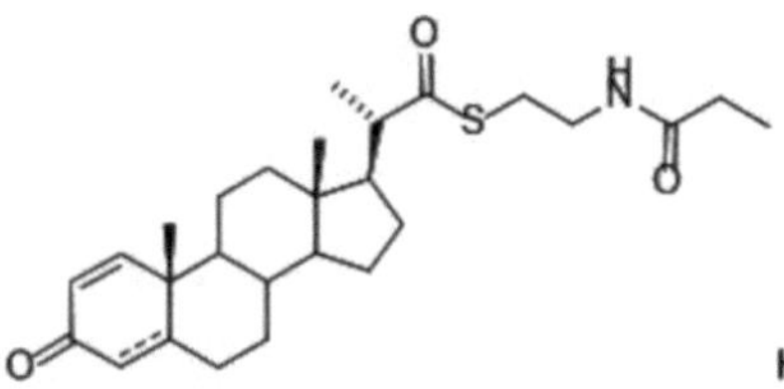

105 Δ4
106

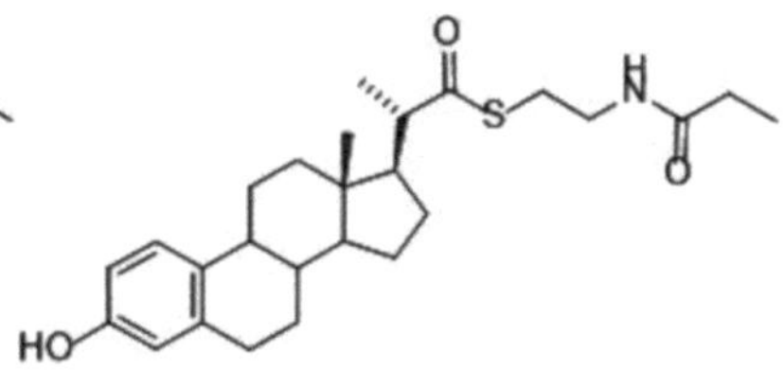

107

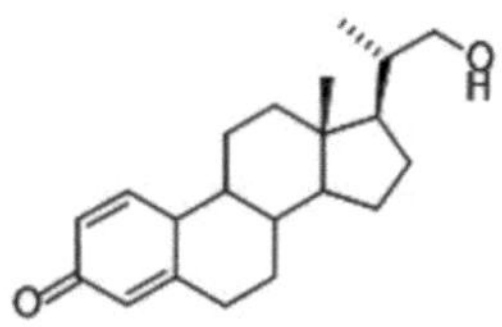

108

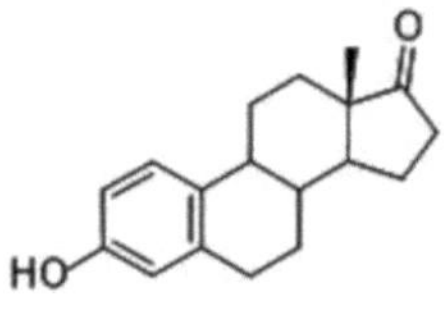

109

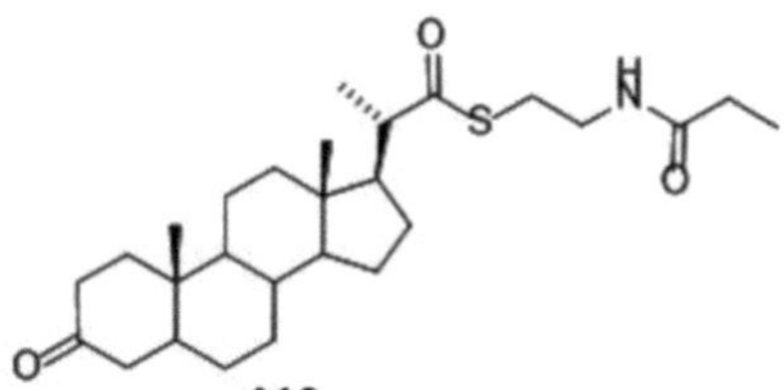

110

2.7. *Lobophytum* sp.

Colonies of *Lobophytum* sp. collected in 2006 in Siladen (North Sulawesi, Indonesia) were extracted with $MeOH:CHCl_3$. The organic extract was chromatographed by MPLC and the obtained fractions were further purified by analytical HPLC to obtain six cembranoids [53]. All compounds were evaluated for cell growth inhibitory activity against three different cell lines:

H9c2 (cardiacmyoblasts), C6 (glioma) and HeLa (epithelial carcinoma). One of the isolated compounds, decaryiol D **111**, showed a significant activity against C6 glioma cell line (IC_{50} 40 μM) compared with the structurally related decaryiol B **112** which was inactive. This fact indicated that the growth inhibitory activity of **111** should be attributed to the presence of the hydroperoxy group in this molecule. Based on the availability of high amounts of decaryiol **113**, it was subjected to several reactions (acetylations, oxidations and epoxidations) to obtain six semi-synthetic derivatives with the purpose of extending the structure-activity relationship knowledge.

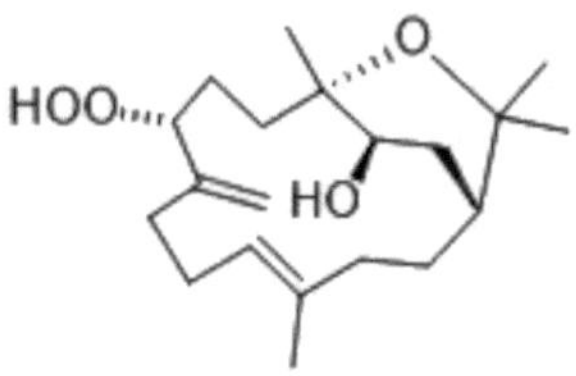

111

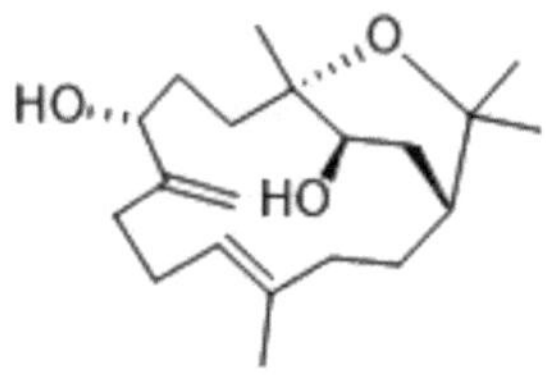

112

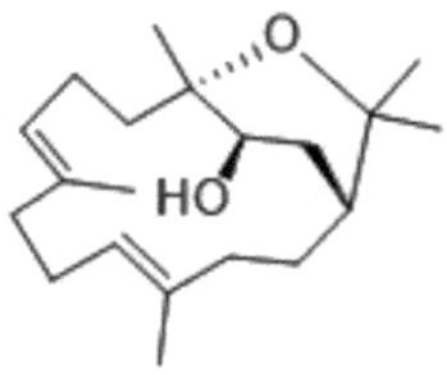

113

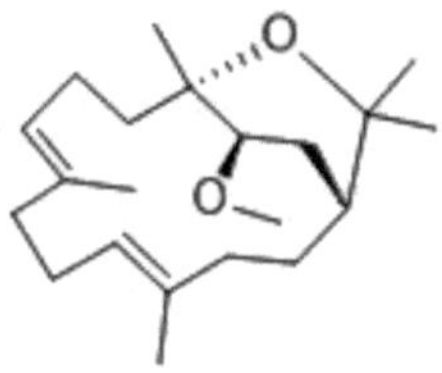

114

The six synthetic cembranoids were also evaluated against the same cell lines and the results showed that the derivative O-methyl decaryiol **114** was more active against C6 glioma cell line (IC_{50} of 8 μM) than the natural compound and it also presented a selectivity as it was inactive against HeLa and practically inactive against the non-tumor H9c2 cell line [53]. These results allowed to establish that minor structural changes on the cembranoid skeleton of decaryiol can radically affect the activity and selectivity as cell growth inhibitors.

2.8. *Sarcophyton glaucum*

Sarcophine **115** is a bioactive cembranoid diterpenes with anticancer activity isolated by Kashman group in 1974 [54, 55] from the Red Sea soft coral *Sarcophyton glaucum*. Continued studies of structure-activity relationship as mentioned in Hassan et al., 2011 [56] suggested the importance of functional groups at C-7/C-8 and the opening of the ring lactone to increase the activity. In addition, later experiments confirmed the importance of macrocyclic double bonds to the mentioned activity, and showed that bromination of sarcophine **115** improved the antiproliferative activity against malignant breast cancer cells. Further experiments through the oxidation of **115**, which resulted in the formation of (+)-sarcophytoxin B exhibiting antiproliferative activity, suggested that C-7/C-8 epoxide is not required for the activity. Thus far, the reported studies clearly demonstrate the need to further optimize the epoxide functionality of sarcophine in relation to its anticancer activity. Then, more experiments [56] were done in the direction of measuring the activity of natural analogs 16-deoxysarcophine **116** and 2-epi-16-deoxysarcophine **117**, and of compounds **118–129** prepared via etherification and carbamoylation of **115** at C-7/C-8. The C-7 ether and carbamate functionalities can be taken as different examples of the introduction of hydrogen bonding donor and hydrogen bonding acceptor and π-aromatic groups at this position to examine their binding affinities toward their molecule target. The analogs thus prepared were subjected to evaluation of their ability to inhibit the proliferation and migration of the human metastatic prostate cancer PC-3 and breast cancer MDA-MB-231 cell lines using MTT and wound healing assays. Most analogs exhibited enhance antimigration activity and lack of cytotoxicity toward the cancer cells.

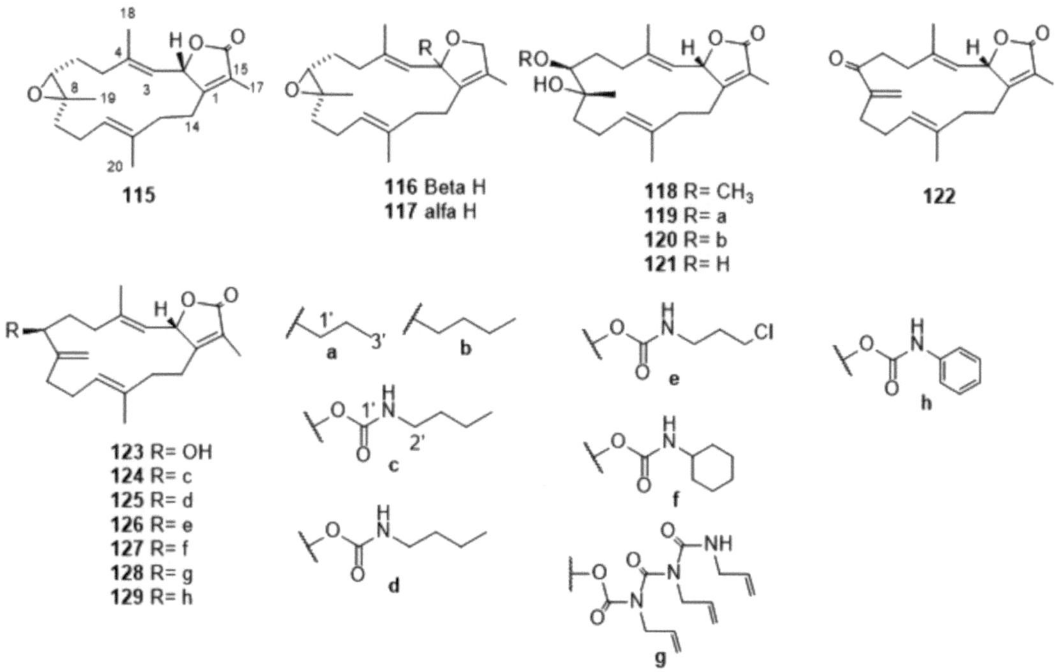

2.9. *Sinularia lochmodes*

An interesting example related to the topic of this chapter is the one published by Tanaka et al. in 2013 [57]. This study mentions lectin SLL-2 isolated from octocoral *Sinularia lochmodes* as an important mediator in the symbiotic relationship of this animal with its zooxanthellae (the symbiotic microalgae *Symbiodinium*) on which the coral depends for energy and nutrients. This lectin SLL-2 influences the transformation of *Symbiodinium* cells into a non-flagellated coccoid form from a flagellated-swimming form. In addition, Forssman antigen pentasaccharide GalNAcα(1,3) GalNAcβ(1,3)Galα(1,4)Galβ(1,4)Glc **130** was also identified as a ligand of lectin SLL-2 [58]. The authors, Tanaka et al. in [57], oriented their work in terms of structure-activity relationship. Thus, the synthesis and biological evaluation of Forssman antigen pentasaccharide and some derivatives obtained by using a one-pot glycosylation and polymer-assisted deprotection were assessed. For the evaluation of the biological activity they used the analysis of the increase or decrease that occurs when oligosaccharides (**131, 132, 133** pentasaccharide derivatives of **130**, protected with 2-trimethylsilylethyl group at the reducing end and protected analogs **134–138**) bind to the fluorescent-labeled lectin SLL-2. The results revealed that the affinity of oligosaccharides for SLL-2 was dependent on the number of sugar units in the oligosaccharide and on the NHAc substituents [57]. Modification of the GalNAc unit to a Gal unit reduced the binding affinity to SLL-2. These results indicated that SLL-2 not only recognized the acetamide group at the non-reducing end of the Forsmann antigen, but also the sugar units at the reducing end. In addition, α-GalNAc **138** showed a stronger affinity than that of β-GalNAc **137**, comparable to that of tetrasaccharide **134**.

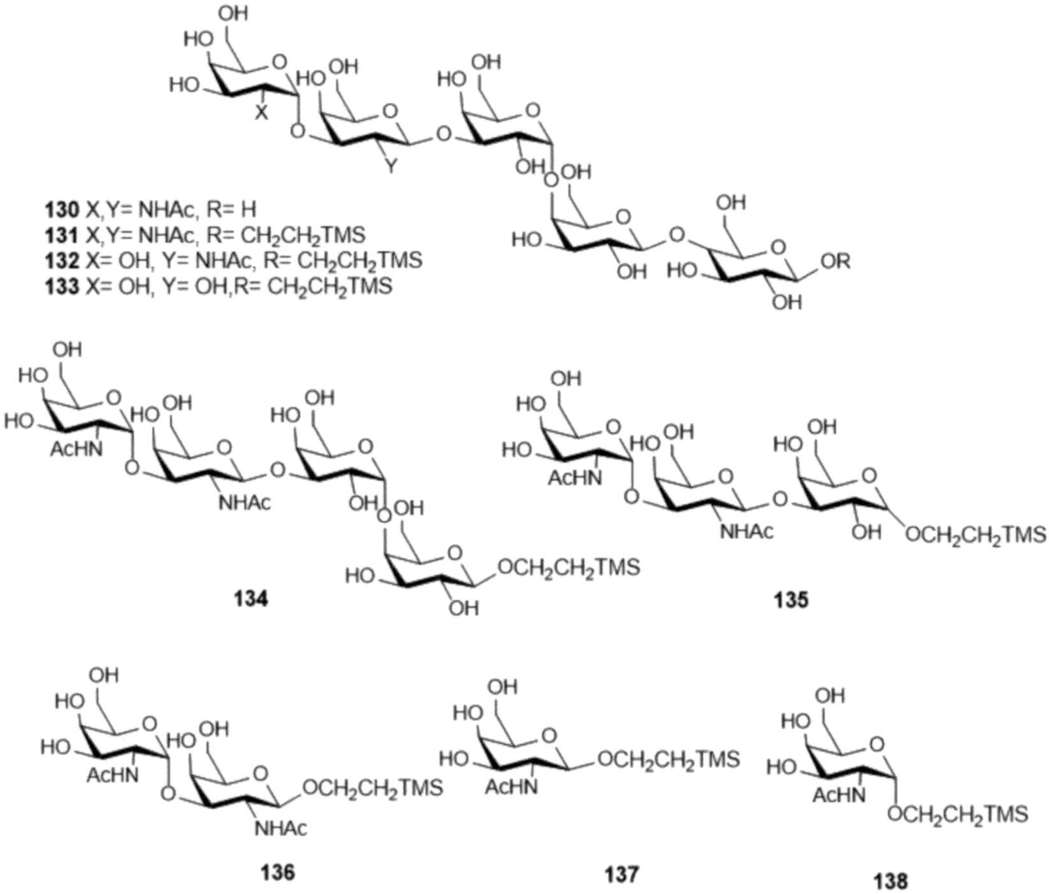

3. Conclusions

As we have shown throughout this chapter, there is no doubt that chemical synthesis plays an important role in the bioprospection of chemical compounds isolated from octocorals, either in the production of the bioactive natural product (supply a natural product), facilitating the further ways of its development as a drug and its subsequent commercialization or in the obtaining of a series of analogs that undoubtedly reveal important features on the interaction of the bioactive molecule and its target, allowing the chemists of marine natural products to change at convenience the activity and toxicity initially detected in the isolated compounds and sometimes even to reach the establishment of the pharmacophore. However, it is also a fact that this strategy has important limitations to consider, among them that the chemical reactions used must be efficient, with the fewest possible steps, economical viable option, easy to perform and to supply products with significant values of biological activity and without by-side toxicity.

The studies using the strategy analyzed in this chapter, suggest in most cases that the natural compounds can be potential scaffolds for the design of potent bioactive leads against different biological targets. In addition, the results indicate that a subtle structural change on the lead compounds can dramatically affect the activity and the selectivity of the structure against the different activities evaluated. The above corroborate that the assessment of the synthetic analogs of this chapter appear to be attractive targets for the development of new anticancer, antiinflammatory, antiviral, antimicrobial, antileishmanial, antiplasmodial and antipathogenic agents. However, this strategy should be accompanied by in silico studies that allow to establish the mechanisms of interaction between the proteins involved in the different biological activities mentioned above with the substrates (natural and synthetic analogs), Thus, the work will be carried out in a more effective way, translating into shorter times and in an adequate investment of resources used in this strategy.

It is important to highlight in this section that for the case of the octocorals *E. knighti* and *P. flagellosa* six natural compounds selected as lead compounds based on their activity values in the QSI and biofilm inhibition assays and on the variety of reactive functional groups present in their cembrane nucleus, selective chemical transformations were used with the purpose of inducing or enhancing antipathogenic properties. As a result, 33 cembrane analogs were obtained, half of them being remarkably more active than the naturals in the bioassays used, without interfering with bacterial growth, which lead us to assume that the inhibition of the phenotype expression is caused by disruption of the bacterial communication (QS) system. The latter is noteworthy, because avoiding the bactericidal effect, there is no selective pressure in the bacteria to develop resistance to this type of compounds.

As for *A. elisabethae*, the work shown above evidenced the importance of the pseudopterosins and *seco*-pseudopterosins isolated from this octocoral collected in the North Caribbean Sea (Bahamas) and in the South Caribbean Sea (Providencia Island), not only in relationship with their novel chemical structure but also for their potent antiinflammatory, cytotoxic and selective antimicrobial activity against Gram-positive bacteria. The efforts of many well-known researchers using the total synthesis of pseudopterosins and the SAR studies described by us in this chapter for natural homologs, allow us to conclude that despite all the research done for about 30 years, the development of these compounds as drugs or as active ingredients in cosmetic creams needs to be continued. Particularly, the supply issue (currently the key point) needs the development of more efficient and commercially viable syntheses or more SAR works aiming the elucidation of the pharmacophore responsible for the activity. In relation to the latter, it is important to emphasize that if we can determine the nature of the pharmacophore there would be no need to synthesize the entire molecule but to achieve by synthesis a partial structure that retains the biological activity, as tried by Fenical and coworkers [40] in their interesting work of the 2010 year.

Acknowledgements

The authors would like to thank Universidad Nacional de Colombia and Universidad de La Sabana for the financial support provided for this project. Also thanks to Prof. Dr. Mónica

Puyana (MP), Prof. Dr. Sven Zea (SZ), PI. Dr. Marcelino Gutierrez (MG), AD. Dr. Lucie Pautet (LP) and PI. Dr. Rogelio Fernandez (RF) for the photos of the octocorals.

Author details

Carmenza Duque[1]*, Leonardo Castellanos[1] and Edisson Tello[2]

*Address all correspondence to: cduqueb@unal.edu.co

1 Universidad Nacional de Colombia, Bogotá, Colombia

2 Universidad de La Sabana, Chía, Colombia

References

[1] Arrieta JM, Arnaud-Haond S, Duarte CM. What lies underneath: Concerning the oceans genetic resources. Proceedings of the National Academy of Sciences of the United States of America. 2010;**107**:18318-18324. DOI: 10.1073/pnas.0911897107

[2] Margulis L, Schwartz KV. Five Kingdoms—An Illustrated Guide to the Phyla of Life on Earth. 3rd ed. W.H. Freeman & Company; 1998

[3] Newman DJ, Cragg GM. Drugs and drug candidates from marine sources. An assessment of the current "state of play". Planta Medica. 2016;**82**:775-789. DOI: 10.1055/s-0042-101353

[4] Gerwick WH, Moore BS. Lessons from the past and charting the future of marine natural products drug discovery and chemical biology. Chemistry & Biology. 2012;**19**:85-98. DOI: 10.1016/j.chembiol.2011.12.014

[5] Mayer AMS, Glaser KB, Cuevas C, Jacobs RS, Kem W, Little RD, McIntosh JM, Newman DJ, Potts BC, Shuster DE. The odyssey of marine pharmaceuticals: A current pipeline perspective. Trends in Pharmacological Sciences. 2010;**31**:255-265. DOI: 10.1016/j.tips.2010.02.005

[6] Faulkner DJ. Marine natural products. Natural Product Reports. 2002;**19**:1-48 and all the revisions of the same author cited in this reference (1977, 1984, 1986-2001)

[7] Blunt JW, Copp BR, Keyzers RA, Munro MHG, Prinsep MR. Marine natural products. Natural Product Reports. 2016;**33**:382-431. DOI: 10.1039/c5np00156k See the other references of these series started and annually published since 2003

[8] Hu Y, Chen J, Hu G, Yu J, Zhu X, Lin Y, Chen S, Yuan J. Statistical research on the bioactivity of new marine natural products discovered during the 28 years from 1985 to 2012. Marine Drugs. 2015;**13**:202-221. DOI: 10.3390/md13010202

[9] Bergmann W, Feeney RJ. Isolation of new thymine pentoside from sponges. Journal of the American Chemical Society. 1950;**72**:2809-2810

[10] Bergmann W, Feeney RJ. XXXII. The nucleosides of sponges. The Journal of Organic Chemistry. 1951;**16**:981-987

[11] Bergmann W, Burke DC. Marine Products. XXXIX. The nucleosides of sponges. III. Spongothymidine and spongouridine. The Journal of Organic Chemistry. 1955;**20**:1501-1507

[12] Bayer FM, Weinheimer AJ, editors. Prostaglandins from Plexaura *homomalla*: Ecology, Utilization and Conservation of a Major Medical Marine Resource. A Symposium. Studios in Tropical Oceanography Book No. 12. 1974

[13] Martins A, Vieira H, Gaspar H, Santos S. Marketed marine natural products in the pharmaceutical and cosmeceutical industries: Tips for success. Marine Drugs. 2014;**12**: 1066-1101. DOI: 10.3390/md12021066

[14] Butler MS, Robertson AAB, Cooper MA. Natural product and natural product derived drugs in clinical triasl. Natural Product Reports. 2014;**31**:1612-1661. DOI: 10.1039/c4np00064a

[15] Daly M, Brugler MR, Cartwright P, Collins AG, Dawson MN, Fautin DG, France SC, McFadden CS, Opresko DM, Rodriguez E, Romano SL, Stake JL. The phylum Cnidaria: A review of phylogenetic patterns and diversity 300 years after Linnaeus. Zootaxa. 2007; **1668**:127-182

[16] McFadden CS, Sánchez JA, France SC. Molecular phylogenetic insights into the evolution of Octocorallia: A review. Integrative and Comparative Biology. 2010;**50**:389-410. DOI: 10.1093/icb/icq056

[17] Burkholder PR, Burkholder LM. Antimicrobial activity of horny corals. Science. 1958;**127**: 1174-1175. DOI: 10.1126/science.127.3307.1174

[18] Coll JC. The chemistry and chemical ecology of Octocorals (Coelenterata, Anthozoa, Octocorallia). Chemical Reviews. 1992;**92**:613-631

[19] Rodríguez AD. The natural products chemistry of West Indian Gorgonian Octocorals. Tetrahedron. 1995;**51**:4571-4618. DOI: 10.1016/0040-4020(95)00216-U

[20] Berrue F, Kerr RG. Diterpenes from gorgoniana corals. Natural Product Reports. 2009;**26**: 681-710. DOI: 10.1039/b821918b

[21] Berrue F, McCulloch MWB, Kerr RG. Marine diterpenes glycosides. Bioorganic & Medicinal Chemistry. 2011;**19**:6702-6719. DOI: 10.1016/j.bmc.2011.06.083

[22] Almeida MTR, Moritz MIG, Capel KCC, Pérez CD, Schenkel EP. Chemical and biological aspects of octocorals from the Brazilian coast. Revista Brasileira de Farmacognosia. 2014; **24**:446-467. DOI: 10.1016/j.bjp.2014.05.002

[23] Lei H. Diterpenoids of gorgonian corals: Chemistry and bioactivity. Chemistry & Biodiversity. 2016;**13**:345-365. DOI: 10.1002/cbdv.201500030

[24] Rodriguez AD, Piña IC, Barnes CL. Synthesis and biological evaluation of Cembranolide analogs containing cyclic ethers. The Journal of Organic Chemistry. 1995;**60**:8096-8100. DOI: 10.1021/jo00129a063

[25] Rodríguez AD, Piña IC, Acosta AL, Ramírez C, Soto JJ. Synthesis of analogues of *Eunicea* γ-cembranolides containing cyclic ethers via saponification. The Journal of Organic Chemistry. 2001;**66**:648-658. DOI: 10.1021/jo001025j

[26] Nieto NI, González N, Rodríguez J, Kerr RG, Jiménez C. New cytotoxic cembranolides: Isolation, biogenetic studies, and synthesis of analogues. Tetrahedron. 2006;**62**:1147-1154. DOI: 10.1016/j.tet.2006.09.033

[27] Pardo-Vargas A, de Barcelos Oliveira I, Stephens P, Cirne-Santos C, de Palmer Paixão I, Ramos F, Jiménez C, Rodríguez J, Resende J, Teixeira V, Castellanos L. Dolabelladienols A–C, new Diterpenes isolated from Brazilian Brown Alga *Dictyota pfaffii*. Marine Drugs. 2014;**12**:4247-4259. DOI: 10.3390/md12074247

[28] Pereira HS, Leão-Ferreira LR, Moussatché N, Teixeira VL, Cavalcanti DN, Costa LJ, Diaz R, Frugulhetti ICPP. Antiviral activity of diterpenes isolated from the Brazilian marine alga *Dictyota menstrualis* against human immunodeficiency virus type 1 (HIV-1). Antiviral Research. 2004;**64**:69-76. DOI: 10.1016/j.antiviral.2004.06.006

[29] Pardo-Vargas A, Ramos FA, Cirne-Santos CC, Stephens PR, Paixao ICP, Teixeira VL, Castellanos L. Semi-synthesis of oxygenated dolabellane diterpenes with highly in vitro anti-HIV-1 activity. Bioorganic & Medicinal Chemistry Letters. 2014;**24**:4381-4383. DOI: 10.1016/j.bmcl.2014.08.019

[30] Duque C, Tello E, Castellanos L, Fernández M, Arévalo-Ferro C. Bacterial sensors in microfouling assays. In: State of the Art in Biosensors/Book 2. Intech Open Science/Open Minds; 2012. DOI: 10.5772/53756. ISBN: 980-953-307-691-6

[31] Blunt J, Buckingham J, Munro M. Taxonomy and marine natural products research. In: Fattorusso E, Gerwick WH, Taglialatela-Scafati O, editors. Handbook of Marine Natural Products. Dordrecht: Springer Netherlands; 2012. pp. 3-54. DOI: 10.1007/978-90-481-3834-0

[32] Tello E, Castellanos L, Arevalo-Ferro C, Duque C. Cembranoid Diterpenes from the Caribbean Sea Whip *Eunicea Knighti*. Journal of Natural Products. 2009;**72**:1595-1602. DOI: 10.1021/np9002492

[33] Tello E, Castellanos L, Arevalo-Ferro C, Rodríguez J, Jiménez C, Duque C. Absolute stereochemistry of antifouling cembranoid epimers at C-8 from the Caribbean octocoral Pseudoplexaura flagellosa. Revised structures of plexaurolones. Tetrahedron. 2011;**67**: 9112-9121. DOI: 10.1016/j.tet.2011.09.094

[34] Tello E, Castellanos L, Arévalo-Ferro C, Duque C. Disruption in quorum-sensing systems and bacterial biofilm inhibition by cembranoid diterpenes isolated from the Octocoral *Eunicea knighti*. Journal of Natural Products. 2012;**75**:1637-1642. DOI: 10.1021/np300313k

[35] Tello E, Castellanos L, Duque C. Synthesis of cembranoid analogues and evaluation of their potential as quorum sensing inhibitors. Bioorganic & Medicinal Chemistry. 2013;**21**: 242-256. DOI: 10.1016/j.bmc.2012.10.022

[36] Tello E. Compuestos tipo cembrano naturales y algunos análogos sintéticos como potenciales agentes antifouling. Doctoral thesis. Bogotá, Colombia: Universidad Nacional de Colombia; 2013

[37] Williams GC, Chen JY. Resurrection of the octocorallian genus Antillogorgia for Caribbean species previously assigned to Pseudopterogorgia, and a taxonomic assessment of the relationship of these genera with Leptogorgia (Cnidaria, Anthozoa, Gorgoniidae). Zootaxa. 2012;**3505**:39-52

[38] Look SA, Fenical W, Jacobs RS, Clardy J. The pseudopterosins: Anti-inflammatory and analgesic natural products from the sea whip *Pseudopterogorgia elisabethae*. Proceedings of the National Academy of Sciences of the United States of America. 1986;**83**:6238-6240

[39] Look SA, Fenical W, Matsumoto GK, Clardy J. The pseudopterosins: A new class of anti-inflammatory and analgesic diterpene pentosides from the marine sea whip *Pseudopterogorgia elisabethae* (Octocorallia). The Journal of Organic Chemistry. 1986;**51**:5140-5145. DOI: 10.1021/jo00376a016

[40] Flachsmann F, Schellhaas K, Moya CE, Jacobs RS, Fenical W. Synthetic pseudopterosin analogues: A novel class of antiinflammatory drug candidates. Bioorganic & Medicinal Chemistry. 2010;**18**:8324-8333. DOI: 10.1016/j.bmc.2010.09.067

[41] Rodríguez II, Shi Y-P, García OJ, Rodríguez AD, Mayer AMS, Sánchez JA, Ortega E, González J. New pseudopterosin and seco-pseudopterosins diterpene glycosides from two Colombian isolates of *Pseudopterogorgia elisabethae* and their diverse biological activities. Journal of Natural Products. 2004;**67**:1672-1680

[42] Caplan SL, Zheng B, Dawson-Scully K, White CA, West LM. Pseudopterosin A: Protection of synaptic function and potential as a Neuromodulatory agent. Marine Drugs. 2016; **14**:55. DOI: 10.3390/md14030055

[43] Puyana M, Narvaez G, Paz A, Osorno O, Duque C. Psudopterosin content variability of the purple sea whip *Pseudopterogorgia elisabethae* at the Islands of San Andrés and Providencia (SW Caribbean). Journal of Chemical Ecology. 2004;**30**:1183-1201

[44] Duque C, Puyana M, Narvaez G, Paz A, Osorno O, Hara N, Fuyimoto Y. Pseudopterosins P-V. New compounds from the gorgonian octocoral *Pseudopterogorgia elisabethae* from Providencia Island, Colombian Caribbean. Tetrahedron. 2004;**60**:10627-10635. DOI: 10.1016/j.tet.2004.09.017

[45] Duque C, Puyana M, Castellanos L, Arias A, Correa H, Osorno O, Asai T, Hara N, Fujimoto Y. Further studies on the constituents of gorgonian octocoral *Pseudopterogorgia elisabethae* collected in San Andrés and Providencia Islands, Colombian Caribbean: Isolation of a putative biosynthetic intermediate leading to erogorgiane. Tetrahedron. 2006;**62**:4205-4213

[46] Correa H, Valenzuela AL, Ospina LF, Duque C. Anti-inflammatory effects of the gorgonian *Pseudopterogorgia elisabethae* collected at the Islands of Providencia and San Andrés (SW Caribbean). Journal of Inflammation. 2009;**6**:5. DOI: 10.1186/1476-9255-6-5

[47] Correa H, Aristizabal F, Duque C, Kerr RG. Cytotoxic and antimicrobial activity of pseudopterosins and seco-pseudopterosins isolated from the octocoral *Pseudopterogorgia elisabethae* of San Andrés and Providencia islands (Southwest Caribbean Sea). Marine Drugs. 2011;**9**:334-344. DOI: 10.3390/md9030334

[48] Correa H, Zorro P, Arevalo-Ferro C, Puyana M, Duque C. Possible ecological role of pseudopterosins G and P-U and seco-pseudopterosins J and K from the gorgonian *Pseudopterogorgia elisabethae* from Providencia island (SW Caribbean) in regulating microbial surface communities. Journal of Chemical Ecology. 2012;**38**:1190-1202. DOI: 10.1007/s10886-012-0182-y

[49] Gutiérrez M, Capson TL, Guzmán HM, González J, Ortega-Barría E, Quiñoá E, Riguera R. Antiplasmodial metabolites isolated from the marine Octocoral *Muricea austera*. Journal of Natural Products. 2006;**69**:1379-1783. DOI: 10.1021/np060007f

[50] D'Ambrosio M, Guerriero A, Pietra F. Arboxeniolide-1, a new, naturally occurring Xeniolide Diterpenoid from the Gorgonian *Paragorgia arborea* of the Crozet Is. (S. Indian Ocean). Zeitschrift für Naturforsch C. 1984;**39**:1180-1183. DOI: 10.1515/ZNC-1984-11-1232

[51] Poza JJ, Fernández R, Reyes F, Rodríguez J, Jiménez C. Isolation, biological significance, synthesis, and cytotoxic evaluation of new natural parathiosteroids A−C and analogues from the soft coral *Paragorgia* sp. The Journal of Organic Chemistry. 2008;**73**:7978-7984. DOI: 10.1021/jo801198u

[52] Birkenmaier A, Holert J, Erdbrink H, Moeller HM, Friemel A, Schoenenberger R, Suter MJF, Klebensberger J, Philipp B. Biochemical and genetic investigation of initial reactions in aerobic degradation of the bile acid cholate in *Pseudomonas* sp. strain Chol1. Journal of Bacteriology. 2007;**189**:7165-7173. DOI: 10.1128/JB.00665-07

[53] Fattorusso E, Romano A, Taglialatela-Scafati O, Irace C, Maffettone C, Bavestrello G, Cerrano C. Oxygenated cembranoids of the decaryiol type from the Indonesian soft coral *Lobophytum* sp. Tetrahedron. 2009;**65**:2898-2904. DOI: 10.1016/j.tet.2009.02.008

[54] Bernstein J, Shmeuli U, Zadock E, Kashman Y, Neeman I. Sarcophine, a new epoxy cembranolide from marine origin. Tetrahedron. 1974;**30**:2817-2824. DOI: 10.1016/S0040-4020(01)97451-4

[55] Néeman I, Fishelson L, Kashman Y. Sarcophine—A new toxin from the soft coral *Sarcophyton glaucum* (Alcyonaria). Toxicon. 1974;**12**:593-598. DOI: 10.1016/0041-0101(74)90192-5

[56] Hassan HM, Sallam AA, Mohammed R, Hifnawy S, DTA Y, El Sayed KA. Semisynthetic analogues of the marine cembranoid sarcophine as prostate and breast cancer migration inhibitors. Bioorganic & Medicinal Chemistry. 2011;**19**:4928-4934. DOI: 10.1016/j.bmc.2011.06.060

[57] Tanaka H, Takeuchi R, Jimbo M, Kuniya N, Takahashi T. Synthesis and biological evaluation of the Forssman antigen Pentasaccharide and derivatives by a one-pot glycosylation procedure. Chemistry—A European Journal. 2013;**19**:3177-3187. DOI: 10.1002/chem.2012 03865

[58] Jimbo M, Yanohara T, Koike K, Sakai R, Muramoto K, Kamiya H. The D-galactose-binding lectin of the octocoral Sinularia lochmodes: Characterization and possible relationship to the symbiotic dinoflagellates. Comparative Biochemistry and Physiology Part B. 2000; **125**:227-236